# Lecture Notes in Mathematics　2033

**Editors:**
J.-M. Morel, Cachan
B. Teissier, Paris

**Subseries:**
École d'Été de Probabilités de Saint-Flour

For further volumes:
http://www.springer.com/series/304

## Saint-Flour Probability Summer School

The Saint-Flour volumes are reflections of the courses given at the Saint-Flour Probability Summer School. Founded in 1971, this school is organised every year by the Laboratoire de Mathématiques (CNRS and Université Blaise Pascal, Clermont-Ferrand, France). It is intended for PhD students, teachers and researchers who are interested in probability theory, statistics, and in their applications.

The duration of each school is 13 days (it was 17 days up to 2005), and up to 70 participants can attend it. The aim is to provide, in three high-level courses, a comprehensive study of some fields in probability theory or Statistics. The lecturers are chosen by an international scientific board. The participants themselves also have the opportunity to give short lectures about their research work.

Participants are lodged and work in the same building, a former seminary built in the 18th century in the city of Saint-Flour, at an altitude of 900 m. The pleasant surroundings facilitate scientific discussion and exchange.

The Saint-Flour Probability Summer School is supported by:

– Université Blaise Pascal
– Centre National de la Recherche Scientifique (C.N.R.S.)
– Ministère délégué à l'Enseignement supérieur et à la Recherche

For more information, see back pages of the book and
http://math.univ-bpclermont.fr/stflour/

Jean Picard
Summer School Chairman
Laboratoire de Mathématiques
Université Blaise Pascal
63177 Aubière Cedex
France

Vladimir Koltchinskii

# Oracle Inequalities in Empirical Risk Minimization and Sparse Recovery Problems

École d'Été de Probabilités
de Saint-Flour XXXVIII-2008

 Springer

Prof. Vladimir Koltchinskii
Georgia Institute of Technology
School of Mathematics
Atlanta
USA
vlad@math.gatech.edu

ISBN 978-3-642-22146-0          e-ISBN 978-3-642-22147-7
DOI 10.1007/978-3-642-22147-7
Springer Heidelberg Dordrecht London New York

Lecture Notes in Mathematics ISSN print edition: 0075-8434
ISSN electronic edition: 1617-9692

Library of Congress Control Number: 2011934366

Mathematics Subject Classification (2010): 62J99, 62H12, 60B20, 60G99

*Cover design*: deblik, Berlin

Printed on acid-free paper

Springer is part of Springer Science+Business Media (www.springer.com)

# Preface

The purpose of these lecture notes is to provide an introduction to the general theory of empirical risk minimization with an emphasis on excess risk bounds and oracle inequalities in penalized problems. In the recent years, there have been new developments in this area motivated by the study of new classes of methods in machine learning such as large margin classification methods (boosting, kernel machines). The main probabilistic tools involved in the analysis of these problems are concentration and deviation inequalities by Talagrand along with other methods of empirical processes theory (symmetrization inequalities, contraction inequality for Rademacher sums, entropy and generic chaining bounds). Sparse recovery based on $\ell_1$-type penalization and low rank matrix recovery based on the nuclear norm penalization are other active areas of research, where the main problems can be stated in the framework of penalized empirical risk minimization, and concentration inequalities and empirical processes tools proved to be very useful.

My interest in empirical processes started in the late 1970s and early 1980s. It was largely influenced by the work of Vapnik and Chervonenkis on Glivenko–Cantelli problem and on empirical risk minimization in pattern recognition, and, especially, by the results of Dudley on uniform central limit theorems. Talagrand's concentration inequality proved in the 1990s was a major result with deep consequences in the theory of empirical processes and related areas of statistics, and it inspired many new approaches in analysis of empirical risk minimization problems.

Over the last years, the work of many people have had a profound impact on my own research and on my view of the subject of these notes. I was lucky to work together with several of them and to have numerous conversations and email exchanges with many others. I am especially thankful to Peter Bartlett, Lucien Birgé, Gilles Blanchard, Stephane Boucheron, Olivier Bousquet, Richard Dudley, Sara van de Geer, Evarist Giné, Gabor Lugosi, Pascal Massart, David Mason, Shahar Mendelson, Dmitry Panchenko, Alexandre Tsybakov, Aad van der Vaart, Jon Wellner and Joel Zinn.

I am thankful to the School of Mathematics, Georgia Institute of Technology and to the Department of Mathematics and Statistics, University of New Mexico where most of my work for the past years have taken place.

The research described in these notes has been supported in part by NSF grants DMS-0304861, MSPA-MPS-0624841, DMS-0906880 and CCF-0808863.

I was working on the initial draft while visiting the Isaac Newton Institute for Mathematical Sciences in Cambridge in 2008. I am thankful to the Institute for its hospitality.

I am also very thankful to Jean-Yves Audibert, Kai Ni and Jon Wellner who provided me with the lists of typos and inconsistencies in the first draft.

I am very grateful to all the students and colleagues who attended the lectures in 2008 and whose questions motivated many of the changes I have made since then.

I am especially thankful to Jean Picard for all his efforts that make The Saint-Flour School truly unique.

Atlanta                                                                    *Vladimir Koltchinskii*
February 2011

# Contents

# Chapter 1
# Introduction

We start with a brief overview of empirical risk minimization problems and of the role of empirical and Rademacher processes in constructing distribution dependent and data dependent excess risk bounds. We then discuss penalized empirical risk minimization and oracle inequalities and conclude with sparse recovery and low rank matrix recovery problems. Many important aspects of empirical risk minimization are beyond the scope of these notes, in particular, the circle of questions related to approximation theory (see well known papers by Cucker and Smale [47], DeVore et al. [49] and references therein).

## 1.1 Abstract Empirical Risk Minimization

Let $X, X_1, \ldots, X_n, \ldots$ be i.i.d. random variables defined on a probability space $(\Omega, \Sigma, \mathbb{P})$ and taking values in a measurable space $(S, \mathscr{A})$ with common distribution $P$. Let $P_n$ denote the empirical measure based on the sample $(X_1, \ldots, X_n)$ of the first $n$ observations:

$$P_n := n^{-1} \sum_{j=1}^{n} \delta_{X_j},$$

where $\delta_x$, $x \in S$ is the Diracs's measure. Let $\mathscr{F}$ be a class of measurable functions $f : S \mapsto \mathbb{R}$. In what follows, the values of a function $f \in \mathscr{F}$ will be interpreted as "losses" associated with certain "actions" and the expectation of $f(X)$,

$$\mathbb{E} f(X) = \int_S f dP = Pf,$$

will be viewed as the risk of a certain "decision rule". We will be interested in the problem of risk minimization

$$Pf \longrightarrow \min, \ f \in \mathscr{F} \tag{1.1}$$

V. Koltchinskii, *Oracle Inequalities in Empirical Risk Minimization and Sparse Recovery Problems*, Lecture Notes in Mathematics 2033, DOI 10.1007/978-3-642-22147-7_1, © Springer-Verlag Berlin Heidelberg 2011

in the cases when the distribution $P$ is unknown and has to be estimated based on the data $(X_1, \ldots, X_n)$. Since the empirical measure $P_n$ is a natural estimator of $P$, the true risk can be estimated by the corresponding empirical risk,

$$n^{-1} \sum_{j=1}^{n} f(X_j) = \int_S f \, dP_n = P_n f,$$

and the risk minimization problem has to be replaced by the empirical risk minimization:

$$P_n f \longrightarrow \min, \ f \in \mathscr{F}. \tag{1.2}$$

Many important methods of statistical estimation such as maximum likelihood and more general $M$-estimation are versions of empirical risk minimization. The general theory of empirical risk minimization has started with seminal paper of Vapnik and Chervonenkis [147] (see Vapnik [146] for more references) although some important ideas go back to much earlier work on asymptotic theory of $M$-estimation. Vapnik and Chervonenkis were motivated by applications of empirical risk minimization in pattern recognition and learning theory that required the development of the theory in a much more general framework than what was common in statistical literature. Their key idea was to relate the quality of the solution of empirical risk minimization problem to the accuracy of approximation of the true distribution $P$ by the empirical distribution $P_n$ uniformly over function classes representing losses of decision rules. Because of this, they have studied general Glivenko–Cantelli problems about convergence of $\| P_n - P \|_{\mathscr{F}}$ to 0, where

$$\| Y \|_{\mathscr{F}} := \sup_{f \in \mathscr{F}} |Y(f)|$$

for $Y : \mathscr{F} \mapsto \mathbb{R}$. Vapnik and Chervonenkis introduced a number of important characteristics of complexity of function classes, such as VC-dimensions and random entropies, that control the accuracy of empirical approximation. These results along with the development of classical limit theorems in Banach spaces in the 1960s and 1970s led to the general theory of empirical processes that started with the pathbreaking paper by Dudley [58] on central limit theorems for empirical measures (see well known books by Dudley [59], Pollard [123], van der Vaart and Wellner [148]).

In the 1990s, Talagrand studied isoperimetric inequalities for product measures and, in particular, he proved a striking uniform version of Bernstein inequality describing concentration of $\| P_n - P \|_{\mathscr{F}}$ around its expectation (see Talagrand [138, 139]). This was a real breakthrough in the theory of empirical processes and empirical risk minimization. At about the same time, a concept of *oracle inequalities* has been developed in nonparametric statistics (see, e.g., Johnstone [74]). In modern statistics, it is common to deal with a multitude of possible models that describe the same data (for instance, a family of models for unknown regression functions of varying complexity). An oracle inequality is a bound on the risk of a statistical estimator that shows that the performance of the estimator is almost (often, up to numerical constants) as good as it would be if the statistician had an access to an oracle that knows what the best model for the target function is. It happened that concentration inequalities provide rather natural probabilistic tools needed to develop oracle inequalities in a number of statistical problems. In particular, Birgé and Massart [23], Barron et al. [12], and, more recently, Massart [106, 107] suggested a general approach to model selection in a variety of statistical problems such as density estimation, regression and classification that is based on penalized empirical risk minimization. They used Talagrand's concentration and deviation inequalities in a systematic way to establish a number of oracle inequalities showing some form of optimality of penalized empirical risk minimization as a model selection tool.

In the recent years, new important classes of algorithms in machine learning have been introduced that are based on empirical risk minimization. In particular, large margin classification algorithms, such as boosting and support vector machines (SVM), can be viewed as empirical risk minimization over infinite dimensional functional spaces with special convex loss functions. In an attempt to understand the nature of these classification methods and to explain their superb generalization performance, there has been another round of work on the abstract theory of empirical risk minimization. One of the main ideas was to use the sup-norms or localized sup-norms of the Rademacher processes indexed by function classes to develop a general approach to measuring the complexities of these classes (see Koltchinskii [81], Bartlett et al. [14], Koltchinskii and Panchenko [92], Bousquet et al. [34], Bartlett et al. [15], Lugosi and Wegkamp [104], Bartlett and Mendelson [17]). This resulted in rather flexible definitions of distribution dependent and data dependent complexities in an abstract framework as well as more specialized complexities reflecting relevant parameters of specific learning machines. Moreover, such complexities have been used as natural penalties in model selection methods. This approach provided a general explanation of fast convergence rates in classification and other learning problems, the phenomenon discovered and studied by several authors, in particular, by Mammen and Tsybakov [105] and in an influential paper by Tsybakov [144].

## 1.2   Excess Risk: Distribution Dependent Bounds

**Definition 1.1.** Let

$$\mathscr{E}(f) := \mathscr{E}_P(f) := \mathscr{E}_P(\mathscr{F}; f) := Pf - \inf_{g \in \mathscr{F}} Pg.$$

This quantity will be called the *excess risk* of $f \in \mathscr{F}$.

Let

$$\hat{f} = \hat{f}_n \in \operatorname{Argmin}_{f \in \mathscr{F}} P_n f$$

be a solution of the empirical risk minimization problem (1.2). The function $\hat{f}_n$ is used as an approximation of the solution of the true risk minimization problem (1.1) and its excess risk $\mathscr{E}_P(\hat{f}_n)$ is a natural measure of accuracy of this approximation.

It is of interest to find tight upper bounds on the excess risk of $\hat{f}_n$ that hold with a high probability. Such bounds usually depend on certain "geometric" properties of the function class $\mathscr{F}$ and on various measures of its "complexity" that determine the accuracy of approximation of the true risk $Pf$ by the empirical risk $P_n f$ in a neighborhood of a proper size of the minimal set of the true risk.

In fact, it is rather easy to describe a general approach to derivation of such bounds in an abstract framework of empirical risk minimization discussed in these notes. This approach does give a correct answer in many specific examples. To be precise, define the $\delta$-*minimal set* of the risk as

$$\mathscr{F}(\delta) := \mathscr{F}_P(\delta) := \{f : \mathscr{E}_P(f) \leq \delta\}.$$

Suppose, for simplicity, that the infimum of the risk $Pf$ is attained at $\bar{f} \in \mathscr{F}$ (the argument can be easily modified if the infimum is not attained in the class). Denote $\hat{\delta} := \mathscr{E}_P(\hat{f})$. Then $\hat{f}, \bar{f} \in \mathscr{F}(\hat{\delta})$ and $P_n\hat{f} \leq P_n\bar{f}$. Therefore,

$$\hat{\delta} = \mathscr{E}_P(\hat{f}) = P(\hat{f} - \bar{f}) \leq P_n(\hat{f} - \bar{f}) + (P - P_n)(\hat{f} - \bar{f}),$$

which implies

$$\hat{\delta} \leq \sup_{f,g \in \mathscr{F}(\hat{\delta})} |(P_n - P)(f - g)|.$$

Imagine there exists a nonrandom upper bound

$$U_n(\delta) \geq \sup_{f,g \in \mathscr{F}(\delta)} |(P_n - P)(f - g)| \tag{1.3}$$

that holds uniformly in $\delta$ with a high probability. Then, with the same probability, the excess risk $\mathscr{E}_P(\hat{f})$ will be bounded by the largest solution of the inequality $\delta \leq U_n(\delta)$. There are many different ways to construct upper bounds on the sup-norms of empirical processes. A very general approach is based on Talagrand's concentration inequalities. Assume for simplicity that functions in the class $\mathscr{F}$ take their values in the interval $[0, 1]$. Based on the $L_2(P)$-diameter $D_P(\mathscr{F}; \delta)$ of the $\delta$-minimal set $\mathscr{F}(\delta)$ and the function

$$\phi_n(\mathscr{F}; \delta) := \mathbb{E} \sup_{f,g \in \mathscr{F}(\delta)} |(P_n - P)(f - g)|,$$

define

$$\bar{U}_n(\delta; t) := K \left( \phi_n(\mathscr{F}; \delta) + D(\mathscr{F}; \delta)\sqrt{\frac{t}{n}} + \frac{t}{n} \right).$$

Talagrand's concentration inequality then implies that with some numerical constant $K > 0$, for all $t > 0$,

$$\mathbb{P}\left\{ \sup_{f,g\in\mathscr{F}(\delta)} |(P_n - P)(f - g)| \geq \bar{U}_n(\delta; t) \right\} \leq e^{-t}.$$

This observation provides an easy way to construct a function $U_n(\delta)$ such that (1.3) holds with a high probability uniformly in $\delta$ (first, by defining such a function at a discrete set of values of $\delta$ and then extending it to all the values by monotonicity). By solving the inequality $\delta \leq U_n(\delta)$, one can construct a bound $\bar{\delta}_n(\mathscr{F})$ such that the probability $\mathbb{P}\{\mathscr{E}_P(\hat{f}_n) \geq \bar{\delta}_n(\mathscr{F})\}$ is small. Thus, constructing an upper bound on the excess risk essentially reduces to solving a fixed point equation of the type $\delta = U_n(\delta)$. Such a fixed point method has been studied, for instance, in Massart [106], Koltchinskii and Panchenko [92], Bartlett et al. [15], Koltchinskii [83] (and in several other papers of these authors).

In the case of $P$-Donsker classes $\mathscr{F}$,

$$\phi_n(\mathscr{F}; \delta) \leq \mathbb{E}\| P_n - P \|_{\mathscr{F}} = O(n^{-1/2}),$$

which implies that

$$\bar{\delta}_n(\mathscr{F}) = O(n^{-1/2}).$$

Moreover, if the diameter $D(\mathscr{F}; \delta)$ of the $\delta$-minimal set tends to 0 as $\delta \to 0$ (which is typically the case if the risk minimization problem (1.1) has a unique solution), then, by asymptotic equicontinuity, we have

$$\lim_{\delta\to 0} \limsup_{n\to\infty} n^{1/2}\phi_n(\mathscr{F}; \delta) = 0,$$

which allows one to conclude that

$$\bar{\delta}_n(\mathscr{F}) = o(n^{-1/2}).$$

It happens that the bound $\bar{\delta}_n(\mathscr{F})$ is of asymptotically correct order as $n \to \infty$ in many specific examples of risk minimization problem in statistics and learning theory.

The bounds of this type are *distribution dependent* (that is, they depend on the unknown distribution $P$).

## 1.3  Rademacher Processes and Data Dependent Bounds on Excess Risk

The next challenge is to construct *data dependent* upper confidence bounds on the excess risk $\mathscr{E}_P(\hat{f})$ of empirical risk minimizers that depend only on the sample $(X_1, \ldots, X_n)$, but do not depend explicitly on the unknown distribution $P$. Such bounds can be used in model selection procedures. Their construction usually requires the development of certain statistical estimates of the quantities involved in the definition of the distribution dependent bound $\bar{\delta}_n(\mathscr{F})$ based on the sample $(X_1, \ldots, X_n)$. Namely, we have to estimate the expectation of the local sup-norm of the empirical process $\phi_n(\mathscr{F}; \delta)$ and the diameter of the $\delta$-minimal set.

A natural way to estimate the empirical process is to replace it by the Rademacher process

$$R_n(f) := n^{-1} \sum_{j=1}^{n} \varepsilon_j f(X_j), \ f \in \mathscr{F},$$

where $\{\varepsilon_j\}$ are i.i.d. Rademacher random variables (that is, they are symmetric Bernoulli random variables taking values $+1$ and $-1$ with probability $1/2$ each) that are also independent of the data $(X_1, \ldots, X_n)$. The process $R_n(f)$, $f \in \mathscr{F}$ depends only on the data (and on the independent sequence of Rademacher random variables that can be simulated). For each $f \in \mathscr{F}$, $R_n(f)$ is essentially the "correlation coefficient" between the values of the function $f$ at data points and independent Rademacher noise. The fact that the sup-norm $\|R_n\|_{\mathscr{F}}$ of the Rademacher process is "large" means that there exists a function $f \in \mathscr{F}$ that fits the Rademacher noise very well. This usually means that the class of functions is too complex for the purposes of statistical estimation and performing empirical risk minimization over such a class is likely to lead to overfitting. Thus, the size of sup-norms or local sup-norms of the Rademacher process provides natural data dependent measures of complexity of function classes used in statistical estimation. Symmetrization inequalities well known in the theory of empirical processes show that the expected sup-norms of Rademacher processes are within a constant from the corresponding sup-norms of the empirical process. Moreover, using concentration inequalities, one can directly relate the sup-norms of these two processes.

The $\delta$-minimal sets (the level sets) of the true risk involved in the construction of the bounds $\bar{\delta}_n(\mathscr{F})$ can be estimated by the level sets of the empirical risk. This is based on *ratio type inequalities* for the excess risk, that is, on bounding the following probabilities

$$\mathbb{P}\left\{ \sup_{f \in \mathscr{F}, \mathscr{E}_P(f) \geq \delta} \left| \frac{\mathscr{E}_{P_n}(f)}{\mathscr{E}_P(f)} - 1 \right| \geq \varepsilon \right\}.$$

This problem is closely related to the study of ratio type empirical processes (see Giné et al. [65], Giné and Koltchinskii [66] and references therein). Finally, the $L_2(P)$-diameter of the $\delta$-minimal sets of $P$ can be estimated by the $L_2(P_n)$-diameter of the $\delta$-minimal sets of $P_n$. Thus, we can estimate all the distribution dependent parameters involved in the construction of $\bar{\delta}_n(\mathscr{F})$ by their empirical versions and, as a result, construct data-dependent upper bounds on the excess risk $\mathscr{E}_P(\hat{f})$ that hold with a guaranteed high probability. The proofs of these facts heavily rely on Talagrand's concentration inequalities for empirical processes.

## 1.4  Penalized Empirical Risk Minimization and Oracle Inequalities

The data-dependent bounds on the excess risk can be used in general model selection techniques in abstract empirical risk minimization problems. In such problems, there is a need to deal with minimizing the risk over a very large class of functions $\mathcal{F}$, and there is a specified family ("a sieve") of subclasses $\{\mathcal{F}_\alpha, \alpha \in A\}$ of varying complexity that are used to approximate functions from $\mathcal{F}$. Often, classes $\mathcal{F}_\alpha$ correspond to different statistical models. Instead of one empirical risk minimization problem (1.2) one has to deal now with a family of problems

$$P_n f \longrightarrow \min, \ f \in \mathcal{F}_\alpha, \ \alpha \in A, \tag{1.4}$$

that has a set of solutions $\{\hat{f}_{n,\alpha} : \alpha \in A\}$. In many cases, there is a natural way to measure the quality of the solution of each of the problems (1.4). For instance, it can be based on distribution dependent upper bounds $\bar{\delta}_n(\alpha) = \bar{\delta}_n(\mathcal{F}_\alpha)$ on the excess risk $\mathcal{E}_P(\mathcal{F}_\alpha; \hat{f}_{n,\alpha})$ discussed above. The goal of model selection is to provide a data driven (adaptive) choice $\hat{\alpha} = \hat{\alpha}(X_1, \ldots, X_n)$ of model index $\alpha$ such that the empirical risk minimization over the class $\mathcal{F}_{\hat{\alpha}}$ results in an estimator $\hat{f} = \hat{f}_{n,\hat{\alpha}}$ with the nearly "optimal" excess risk $\mathcal{E}_P(\mathcal{F}; \hat{f})$. One of the most important approaches to model selection is based on penalized empirical risk minimization, that is, on solving the following problem

$$\hat{\alpha} := \text{argmin}_{\alpha \in A} \left[ \min_{f \in \mathcal{F}_\alpha} P_n f + \hat{\pi}_n(\alpha) \right], \tag{1.5}$$

where $\hat{\pi}_n(\alpha), \alpha \in A$ are properly chosen complexity penalties. Often, $\hat{\pi}_n(\alpha)$ is designed as a data dependent upper bound on $\bar{\delta}_n(\alpha)$, the "desired accuracy" of empirical risk minimization for the class $\mathcal{F}_\alpha$. This approach has been developed under several different names (Vapnik–Chervonenkis structural risk minimization, method of sieves, etc.). Sometimes, it is convenient to write penalized empirical risk minimization problem in the following form

$$\hat{f} := \text{argmin}_{f \in \mathcal{F}} \left[ P_n f + \text{pen}(n; f) \right],$$

where $\text{pen}(n; \cdot)$ is a real valued complexity penalty defined on $\mathcal{F}$. Denoting, for each $\alpha \in \mathbb{R}$,

$$\mathcal{F}_\alpha := \{f \in \mathcal{F} : \text{pen}(n; f) = \alpha\}$$

and defining $\hat{\pi}_n(\alpha) = \alpha$, the problem can be again rewritten as (1.5).

The bounds on the excess risk of $\hat{f} = \hat{f}_{n,\hat{\alpha}}$ of the following type (with some constant $C$)

$$\mathscr{E}_P(\mathscr{F}; \hat{f}) \leq C \inf_{\alpha \in A} \left[ \inf_{f \in \mathscr{F}_\alpha} \mathscr{E}_P(f) + \bar{\delta}_n(\alpha) \right] \qquad (1.6)$$

that hold with a high probability are often used to express the optimality properties of model selection. The meaning of these inequalities can be explained as follows. Imagine that the minimum of the true risk in the class $\mathscr{F}$ is attained in a subclass $\mathscr{F}_\alpha$ for some $\alpha = \alpha(P)$. If there were an oracle that knew the model index $\alpha(P)$, then with the help of the oracle one could achieve the excess risk at least as small as $\bar{\delta}_n(\alpha(P))$. The model selection method for which the inequality (1.6) holds is not using the help of the oracle. However, it follows from (1.6) that the excess risk of the resulting estimator is upper bounded by $C\bar{\delta}_n(\alpha(P))$ (which is within a constant of the performance of the oracle).

## 1.5   Concrete Empirical Risk Minimization Problems

*Density estimation.* The most popular method of statistical estimation, the maximum likelihood method, can be viewed as a special case of empirical risk minimization. Let $\mu$ be a $\sigma$-finite measure on $(S, \mathscr{A})$ and let $\mathscr{P}$ be a statistical model, that is, $\mathscr{P}$ is a family of probability densities with respect to $\mu$. In particular, $\mathscr{P}$ can be a parametric model with a parameter set $\Theta$, $\mathscr{P} = \{p(\theta, \cdot) : \theta \in \Theta\}$. A maximum likelihood estimator of unknown density $p_* \in \mathscr{P}$ based on i.i.d. observations $X_1, \ldots, X_n$ sampled from $p_*$ is a solution of the following empirical risk minimization problem

$$n^{-1} \sum_{j=1}^{n} \left( -\log p(X_j) \right) \longrightarrow \min, \ p \in \mathscr{P}. \qquad (1.7)$$

Another popular approach to density estimation is based on a penalized empirical risk minimization problem

$$-\frac{2}{n} \sum_{j=1}^{n} p(X_j) + \|p\|_{L_2(\mu)}^2 \longrightarrow \min, \ p \in \mathscr{P}. \qquad (1.8)$$

This approach can be explained as follows. The best $L_2(\mu)$-approximation of the density $p_*$ is obtained by solving

$$\|p - p_*\|_{L_2(\mu)}^2 = -2 \int_S pp_* d\mu + \|p\|_{L_2(\mu)}^2 + \|p_*\|_{L_2(\mu)}^2 \longrightarrow \min, \ p \in \mathscr{P}.$$

The integral $\int_S pp_* d\mu = \mathbb{E}p(X)$ can be estimated by $n^{-1} \sum_{j=1}^{n} p(X_j)$, leading to problem (1.8). Of course, in the case of complex enough models $\mathscr{P}$, there might be a need in complexity penalization in (1.7) and (1.8).

*Prediction problems.* Empirical risk minimization is especially useful in a variety of prediction problems. In these problems, the data consists of i.i.d. couples $(X_1, Y_1), \ldots (X_n, Y_n)$ in $S \times T$ with common distribution $P$. Assume that $T \subset \mathbb{R}$. Given another couple $(X, Y)$ sampled from $P$, the goal is to predict $Y$ based on an observation of $X$. To formalize this problem, introduce a loss function $\ell : T \times \mathbb{R} \mapsto \mathbb{R}_+$. Given $g : S \mapsto \mathbb{R}$, denote $(\ell \bullet g)(x, y) := \ell(y, g(x))$, which will be interpreted as a loss suffered as a result of using $g(x)$ to predict $y$. Then the risk associated with an "action" $g$ is defined as

$$P(\ell \bullet g) = \mathbb{E}\ell(Y, g(X)).$$

Given a set $\mathscr{G}$ of possible actions $g$, we want to minimize the risk:

$$P(\ell \bullet g) \longrightarrow \min, \ g \in \mathscr{G}.$$

The risk can be estimated based on the data $(X_1, Y_1), \ldots, (X_n, Y_n)$, which leads to the following empirical risk minimization problem:

$$P_n(\ell \bullet g) = n^{-1} \sum_{j=1}^{n} \ell(Y_j, g(X_j)) \longrightarrow \min, \ g \in \mathscr{G}.$$

Introducing the notation $f := \ell \bullet g$ and setting $\mathscr{F} := \{\ell \bullet g : g \in \mathscr{G}\}$, one can rewrite the problems in the form (1.1), (1.2).

Regression and classification are two most common examples of prediction problems. In regression problems, the loss function is usually defined as $\ell(y; u) = \phi(y - u)$, where $\phi$ is, most often, nonnegative, even and convex function with $\phi(0) = 0$. The empirical risk minimization becomes

$$n^{-1} \sum_{j=1}^{n} \phi(Y_j - g(X_j)) \longrightarrow \min, \ g \in \mathscr{G}.$$

The choice $\phi(u) = u^2$ is, by far, the most popular and it means fitting the regression model using the least squares method.

In the case of *binary classification* problems, $T := \{-1, 1\}$ and it is natural to consider a class $\mathscr{G}$ of binary functions (classifiers) $g : S \mapsto \{-1, 1\}$ and to use the binary loss $\ell(y; u) = I(y \neq u)$. The risk of a classifier $g$ with respect to the binary loss

$$P(\ell \bullet g) = \mathbb{P}\{Y \neq g(X)\}$$

is just the probability of misclassification and, in learning theory, it is known as *the generalization error*. A binary classifier that minimizes the generalization error over all measurable binary functions is called *the Bayes classifier* and its generalization error is called *the Bayes risk*. The corresponding empirical risk

$$P_n(\ell \bullet g) = n^{-1} \sum_{j=1}^{n} I(Y_j \neq g(X_j))$$

is known as *the training error*. Minimizing the training error over $\mathscr{G}$

$$n^{-1} \sum_{j=1}^{n} I(Y_j \neq g(X_j)) \longrightarrow \min, \ g \in \mathscr{G}$$

is, usually, a computationally intractable problem (with an exception of very simple families of classifiers $\mathscr{G}$) since the functional to be minimized lacks convexity, smoothness or any other form of regularity.

*Large margin classification.* Large margin classification methods are based on the idea of considering real valued classifiers $g : S \mapsto \mathbb{R}$ instead of binary classifiers and replacing the binary loss by a convex "surrogate loss". A real valued classifier $g$ can be easily transformed into binary: $g \mapsto \text{sign}(g)$. Define $\ell(y, u) := \phi(yu)$, where $\phi : \mathbb{R} \mapsto \mathbb{R}_+$ is a convex nonincreasing function such that $\phi(u) \geq I_{(-\infty, 0]}(u), u \in \mathbb{R}$. The product $Yg(X)$ is called *the margin* of classifier $g$ on the training example $(X, Y)$. If $Yg(X) \geq 0$, $(X, Y)$ is correctly classified by $g$, otherwise the example is misclassified. Given a convex set $\mathscr{G}$ of classifiers $g : S \mapsto \mathbb{R}$ the risk minimization problem becomes

$$P(\ell \bullet g) = \mathbb{E}\phi(Yg(X)) \longrightarrow \min, \ g \in \mathscr{G}$$

and its empirical version is

$$P_n(\ell \bullet g) = n^{-1} \sum_{j=1}^{n} \phi(Y_j g(X_j)) \longrightarrow \min, \ g \in \mathscr{G}, \qquad (1.9)$$

which are convex optimization problems.

It is well known that, under very mild conditions on the "surrogate loss" $\phi$ (so called *classification calibration*, see, e.g., [16]) the solution $g_*$ of the problem

$$P(\ell \bullet g) = \mathbb{E}\phi(Yg(X)) \longrightarrow \min, \ g : S \mapsto \mathbb{R}$$

possesses the property that $\text{sign}(g_*)$ is the Bayes classifier. Thus, it becomes plausible that the empirical risk minimization problem (1.9) with a large enough and properly chosen convex function class $\mathscr{G}$ would have a solution $\hat{g}$ such that the generalization error of the binary classifier $\text{sign}(\hat{g})$ is close enough to the Bayes risk. Because of the nature of the loss function (heavy penalization for negative and even small positive margins), the solution $\hat{g}$ tends to be a classifier with most of the margins on the training data positive and large, which explains the name "large margin classifiers".

Among common choices of the surrogate loss function are $\phi(u) = e^{-u}$ (the exponential loss), $\phi(u) = \log_2(1 + e^{-u})$ (the logit loss) and $\phi(u) = (1 - u) \vee 0$ (the hinge loss).

A possible choice of class $\mathcal{G}$ is

$$\mathcal{G} := \mathrm{conv}(\mathcal{H}) := \left\{ \sum_{j=1}^{N} \lambda_j h_j, N \geq 1, \lambda_j \geq 0, \sum_{j=1}^{N} \lambda_j h_j, h_j \in \mathcal{H} \right\},$$

where $\mathcal{H}$ is a given *base class* of classifiers. Usually, $\mathcal{H}$ consists of binary classifiers and it is a rather simple class such that the direct minimization of the training error over $\mathcal{H}$ is computationally tractable. The problem (1.9) is then solved by a version of gradient descent algorithm in a functional space. This leads to a family of classification methods called *boosting* (also, voting methods, ensemble methods, etc). Classifiers output by boosting are convex combinations of base classifiers and the whole method is often interpreted in machine learning literature as a way to combine simple base classifiers into more complex and powerful classifiers with a much better generalization performance.

Another popular approach is based on penalized empirical risk minimization in a reproducing kernel Hilbert space (RKHS) $\mathcal{H}_K$ generated by a symmetric nonnegatively definite kernel $K : S \times S \mapsto \mathbb{R}$. For instance, using the square of the norm as a penalty results in the following problem:

$$n^{-1} \sum_{j=1}^{n} \phi(Y_j g(X_j)) + \varepsilon \|g\|_{\mathcal{H}_K}^2 \longrightarrow \min, \ g \in \mathcal{H}_K, \tag{1.10}$$

where $\varepsilon > 0$ is a regularization parameter. In the case of hinge loss $\phi(u) = (1-u)\vee 0$ the method is called *support vector machine* (SVM). By the basic properties of RKHS, a function $g \in \mathcal{H}_K$ can be represented as $g(x) = \langle g, K(x, \cdot)\rangle_{\mathcal{H}_K}$. Because of this, it is very easy to conclude that the solution $\hat{g}$ of (1.10) must be in the linear span of functions $K(X_1, \cdot), \ldots, K(X_n, \cdot)$. Thus, the problem (1.10) is essentially a finite dimensional convex problem (in the case of hinge loss, it becomes a quadratic programming problem).

## 1.6   Sparse Recovery Problems

Let $\mathcal{H} = \{h_1, \ldots, h_N\}$ be a given set of functions from $S$ into $\mathbb{R}$ called *a dictionary*. Given $\lambda \in \mathbb{R}^N$, denote $f_\lambda = \sum_{j=1}^{N} \lambda_j h_j$. Suppose that a function $f_* \in \mathrm{l.s.}(\mathcal{H})$ is observed at random points $X_1, \ldots, X_n$ with common distribution $\Pi$,

$$Y_j = f_*(X_j), \ j = 1, \ldots, n$$

being the observations. The goal is to find a representation of $f_*$ in the dictionary, that is, to find $\lambda \in \mathbb{R}^N$ such that

$$f_\lambda(X_j) = Y_j, \; j = 1, \ldots, n. \tag{1.11}$$

In the case when the functions in the dictionary are not linearly independent, such a representation does not have to be unique. Moreover, if $N > n$, the system of linear equations (1.11) is underdetermined and the set

$$L := \{\lambda \in \mathbb{R}^N : f_\lambda(X_j) = Y_j, j = 1, \ldots, n\}$$

is a nontrivial affine subspace of $\mathbb{R}^N$. However, even in this case, the following problem still makes sense:

$$\|\lambda\|_{\ell_0} = \sum_{j=1}^{N} I(\lambda_j \neq 0) \longrightarrow \min, \lambda \in L. \tag{1.12}$$

In other words, the goal is to find *the sparsest solution* of the linear system (1.11). In general, the sparse recovery problem (1.12) is not computationally tractable since solving such a nonconvex optimization problem essentially requires searching through all $2^N$ coordinate subspaces of $\mathbb{R}^N$ and then solving the corresponding linear systems. However, the following problem

$$\|\lambda\|_{\ell_1} = \sum_{j=1}^{N} |\lambda_j| \longrightarrow \min, \lambda \in L. \tag{1.13}$$

is convex, and, moreover, it is a linear programming problem. It happens that for some dictionaries $\mathcal{H}$ and distributions $\Pi$ of design variables the solution of problem (1.13) is unique and coincides with the sparsest solution $\lambda^*$ of problem (1.12) (provided that $\|\lambda^*\|_{\ell_0}$ is sufficiently small). This fact is closely related to some problems in convex geometry concerning the neighborliness of convex polytopes.

More generally, one can study sparse recovery problems in the case when $f_*$ does not necessarily belong to the linear span of the dictionary $\mathcal{H}$ and it is measured at random locations $X_j$ with some errors. Given i.i.d. sample $(X_1, Y_1), \ldots, (X_n, Y_n)$ and a loss function $\ell$, this naturally leads to the study of the following penalized empirical risk minimization problem

$$\hat{\lambda}^\varepsilon := \mathrm{argmin}_{\lambda \in \mathbb{R}^N} \left[ P_n(\ell \bullet f_\lambda) + \varepsilon \|\lambda\|_{\ell_1} \right] \tag{1.14}$$

which is an empirical version of the problem

$$\lambda^\varepsilon := \mathrm{argmin}_{\lambda \in \mathbb{R}^N} \Big[ P(\ell \bullet f_\lambda) + \varepsilon \|\lambda\|_{\ell_1} \Big], \tag{1.15}$$

where $\varepsilon > 0$ is a regularization parameter. It is assumed that the loss function $\ell(y; u)$ is convex with respect to $u$ which makes the optimization problems (1.14) and (1.15) convex. This framework includes sparse recovery in both regression and large margin classification contexts. In the case of regression with quadratic loss $\ell(y, u) = (y - u)^2$, this penalization method has been called LASSO in statistical literature. The sparse recovery algorithm (1.13) can be viewed as a version of (1.14) with quadratic loss and with $\varepsilon = 0$.

Another popular method of sparse recovery, introduced recently by Candes and Tao [44] and called *the Dantzig selector*, is based on solving the following linear programming problem

$$\hat{\lambda}^\varepsilon \in \mathrm{Argmin}_{\lambda \in \hat{\Lambda}^\varepsilon} \|\lambda\|_{\ell_1},$$

where

$$\hat{\Lambda}^\varepsilon := \left\{ \lambda \in \mathbb{R}^N : \max_{1 \le k \le N} \left| n^{-1} \sum_{j=1}^n (f_\lambda(X_j) - Y_j) h_k(X_j) \right| \le \varepsilon/2 \right\}.$$

Note that the conditions defining the set $\hat{\Lambda}^\varepsilon$ are just necessary conditions of extremum in the LASSO-optimization problem

$$n^{-1} \sum_{j=1}^n (Y_j - f_\lambda(X_j))^2 + \varepsilon \|\lambda\|_{\ell_1} \longrightarrow \min, \ \lambda \in \mathbb{R}^N,$$

so, the Dantzig selector is closely related to the LASSO.

We will also study some other types of penalties that can be used in sparse recovery problems such as, for instance, the entropy penalty $\sum_{j=1}^N \lambda_j \log \lambda_j$ for sparse recovery problems in the convex hull of the dictionary $\mathscr{H}$.

Our goal will be to establish oracle inequalities showing that the methods of this type allow one to find a sparse approximation of the target function (when it exists).

## 1.7   Recovering Low Rank Matrices

Let $A \in \mathbb{M}_{m_1,m_2}(\mathbb{R})$[1] be an unknown $m_1 \times m_2$ matrix and $X_1, \dots, X_n \in \mathbb{M}_{m_1,m_2}(\mathbb{R})$ be given matrices. The goal is to recover $A$ based on its measurements

$$Y_j = \langle A, X_j \rangle = \mathrm{tr}(AX_j^*), \ j = 1, \dots, n. \tag{1.16}$$

---

[1] In this section, we are using the notations of linear algebra introduced in Sect. A.4.

In the case when $A$ is a large matrix, but its rank $\text{rank}(A)$ is relatively small, it is of interest to recover $A$ based on a relatively small number of linear measurements (1.16) with $n$ of the order $(m_1 \vee m_2)\text{rank}(A)$ (up to constants and logarithmic factors). This noncommutative generalization of sparse recovery problems has been intensively studied in the recent years, see [41,45,70,71,124] and references therein. As in the case of sparse recovery, the main methods of low rank matrix recovery are based on convex relaxation of a rank minimization problem

$$\text{rank}(S) \longrightarrow \min, \ S \in \mathscr{L}, \ \mathscr{L} := \Big\{ S : \langle S, X_j \rangle = Y_j, \ j = 1, \ldots, n \Big\}, \quad (1.17)$$

which is not computationally tractable. The most popular algorithm is based on nuclear norm minimization:

$$\|S\|_1 \longrightarrow \min, \ S \in \mathscr{L}, \ \mathscr{L} = \Big\{ S : \langle S, X_j \rangle = Y_j, \ j = 1, \ldots, n \Big\}. \quad (1.18)$$

Of course, similar problems can be also considered under further constraints on the set of matrices in question (for instance, when the matrices are Hermitian, nonnegatively definite, etc).

*Matrix completion*, in which $Y_j, j = 1, \ldots, n$ are noiseless observations of randomly picked entries of the target matrix $A$, is a typical example of matrix recovery problems that has been studied in a great detail. It can be viewed as a special case of *sampling from an orthonormal basis*. Let $E_i, i = 1, \ldots, m_1 m_2$ be an orthonormal basis of $\mathbb{M}_{m_1,m_2}(\mathbb{C})$ with respect to the Hilbert–Schmidt inner product and let $X_j, j = 1, \ldots, n$ be i.i.d. random variables sampled from a distribution $\Pi$ on the set $\{E_1, \ldots, E_{m^2}\}$. Most often, $\Pi$ is the uniform distribution that assigns probability $\frac{1}{m_1 m_2}$ to each basis matrix $E_i$. Note that

$$\mathbb{E}|\langle A, X \rangle|^2 = \frac{1}{m_1 m_2} \|A\|_2^2, \ A \in \mathbb{M}_{m_1,m_2}(\mathbb{R}).$$

In the case of matrix completion problems, $\{E_i : i = 1, \ldots, m_1 m_2\}$ is *the matrix completion basis*

$$\Big\{ e_i^{m_1} \otimes e_j^{m_2} : 1 \leq i \leq m_1, 1 \leq j \leq m_2 \Big\},$$

where $\{e_i^{m_1} : i = 1, \ldots, m_1\}$, $\{e_j^{m_2} : j = 1, \ldots, m_2\}$ are the canonical bases of $\mathbb{R}^{m_1}, \mathbb{R}^{m_2}$, respectively. Clearly, the Fourier coefficients $\langle A, e_i^{m_1} \otimes e_j^{m_2} \rangle$ coincide with the entries of matrix $A$. We will discuss only the case when the matrices $X_1, \ldots, X_n$ are i.i.d. with uniform distribution in the matrix completion basis, which corresponds to sampling the entries of the target matrix with replacement (although it is even more natural to study the sampling without replacement, and it is often done in the literature).

Another example of sampling from an orthonormal basis is related to *quantum state tomography*, an important problem in quantum statistics (see [70, 71]). The goal is to estimate *the density matrix* $\rho \in \mathbb{H}_m(\mathbb{C})$ of a quantum system, which is a Hermitian nonnegatively definite matrix of trace 1. The estimation is based on measurements of $n$ observables $X_1, \ldots, X_n \in \mathbb{H}_m(\mathbb{C})$ under the assumption that, for each measurement, the system is prepared in state $\rho$. In the noiseless case, $\rho$ has to be recovered based on the outcomes of the measurements

$$Y_j = \langle \rho, X_j \rangle = \mathrm{tr}(\rho X_j), \quad j = 1, \ldots, n \qquad (1.19)$$

and the following version of (1.18) can be used:

$$\|S\|_1 \longrightarrow \min, \quad S \in \mathscr{S}, \; \langle S, X_j \rangle = Y_j, \; j = 1, \ldots, n \qquad (1.20)$$

where

$$\mathscr{S} = \left\{ S \in \mathbb{H}_m(\mathbb{C}) : S \geq 0, \mathrm{tr}(S) = 1 \right\}$$

is the set of all density matrices. As an example of an interesting design $\{X_j\}$, let $m = 2^k$ and consider the *Pauli basis* in the space of $2 \times 2$ matrices $\mathbb{M}_2(\mathbb{C})$: $W_i := \frac{1}{\sqrt{2}} \sigma_i$, where

$$\sigma_1 := \begin{pmatrix} 0 & 1 \\ 1 & 0 \end{pmatrix}, \quad \sigma_2 := \begin{pmatrix} 0 & -i \\ i & 0 \end{pmatrix}, \quad \sigma_3 := \begin{pmatrix} 1 & 0 \\ 0 & -1 \end{pmatrix} \text{ and } \sigma_4 := \begin{pmatrix} 1 & 0 \\ 0 & 1 \end{pmatrix}$$

are the *Pauli matrices* (they are both Hermitian and unitary). The Pauli basis in the space $\mathbb{M}_m(\mathbb{C})$ can be now defined by tensorizing the Pauli basis in $\mathbb{M}_2(\mathbb{C})$ : it consists of all tensor products $W_{i_1} \otimes \cdots \otimes W_{i_k}$, $(i_1, \ldots, i_k) \in \{1, 2, 3, 4\}^k$. As in the case of matrix completion, $X_1, \ldots, X_n$ are i.i.d. random variables sampled from the uniform distribution in the Pauli basis and the state $\rho$ has to be recovered based on the outcomes of $n$ measurements (1.19). Such a measurement model for a $k$ qubit system is relatively standard in quantum information, in particular, in quantum state and quantum process tomography (see Nielsen and Chuang [120], Sect. 8.4.2).

One more example of a random design in matrix recovery problems is *subgaussian design* (which is similar to the design of dictionaries in sparse recovery and compressed sensing). Assume again that the matrix $A \in \mathbb{H}_m(\mathbb{C})$ to be recovered is Hermitian and let $X, X_1, \ldots, X_n$ be i.i.d. random matrices in $\mathbb{H}_m(\mathbb{C})$. Suppose that $\langle A, X \rangle$ is a subgaussian random variable for each $A \in \mathbb{H}_m(\mathbb{C})$ (see Sect. 3.1). One specific example is the *Gaussian design*, where $X$ is a symmetric random matrix with real entries such that $\{X_{ij} : 1 \leq i \leq j \leq m\}$ are independent centered normal random variables with $\mathbb{E}X_{ii}^2 = 1$, $i = 1, \ldots, m$ and $\mathbb{E}X_{ij}^2 = \frac{1}{2}$, $i < j$. Another example is the *Rademacher design*, where $X_{ii} = \varepsilon_{ii}$, $i = 1, \ldots, m$ and $X_{ij} = \frac{1}{\sqrt{2}} \varepsilon_{ij}$, $i < j$, $\{\varepsilon_{ij} : 1 \leq i \leq j \leq m\}$ being i.i.d. Rademacher random variables (that is, random variables taking values $+1$ or $-1$ with probability $1/2$ each). In both cases, $\mathbb{E}|\langle A, X \rangle|^2 = \|A\|_2^2$, $A \in \mathbb{M}_m(\mathbb{C})$, which means that $X$

is an *isotropic* random matrix, and $\langle A, X \rangle$ is a subgaussian random variable with subgaussian parameter $\|A\|_2$ (up to a constant).

In the case of *matrix regression* model

$$Y_j = \langle A, X_j \rangle + \xi_j, \ j = 1, \ldots, n, \tag{1.21}$$

where $A \in \mathbb{M}_{m_1, m_2}(\mathbb{R})$ is an unknown target matrix and $\xi_j, \ j = 1, \ldots, n$ are i.i.d. mean zero random variables (random noise), one can replace the nuclear norm minimization algorithm (1.18) by the following version of penalized empirical risk minimization:

$$\hat{A}^\varepsilon := \mathrm{argmin}_{S \in \mathbb{M}_{m_1, m_2}(\mathbb{R})} \left[ n^{-1} \sum_{j=1}^{n} \left( Y_j - \langle S, X_j \rangle \right)^2 + \varepsilon \|S\|_1 \right],$$

where $\varepsilon > 0$ is a regularization parameter. Such problems have been studied in [40, 90, 127] and they will be discussed in Chap. 9 (for some other penalization methods, for instance, von Neumann entropy penalization in density matrix estimation problem, see also [88]). The main goal will be to establish oracle inequalities for the error of matrix estimators that show how it depends on the rank of the target matrix $A$, or, more generally, on the rank of oracles approximating $A$.

# Chapter 2
# Empirical and Rademacher Processes

The empirical process is defined as

$$Z_n := n^{1/2}(P_n - P)$$

and it can be viewed as a random measure. However, more often, it has been viewed as a stochastic process indexed by a function class $\mathscr{F}$ :

$$Z_n(f) = n^{1/2}(P_n - P)(f), f \in \mathscr{F}$$

(see Dudley [59] or van der Vaart and Wellner [148]).

The Rademacher process indexed by a class $\mathscr{F}$ was defined in Sect. 1.3 as

$$R_n(f) := n^{-1} \sum_{i=1}^{n} \varepsilon_i f(X_i), \ f \in \mathscr{F},$$

$\{\varepsilon_i\}$ being i.i.d. Rademacher random variables (that is, $\varepsilon_i$ takes the values $+1$ and $-1$ with probability $1/2$ each) independent of $\{X_i\}$.

It should be mentioned that certain measurability assumptions are required in the study of empirical and Rademacher processes. In particular, under these assumptions, such quantities as $\|P_n - P\|_{\mathscr{F}}$ are properly measurable random variables. We refer to the books of Dudley [59], Chap. 5 and van der Vaart and Wellner [148], Sect. 1.7 for precise formulations of these measurability assumptions. Some of the bounds derived and used below hold even without the assumptions of this nature, if the expectation is replaced by the outer expectation, as it is often done, for instance, in [148]. Another option is to "define"

$$\mathbb{E}\|P_n - P\|_{\mathscr{F}} := \sup\left\{\mathbb{E}\|P_n - P\|_{\mathscr{G}} : \mathscr{G} \subset \mathscr{F}, \mathscr{G} \text{ is finite}\right\},$$

V. Koltchinskii, *Oracle Inequalities in Empirical Risk Minimization and Sparse Recovery Problems*, Lecture Notes in Mathematics 2033, DOI 10.1007/978-3-642-22147-7_2, © Springer-Verlag Berlin Heidelberg 2011

which provides a simple way to get around the measurability difficulties. Such an approach has been frequently used by Talagrand (see, e.g., [140]). In what follows, it will be assumed that the measurability problems have been resolved in one of these ways.

## 2.1  Symmetrization Inequalities

The following important inequality reveals close relationships between empirical and Rademacher processes.

**Theorem 2.1.** *For any class $\mathscr{F}$ of $P$-integrable functions and for any convex function $\Phi : \mathbb{R}_+ \mapsto \mathbb{R}_+$*

$$\mathbb{E}\Phi\left(\frac{1}{2}\|R_n\|_{\mathscr{F}_c}\right) \le \mathbb{E}\Phi\left(\|P_n - P\|_{\mathscr{F}}\right) \le \mathbb{E}\Phi\left(2\|R_n\|_{\mathscr{F}}\right),$$

*where $\mathscr{F}_c := \{f - Pf : f \in \mathscr{F}\}$. In particular,*

$$\frac{1}{2}\mathbb{E}\|R_n\|_{\mathscr{F}_c} \le \mathbb{E}\|P_n - P\|_{\mathscr{F}} \le 2\mathbb{E}\|R_n\|_{\mathscr{F}}.$$

*Proof.* Assume that the random variables $X_1, \dots X_n$ are defined on a probability space $(\bar{\Omega}, \bar{\Sigma}, \bar{\mathbb{P}})$. We will also need two other probability spaces: $(\tilde{\Omega}, \tilde{\Sigma}, \tilde{\mathbb{P}})$ and $(\Omega_\varepsilon, \Sigma_\varepsilon, \mathbb{P}_\varepsilon)$. The main probability space on which all the random variables are defined will be denoted $(\Omega, \Sigma, \mathbb{P})$ and it will be the product space

$$(\Omega, \Sigma, \mathbb{P}) = (\bar{\Omega}, \bar{\Sigma}, \bar{\mathbb{P}}) \times (\tilde{\Omega}, \tilde{\Sigma}, \tilde{\mathbb{P}}) \times (\Omega_\varepsilon, \Sigma_\varepsilon, \mathbb{P}_\varepsilon).$$

The corresponding expectations will be denoted by $\bar{\mathbb{E}}, \tilde{\mathbb{E}}, \mathbb{E}_\varepsilon$ and $\mathbb{E}$. Let $(\tilde{X}_1, \dots, \tilde{X}_n)$ be an independent copy of $(X_1, \dots, X_n)$. Think of random variables $\tilde{X}_1, \dots, \tilde{X}_n$ as being defined on $(\tilde{\Omega}, \tilde{\Sigma}, \tilde{\mathbb{P}})$. Denote $\tilde{P}_n$ the empirical measure based on $(\tilde{X}_1, \dots, \tilde{X}_n)$ (it is an independent copy of $P_n$). Then $\tilde{\mathbb{E}}\tilde{P}_n f = Pf$ and, using Jensen's inequality,

$$\mathbb{E}\Phi\left(\|P_n - P\|_{\mathscr{F}}\right) = \bar{\mathbb{E}}\Phi\left(\|P_n - \tilde{\mathbb{E}}\tilde{P}_n\|_{\mathscr{F}}\right) = \bar{\mathbb{E}}\Phi\left(\|\tilde{\mathbb{E}}(P_n - \tilde{P}_n)\|_{\mathscr{F}}\right)$$

$$\le \bar{\mathbb{E}}\tilde{\mathbb{E}}\Phi\left(\|P_n - \tilde{P}_n\|_{\mathscr{F}}\right) = \bar{\mathbb{E}}\tilde{\mathbb{E}}\Phi\left(\left\|n^{-1}\sum_{j=1}^{n}(\delta_{X_j} - \delta_{\tilde{X}_j})\right\|_{\mathscr{F}}\right).$$

Since $X_1, \dots, X_n, \tilde{X}_1, \dots, \tilde{X}_n$ are i.i.d., the distribution of $(X_1, \dots, X_n, \tilde{X}_1, \dots, \tilde{X}_n)$ is invariant with respect to all permutations of the components. In particular, one can switch any couple $X_j, \tilde{X}_j$. Because of this,

$$\bar{\mathbb{E}}\tilde{\mathbb{E}}\varPhi\left(\left\|n^{-1}\sum_{j=1}^{n}(\delta_{X_j}-\delta_{\tilde{X}_j})\right\|_{\mathscr{F}}\right) = \bar{\mathbb{E}}\tilde{\mathbb{E}}\varPhi\left(\left\|n^{-1}\sum_{j=1}^{n}\sigma_j(\delta_{X_j}-\delta_{\tilde{X}_j})\right\|_{\mathscr{F}}\right),$$

for an arbitrary choice of $\sigma_j = +1$ or $\sigma_j = -1$. Define now i.i.d. Rademacher random variables on $(\Omega_\varepsilon, \Sigma_\varepsilon, \mathbb{P}_\varepsilon)$ (thus, independent of $(X_1, \ldots, X_n, \tilde{X}_1, \ldots, \tilde{X}_n)$). Then, we have

$$\bar{\mathbb{E}}\tilde{\mathbb{E}}\varPhi\left(\left\|n^{-1}\sum_{j=1}^{n}(\delta_{X_j}-\delta_{\tilde{X}_j})\right\|_{\mathscr{F}}\right) = \mathbb{E}_\varepsilon\bar{\mathbb{E}}\tilde{\mathbb{E}}\varPhi\left(\left\|n^{-1}\sum_{j=1}^{n}\varepsilon_j(\delta_{X_j}-\delta_{\tilde{X}_j})\right\|_{\mathscr{F}}\right)$$

and the proof can be completed as follows:

$$\mathbb{E}\varPhi\left(\|P_n - P\|_{\mathscr{F}}\right) \le \mathbb{E}_\varepsilon\bar{\mathbb{E}}\tilde{\mathbb{E}}\varPhi\left(\left\|n^{-1}\sum_{j=1}^{n}\varepsilon_j(\delta_{X_j}-\delta_{\tilde{X}_j})\right\|_{\mathscr{F}}\right)$$

$$\le \frac{1}{2}\mathbb{E}_\varepsilon\bar{\mathbb{E}}\varPhi\left(2\left\|n^{-1}\sum_{j=1}^{n}\varepsilon_j\delta_{X_j}\right\|_{\mathscr{F}}\right) + \frac{1}{2}\mathbb{E}_\varepsilon\tilde{\mathbb{E}}\varPhi\left(2\left\|n^{-1}\sum_{j=1}^{n}\varepsilon_j\delta_{\tilde{X}_j}\right\|_{\mathscr{F}}\right)$$

$$= \mathbb{E}\varPhi\left(2\|R_n\|_{\mathscr{F}}\right).$$

The proof of the lower bound is similar.                                         □

The upper bound is called the *symmetrization inequality* and the lower bound is sometimes called the *desymmetrization inequality*. The desymmetrization inequality is often used together with the following elementary lower bound (in the case of $\varPhi(u) = u$)

$$\mathbb{E}\|R_n\|_{\mathscr{F}_c} \ge \mathbb{E}\|R_n\|_{\mathscr{F}} - \sup_{f\in\mathscr{F}}|Pf|\,\mathbb{E}|R_n(1)| \ge$$

$$\ge \mathbb{E}\|R_n\|_{\mathscr{F}} - \sup_{f\in\mathscr{F}}|Pf|\,\mathbb{E}^{1/2}|n^{-1}\sum_{j=1}^{n}\varepsilon_j|^2 = \mathbb{E}\|R_n\|_{\mathscr{F}} - \frac{\sup_{f\in\mathscr{F}}|Pf|}{\sqrt{n}}.$$

## 2.2  Comparison Inequalities for Rademacher Sums

Given a set $T \subset \mathbb{R}^n$ and i.i.d. Rademacher variables $\varepsilon_i, i = 1, 2, \ldots$, it is of interest to know how the expected value of the sup-norm of Rademacher sums indexed by $T$

$$R_n(T) := \mathbb{E}\sup_{t\in T}\left|\sum_{i=1}^{n}t_i\varepsilon_i\right|$$

depends on the geometry of the set $T$. The following beautiful *comparison inequality* for Rademacher sums is due to Talagrand and it is often used to control $R_n(T)$ for more complex sets $T$ in terms of similar quantities for simpler sets.

**Theorem 2.2.** *Let $T \subset \mathbb{R}^n$ and let $\varphi_i : \mathbb{R} \mapsto \mathbb{R}$, $i = 1, \ldots, n$ be functions such that $\varphi_i(0) = 0$ and*

$$|\varphi_i(u) - \varphi_i(v)| \leq |u - v|, \ u, v \in \mathbb{R}$$

*(that is, $\varphi_i$ are contractions). For all convex nondecreasing functions $\Phi : \mathbb{R}_+ \mapsto \mathbb{R}_+$,*

$$\mathbb{E}\Phi\left(\frac{1}{2}\sup_{t \in T}\left|\sum_{i=1}^n \varphi_i(t_i)\varepsilon_i\right|\right) \leq \mathbb{E}\Phi\left(\sup_{t \in T}\left|\sum_{i=1}^n t_i\varepsilon_i\right|\right).$$

*Proof.* First, we prove that for a nondecreasing convex function $\Phi : \mathbb{R} \mapsto \mathbb{R}_+$ and for an arbitrary $A : T \mapsto \mathbb{R}$

$$\mathbb{E}\Phi\left(\sup_{t \in T}\left[A(t) + \sum_{i=1}^n \varphi_i(t_i)\varepsilon_i\right]\right) \leq \mathbb{E}\Phi\left(\sup_{t \in T}\left[A(t) + \sum_{i=1}^n t_i\varepsilon_i\right]\right). \qquad (2.1)$$

We start with the case $n = 1$. Then, the bound is equivalent to the following

$$\mathbb{E}\Phi\left(\sup_{t \in T}[t_1 + \varepsilon\varphi(t_2)]\right) \leq \mathbb{E}\Phi\left(\sup_{t \in T}[t_1 + \varepsilon t_2]\right)$$

for an arbitrary set $T \subset \mathbb{R}^2$ and an arbitrary contraction $\varphi$. One can rewrite it as

$$\frac{1}{2}\left(\Phi\left(\sup_{t \in T}[t_1 + \varphi(t_2)]\right) + \Phi\left(\sup_{t \in T}[t_1 - \varphi(t_2)]\right)\right)$$
$$\leq \frac{1}{2}\left(\Phi\left(\sup_{t \in T}[t_1 + t_2]\right) + \Phi\left(\sup_{t \in T}[t_1 - t_2]\right)\right).$$

If now $(t_1, t_2) \in T$ denotes a point where $\sup_{t \in T}[t_1 + \varphi(t_2)]$ is attained and $(s_1, s_2) \in T$ is a point where $\sup_{t \in T}[t_1 - \varphi(t_2)]$ is attained, then it is enough to show that

$$\Phi\left(t_1 + \varphi(t_2)\right) + \Phi\left(s_1 - \varphi(s_2)\right) \leq \Phi\left(\sup_{t \in T}[t_1 + t_2]\right) + \Phi\left(\sup_{t \in T}[t_1 - t_2]\right)$$

(if the suprema are not attained, one can easily modify the argument). Clearly, we have the following conditions:

$$t_1 + \varphi(t_2) \geq s_1 + \varphi(s_2) \text{ and } t_1 - \varphi(t_2) \leq s_1 - \varphi(s_2).$$

First consider the case when $t_2 \geq 0, s_2 \geq 0$ and $t_2 \geq s_2$. In this case, we will prove that

$$\Phi\left(t_1 + \varphi(t_2)\right) + \Phi\left(s_1 - \varphi(s_2)\right) \leq \Phi\left(t_1 + t_2\right) + \Phi\left(s_1 - s_2\right), \qquad (2.2)$$

which would imply the bound. Indeed, for

$$a := t_1 + \varphi(t_2), b := t_1 + t_2, c := s_1 - s_2, d := s_1 - \varphi(s_2),$$

we have $a \le b$ and $c \le d$ since $\varphi(t_2) \le t_2$, $\varphi(s_2) \le s_2$ (by the assumption that $\varphi$ is a contraction and $\varphi(0) = 0$). We also have that

$$b - a = t_2 - \varphi(t_2) \ge s_2 - \varphi(s_2) = d - c,$$

because again $\varphi$ is a contraction and $t_2 \ge s_2$. Finally, we have

$$a = t_1 + \varphi(t_2) \ge s_1 + \varphi(s_2) \ge s_1 - s_2 = c.$$

Since the function $\Phi$ is nondecreasing and convex, its increment over the interval $[a, b]$ is larger than its increment over the interval $[c, d]$ ($[a, b]$ is longer than $[c, d]$ and $a \ge c$), which is equivalent to (2.2).

If $t_2 \ge 0, s_2 \ge 0$ and $s_2 \ge t_2$, it is enough to use the change of notations $(t, s) \mapsto (s, t)$ and to replace $\varphi$ with $-\varphi$.

The case $t_2 \le 0, s_2 \le 0$ can be now handled by using the transformation $(t_1, t_2) \mapsto (t_1, -t_2)$ and changing the function $\varphi$ accordingly.

We have to consider the case $t_2 \ge 0, s_2 \le 0$ (the only remaining case $t_2 \le 0$, $s_2 \ge 0$ would again follow by switching the names of $t$ and $s$ and replacing $\varphi$ with $-\varphi$). In this case, we have $\varphi(t_2) \le t_2$, $-\varphi(s_2) \le -s_2$, which, in view of monotonicity of $\Phi$, immediately implies

$$\Phi\Big(t_1 + \varphi(t_2)\Big) + \Phi\Big(s_1 - \varphi(s_2)\Big) \le \Phi\Big(t_1 + t_2\Big) + \Phi\Big(s_1 - s_2\Big).$$

This completes the proof of (2.1) in the case $n = 1$.

In the general case, we have

$$\mathbb{E}\Phi\left(\sup_{t \in T}\Big[A(t) + \sum_{i=1}^{n} \varphi_i(t_i)\varepsilon_i\Big]\right)$$

$$= \mathbb{E}_{\varepsilon_1,\dots,\varepsilon_{n-1}} \mathbb{E}_{\varepsilon_n} \Phi\left(\sup_{t \in T}\Big[A(t) + \sum_{i=1}^{n-1} \varphi_i(t_i)\varepsilon_i + \varepsilon_n\varphi(t_n)\Big]\right).$$

The expectation $\mathbb{E}_{\varepsilon_n}$ (conditional on $\varepsilon_1, \dots, \varepsilon_{n-1}$) can be bounded using the result in the case $n = 1$. This yields (after changing the order of integration)

$$\mathbb{E}\Phi\left(\sup_{t \in T}\Big[A(t) + \sum_{i=1}^{n} \varphi_i(t_i)\varepsilon_i\Big]\right) \le \mathbb{E}_{\varepsilon_n} \mathbb{E}_{\varepsilon_1,\dots,\varepsilon_{n-1}} \Phi\left(\sup_{t \in T}\Big[A(t) + \varepsilon_n t_n + \sum_{i=1}^{n-1} \varphi_i(t_i)\varepsilon_i\Big]\right).$$

The proof of (2.1) can now be completed by an induction argument.

Finally, to prove the inequality of the theorem, it is enough to write

$$
\mathbb{E}\Phi\left(\frac{1}{2}\sup_{t\in T}\left|\sum_{i=1}^{n}\varphi_i(t_i)\varepsilon_i\right|\right)
$$

$$
= \mathbb{E}\Phi\left(\frac{1}{2}\left[\left(\sup_{t\in T}\sum_{i=1}^{n}\varphi_i(t_i)\varepsilon_i\right)_+ + \left(\sup_{t\in T}\sum_{i=1}^{n}\varphi_i(t_i)(-\varepsilon_i)\right)_+\right]\right)
$$

$$
\le \frac{1}{2}\left[\mathbb{E}\Phi\left(\left(\sup_{t\in T}\sum_{i=1}^{n}\varphi_i(t_i)\varepsilon_i\right)_+\right) + \mathbb{E}\Phi\left(\left(\sup_{t\in T}\sum_{i=1}^{n}\varphi_i(t_i)(-\varepsilon_i)\right)_+\right)\right],
$$

where $a_+ := a \vee 0$. Applying the inequality (2.1) to the function $u \mapsto \Phi(u_+)$, which is convex and nondecreasing, completes the proof.                               □

We will frequently use a corollary of the above comparison inequality that provides upper bounds on the moments of the sup-norm of Rademacher process $R_n$ on the class

$$
\varphi \circ \mathscr{F} := \{\varphi \circ f : f \in \mathscr{F}\}
$$

in terms of the corresponding moments of the sup-norm of $R_n$ on $\mathscr{F}$ and Lipschitz constant of function $\varphi$.

**Theorem 2.3.** *Let* $\varphi : \mathbb{R} \mapsto \mathbb{R}$ *be a contraction satisfying the condition* $\varphi(0) = 0$. *For all convex nondecreasing functions* $\Phi : \mathbb{R}_+ \mapsto \mathbb{R}_+$,

$$
\mathbb{E}\Phi\left(\frac{1}{2}\|R_n\|_{\varphi\circ\mathscr{F}}\right) \le \mathbb{E}\Phi\left(\|R_n\|_{\mathscr{F}}\right).
$$

*In particular,*

$$
\mathbb{E}\|R_n\|_{\varphi\circ\mathscr{F}} \le 2\mathbb{E}\|R_n\|_{\mathscr{F}}.
$$

The inequality of Theorem 2.3 will be called *the contraction inequality* for Rademacher processes.

A simple rescaling of the class $\mathscr{F}$ allows one to use the contraction inequality in the case of an arbitrary function $\varphi$ satisfying the Lipschitz condition

$$
|\varphi(u) - \varphi(v)| \le L|u - v|
$$

on an arbitrary interval $(a, b)$ that contains the ranges of all the functions in $\mathscr{F}$. In this case, the last bound of Theorem 2.3 takes the form

$$
\mathbb{E}\|R_n\|_{\varphi\circ\mathscr{F}} \le 2L\mathbb{E}\|R_n\|_{\mathscr{F}}.
$$

This implies, for instance, that

$$\mathbb{E} \sup_{f \in \mathscr{F}} \left| n^{-1} \sum_{i=1}^{n} \varepsilon_i f^2(X_i) \right| \leq 4 U \mathbb{E} \sup_{f \in \mathscr{F}} \left| n^{-1} \sum_{i=1}^{n} \varepsilon_i f(X_i) \right| \qquad (2.3)$$

provided that the functions in the class $\mathscr{F}$ are uniformly bounded by a constant $U$.

## 2.3  Concentration Inequalities

A well known, simple and useful concentration inequality for functions

$$Z = g(X_1, \ldots, X_n)$$

of independent random variables with values in arbitrary spaces is valid under so called *bounded difference condition* on $g$ : there exist constants $c_j, j = 1, \ldots, n$ such that for all $j = 1, \ldots, n$ and all $x_1, x_2, \ldots, x_j, x'_j, \ldots, x_n$

$$\left| g(x_1, \ldots, x_{j-1}, x_j, x_{j+1}, \ldots, x_n) - g(x_1, \ldots, x_{j-1}, x'_j, x_{j+1}, \ldots, x_n) \right| \leq c_j. \qquad (2.4)$$

**Theorem 2.4 (Bounded difference inequality).** *Under the condition (2.4),*

$$\mathbb{P}\{Z - \mathbb{E}Z \geq t\} \leq \exp\left\{ -\frac{2t^2}{\sum_{j=1}^{n} c_j^2} \right\}$$

*and*

$$\mathbb{P}\{Z - \mathbb{E}Z \leq -t\} \leq \exp\left\{ -\frac{2t^2}{\sum_{j=1}^{n} c_j^2} \right\}.$$

A standard proof of this inequality is based on bounding the exponential moment $\mathbb{E}e^{\lambda(Z-\mathbb{E}Z)}$, using the following martingale difference representation

$$Z - \mathbb{E}Z = \sum_{j=1}^{n} \left[ \mathbb{E}(Z|X_1, \ldots, X_j) - \mathbb{E}(Z|X_1, \ldots, X_{j-1}) \right],$$

then using Markov inequality and optimizing the resulting bound with respect to $\lambda > 0$.

In the case when $Z = X_1 + \cdots + X_n$, the bounded difference inequality coincides with Hoeffding inequality for sums of bounded independent random variables (see Sect. A.2).

For a class $\mathscr{F}$ of functions uniformly bounded by a constant $U$, the bounded difference inequality immediately implies the following bounds for $\|P_n - P\|_{\mathscr{F}}$, providing a uniform version of Hoeffding inequality.

**Theorem 2.5.** *For all $t > 0$,*

$$\mathbb{P}\left\{\|P_n - P\|_{\mathscr{F}} \geq \mathbb{E}\|P_n - P\|_{\mathscr{F}} + \frac{tU}{\sqrt{n}}\right\} \leq \exp\{-t^2/2\}$$

*and*

$$\mathbb{P}\left\{\|P_n - P\|_{\mathscr{F}} \leq \mathbb{E}\|P_n - P\|_{\mathscr{F}} - \frac{tU}{\sqrt{n}}\right\} \leq \exp\{-t^2/2\}.$$

Developing uniform versions of Bernstein's inequality (see Sect. A.2) happened to be a much harder problem that was solved in the famous papers by Talagrand [138, 139] on concentration inequalities for product measures and empirical processes.

**Theorem 2.6 (Talagrand's inequality).** *Let $X_1, \ldots, X_n$ be independent random variables in $S$. For any class of functions $\mathscr{F}$ on $S$ that is uniformly bounded by a constant $U > 0$ and for all $t > 0$*

$$\mathbb{P}\left\{\left|\left\|\sum_{i=1}^{n} f(X_i)\right\|_{\mathscr{F}} - \mathbb{E}\left\|\sum_{i=1}^{n} f(X_i)\right\|_{\mathscr{F}}\right| \geq t\right\} \leq K \exp\left\{-\frac{1}{K}\frac{t}{U}\log\left(1 + \frac{tU}{V}\right)\right\},$$

*where $K$ is a universal constant and $V$ is any number satisfying*

$$V \geq \mathbb{E}\sup_{f \in \mathscr{F}}\sum_{i=1}^{n} f^2(X_i).$$

Using symmetrization inequality and contraction inequality for the square (2.3), it is easy to show that in the case of i.i.d. random variables $X_1, \ldots, X_n$ with distribution $P$

$$\mathbb{E}\sup_{f \in \mathscr{F}}\sum_{i=1}^{n} f^2(X_i) \leq n \sup_{f \in \mathscr{F}} Pf^2 + 8U\mathbb{E}\left\|\sum_{i=1}^{n}\varepsilon_i f(X_i)\right\|_{\mathscr{F}}. \qquad (2.5)$$

The right hand side of this bound is a common choice of the quantity $V$ involved in Talagrand's inequality. Moreover, in the case when $\mathbb{E}f(X) = 0$, the desymmetrization inequality yields

$$\mathbb{E}\left\|\sum_{i=1}^{n}\varepsilon_i f(X_i)\right\|_{\mathscr{F}} \leq 2\mathbb{E}\left\|\sum_{i=1}^{n} f(X_i)\right\|_{\mathscr{F}}.$$

As a result, one can use Talagrand's inequality with

$$V = n \sup_{f \in \mathscr{F}} Pf^2 + 16U\mathbb{E}\left\|\sum_{i=1}^{n} f(X_i)\right\|$$

and the size of $\left\|\sum_{i=1}^{n} f(X_i)\right\|_{\mathscr{F}}$ is now controlled it terms of its expectation only.

This form of Talagrand's inequality is especially convenient and there have been considerable efforts to find explicit and sharp values of the constants in such inequalities. In particular, we will frequently use the bounds proved by Bousquet [33] and Klein [77] (in fact, Klein and Rio [78] provide an improved version of this inequality). Namely, for a class $\mathscr{F}$ of measurable functions from $S$ into $[0, 1]$ (by a simple rescaling $[0, 1]$ can be replaced by any bounded interval) the following bounds hold for all $t > 0$ :

**Bousquet bound**

$$\mathbb{P}\left\{\|P_n - P\|_{\mathscr{F}} \geq \mathbb{E}\|P_n - P\|_{\mathscr{F}} + \sqrt{2\frac{t}{n}\left(\sigma_P^2(\mathscr{F}) + 2\mathbb{E}\|P_n - P\|_{\mathscr{F}}\right)} + \frac{t}{3n}\right\} \leq e^{-t}$$

**Klein-Rio bound**

$$\mathbb{P}\left\{\|P_n - P\|_{\mathscr{F}} \leq \mathbb{E}\|P_n - P\|_{\mathscr{F}} - \sqrt{2\frac{t}{n}\left(\sigma_P^2(\mathscr{F}) + 2\mathbb{E}\|P_n - P\|_{\mathscr{F}}\right)} - \frac{t}{n}\right\} \leq e^{-t}.$$

Here

$$\sigma_P^2(\mathscr{F}) := \sup_{f \in \mathscr{F}} \left(Pf^2 - (Pf)^2\right).$$

We will also need a version of Talagrand's inequality for unbounded classes of functions. Given a class $\mathscr{F}$ of measurable functions $f : S \mapsto \mathbb{R}$, denote by $F$ an envelope of $\mathscr{F}$, that is, a measurable function such that $|f(x)| \leq F(x), x \in S$, $f \in \mathscr{F}$. The next bounds follow from Theorem 4 of Adamczak [1]: for all $\alpha \in (0, 1]$ there exists a constant $K = K(\alpha)$ such that

**Adamczak bound**

$$\mathbb{P}\left\{\|P_n - P\|_{\mathscr{F}} \geq K\left[\mathbb{E}\|P_n - P\|_{\mathscr{F}} + \sigma_P(\mathscr{F})\sqrt{\frac{t}{n}} + \left\|\max_{1 \leq j \leq n} F(X_j)\right\|_{\psi_\alpha} \frac{t^{1/\alpha}}{n}\right]\right\} \leq e^{-t}$$

and

$$\mathbb{P}\left\{\mathbb{E}\|P_n - P\|_{\mathscr{F}} \geq K\left[\|P_n - P\|_{\mathscr{F}} + \sigma_P(\mathscr{F})\sqrt{\frac{t}{n}} + \left\|\max_{1 \leq j \leq n} F(X_j)\right\|_{\psi_\alpha} \frac{t^{1/\alpha}}{n}\right]\right\} \leq e^{-t}.$$

Concentration inequalities can be also applied to the Rademacher process which can be viewed as an empirical process based on the sample $(X_1, \varepsilon_1), \ldots, (X_n, \varepsilon_n)$ in the space $S \times \{-1, 1\}$ and indexed by the class of functions $\tilde{\mathscr{F}} := \{\tilde{f} : f \in \mathscr{F}\}$, where $\tilde{f}(x, u) := f(x)u, (x, u) \in S \times \{-1, 1\}$.

## 2.4  Exponential Bounds for Sums of Independent Random Matrices

In this section, we discuss very simple, but powerful noncommutative Bernstein type inequalities that go back to Ahlswede and Winter [4]. The goal is to bound the tail probability $\mathbb{P}\{\|X_1 + \cdots + X_n\| \geq t\}$, where $X_1, \ldots, X_n$ are independent Hermitian random $m \times m$ matrices with $\mathbb{E}X_j = 0$ and $\| \cdot \|$ is the operator norm.[1] The proofs of such inequalities are based on a matrix extension of the classical proof of Bernstein's inequality for real valued random variables, but they also rely on important matrix inequalities that have many applications in mathematical physics. In the case of sums of i.i.d. random matrices, it is enough to use the following well known *Golden-Thompson inequality* (see, e.g., Simon [133], p. 94):

**Proposition 2.1.** *For arbitrary Hermitian $m \times m$ matrices $A, B$*

$$\operatorname{tr}(e^{A+B}) \leq \operatorname{tr}(e^A e^B).$$

It is needed to control the matrix moment generating function

$$\mathbb{E}\operatorname{tr}\exp\{\lambda(X_1 + \cdots + X_n)\}.$$

This approach was used in the original paper by Ahlswede and Winter [4], but also in [70, 88, 124]. However, it does not seem to provide the correct form of "variance parameter" in the non i.i.d. case. We will use below another approach suggested by Tropp [142] that is based on the following classical result by Lieb [102] (Theorem 6).

**Proposition 2.2.** *For all Hermitian matrices $A$, the function*

$$G_A(S) := \operatorname{tr}\exp\{A + \log S\}$$

*is concave on the cone of Hermitian positively definite matrices.*

Given independent Hermitian random $m \times m$ matrices $X_1, \ldots, X_n$ with $\mathbb{E}X_j = 0$, denote

$$\sigma^2 := n^{-1} \left\| \mathbb{E}(X_1^2 + \cdots + X_n^2) \right\|.$$

**Theorem 2.7.** *1. Suppose that, for some $U > 0$ and for all $j = 1, \ldots, n$, $\|X_j\| \leq U$. Then*

$$\mathbb{P}\left\{ \|X_1 + \cdots + X_n\| \geq t \right\} \leq 2m \exp\left\{ -\frac{t^2}{2\sigma^2 n + 2Ut/3} \right\}. \qquad (2.6)$$

*2. Let $\alpha \geq 1$ and suppose that, for some $U^{(\alpha)} > 0$ and for all $j = 1, \ldots, n$,*

---

[1] For the notations used in this section, see Sect. A.4.

$$\left\| \|X_j\| \right\|_{\psi_\alpha} \vee 2\mathbb{E}^{1/2}\|X_j\|^2 \le U^{(\alpha)}.$$

*Then, there exists a constant $K > 0$ such that*

$$\mathbb{P}\{\|X_1 + \cdots + X_n\| \ge t\} \le 2m \exp\left\{-\frac{1}{K}\frac{t^2}{n\sigma^2 + tU^{(\alpha)}\log^{1/\alpha}(U^{(\alpha)}/\sigma)}\right\}. \quad (2.7)$$

Inequality (2.6) is a direct noncommutative extension of classical Bernstein's inequality for sums of independent random variables. It is due to Ahlswede and Winter [4] (see also [70, 124, 142]). In inequality (2.7), the $L_\infty$-bound $U$ on $\|X_j\|$ is replaced by a weaker $\psi_\alpha$-norm. This inequality was proved in [88] in the i.i.d. case and in [89] in the general case. We follow the last paper below. Note that, when $\alpha \to \infty$, (2.7) coincides with (2.6) (up to constants).

*Proof.* Denote $Y_n := X_1 + \cdots + X_n$ and observe that $\|Y_n\| < t$ if and only if $-tI_m < Y_n < tI_m$. It follows that

$$\mathbb{P}\{\|Y_n\| \ge t\} \le \mathbb{P}\{Y_n \not< tI_m\} + \mathbb{P}\{Y_n \not> -tI_m\}. \quad (2.8)$$

The next bounds are based on a simple matrix algebra:

$$\mathbb{P}\{Y_n \not< tI_m\} = \mathbb{P}\{e^{\lambda Y_n} \not< e^{\lambda t I_m}\} \le \mathbb{P}\left\{\mathrm{tr}\left(e^{\lambda Y_n}\right) \ge e^{\lambda t}\right\} \le e^{-\lambda t}\mathbb{E}\mathrm{tr}(e^{\lambda Y_n}). \quad (2.9)$$

To bound the matrix moment generating function $\mathbb{E}\mathrm{tr}(e^{\lambda Y_n})$, observe that

$$\mathbb{E}\mathrm{tr}(e^{\lambda Y_n}) = \mathbb{E}\mathbb{E}_n \mathrm{tr}\exp\{\lambda Y_{n-1} + \log e^{\lambda X_n}\} = \mathbb{E}\mathbb{E}_n G_{\lambda Y_{n-1}}(e^{\lambda X_n}).$$

where $\mathbb{E}_n$ denotes the conditional expectation given $X_1, \ldots, X_{n-1}$. Using Lieb's theorem (see Proposition 2.2), Jensen's inequality for the expectation $\mathbb{E}_n$ and the independence of random matrices $X_j$, we get

$$\mathbb{E}\mathrm{tr}(e^{\lambda Y_n}) \le \mathbb{E}G_{\lambda Y_{n-1}}(\mathbb{E}e^{\lambda X_n}) = \mathbb{E}\mathrm{tr}\exp\{\lambda Y_{n-1} + \log \mathbb{E}e^{\lambda X_n}\}.$$

Using the same conditioning trick another time, we get

$$\mathbb{E}\mathrm{tr}(e^{\lambda Y_n}) \le \mathbb{E}\mathrm{tr}\exp\{\lambda Y_{n-1} + \log \mathbb{E}e^{\lambda X_n}\}$$
$$= \mathbb{E}\mathbb{E}_{n-1}\mathrm{tr}\exp\{\lambda Y_{n-2} + \log \mathbb{E}e^{\lambda X_n} + \log e^{\lambda X_{n-1}}\} = \mathbb{E}\mathbb{E}_n G_{\lambda Y_{n-2}+\log \mathbb{E}e^{\lambda X_n}}(e^{\lambda X_{n-1}})$$

and another application of Lieb's theorem and Jensen's inequality yields

$$\mathbb{E}\mathrm{tr}(e^{\lambda Y_n}) \le \mathbb{E}\mathrm{tr}\exp\{\lambda Y_{n-2} + \log \mathbb{E}e^{\lambda X_{n-1}} + \log \mathbb{E}e^{\lambda X_n}\}.$$

Iterating this argument, we get

$$\mathbb{E}\mathrm{tr}(e^{\lambda Y_n}) \leq \mathrm{tr}\exp\{\log\mathbb{E}e^{\lambda X_1} + \log\mathbb{E}e^{\lambda X_2} + \cdots + \log\mathbb{E}e^{\lambda X_n}\}. \qquad (2.10)$$

Next we have to bound $\mathbb{E}e^{\lambda X}$ for an arbitrary Hermitian random matrix with $\mathbb{E}X = 0$ and $\|X\| \leq U$. To this end, we use the Taylor expansion:

$$\begin{aligned}
\mathbb{E}e^{\lambda X} &= I_m + \mathbb{E}\lambda^2 X^2 \left[\frac{1}{2!} + \frac{\lambda X}{3!} + \frac{\lambda^2 X^2}{4!} + \cdots\right] \\
&\leq I_m + \lambda^2 \mathbb{E}X^2 \left[\frac{1}{2!} + \frac{\lambda\|X\|}{3!} + \frac{\lambda^2\|X\|^2}{4!} + \cdots\right] \\
&= I_m + \lambda^2 \mathbb{E}X^2 \left[\frac{e^{\lambda\|X\|} - 1 - \lambda\|X\|}{\lambda^2\|X\|^2}\right].
\end{aligned}$$

Under the assumption $\|X\| \leq U$, this yields

$$\mathbb{E}e^{\lambda X} \leq I_m + \lambda^2 \mathbb{E}X^2 \left[\frac{e^{\lambda U} - 1 - \lambda U}{\lambda^2 U^2}\right].$$

Denoting $\phi(u) := \frac{e^u - 1 - u}{u^2}$, we easily get

$$\log\mathbb{E}e^{\lambda X} \leq \lambda^2 \mathbb{E}X^2 \phi(\lambda U).$$

We will use this bound for each random matrix $X_j$ and substitute the result in (2.10) to get

$$\begin{aligned}
\mathbb{E}\mathrm{tr}(e^{\lambda Y_n}) &\leq \mathrm{tr}\exp\left\{\lambda^2 \mathbb{E}(X_1^2 + \cdots + X_n^2)\phi(\lambda U)\right\} \\
&\leq m\exp\left\{\lambda^2 \|\mathbb{E}(X_1^2 + \cdots + X_n^2)\|\phi(\lambda U)\right\}.
\end{aligned}$$

In view of (2.9), it remains to follow the usual proof of Bernstein–Bennett type inequalities to obtain (2.6).

To prove (2.7), we bound $\mathbb{E}e^{\lambda X}$ in a slightly different way. We do it for an arbitrary Hermitian random matrix with $\mathbb{E}X = 0$ and

$$\left\|\|X\|\right\|_{\psi_\alpha} \vee 2\mathbb{E}^{1/2}\|X\|^2 \leq U^{(\alpha)}.$$

For all $\tau > 0$, we get

$$\mathbb{E}e^{\lambda X} \le I_m + \lambda^2 \mathbb{E} X^2 \left[ \frac{e^{\lambda \|X\|} - 1 - \lambda \|X\|}{\lambda^2 \|X\|^2} \right]$$

$$\le I_m + \lambda^2 \mathbb{E} X^2 \left[ \cdot \frac{e^{\lambda \tau} - 1 - \lambda \tau}{\lambda^2 \tau^2} \right] + I_m \lambda^2 \mathbb{E} \|X\|^2 \left[ \frac{e^{\lambda \|X\|} - 1 - \lambda \|X\|}{\lambda^2 \|X\|^2} \right] I(\|X\| \ge \tau).$$

Take $M := 2(\log 2)^{1/\alpha} U^{(\alpha)}$ and assume that $\lambda \le 1/M$. It follows that

$$\mathbb{E} \|X\|^2 \left[ \frac{e^{\lambda \|X\|} - 1 - \lambda \|X\|}{\lambda^2 \|X\|^2} \right] I(\|X\| \ge \tau) \le M^2 \mathbb{E} e^{\|X\|/M} I(\|X\| \ge \tau) \le$$

$$M^2 \mathbb{E}^{1/2} e^{2\|X\|/M} \mathbb{P}^{1/2} \{\|X\| \ge \tau\}.$$

Since, for $\alpha \ge 1$,

$$M = 2(\log 2)^{1/\alpha} U^{(\alpha)} \ge 2 \Big\| \|X\| \Big\|_{\psi_1}$$

(see Sect. A.1), we get $\mathbb{E} e^{2\|X\|/M} \le 2$ and also

$$\mathbb{P}\{\|X\| \ge \tau\} \le \exp\left\{ -2^\alpha \log 2 \left( \frac{\tau}{M} \right)^\alpha \right\}.$$

Therefore, the following bound holds:

$$\mathbb{E}e^{\lambda X} \le I_m + \lambda^2 \mathbb{E} X^2 \left[ \frac{e^{\lambda \tau} - 1 - \lambda \tau}{\lambda^2 \tau^2} \right] + 2^{1/2} \lambda^2 M^2 \exp\left\{ -2^{\alpha-1} \log 2 \left( \frac{\tau}{M} \right)^\alpha \right\} I_m.$$

Take now $\tau := M \frac{2^{1/\alpha - 1}}{(\log 2)^{1/\alpha}} \log^{1/\alpha} \frac{M^2}{\sigma^2}$ and suppose that $\lambda$ satisfies the condition $\lambda \tau \le 1$. This yields the following bound

$$\mathbb{E}e^{\lambda X} \le I_m + \frac{C_1}{2} \lambda^2 (\mathbb{E} X^2 + \sigma^2 I_m),$$

which implies that

$$\log \mathbb{E}e^{\lambda X} \le \frac{C_1}{2} \lambda^2 (\mathbb{E} X^2 + \sigma^2 I_m)$$

with some constant $C_1 > 0$. We use the last bound for each random matrix $X_j$, $j = 1, \ldots, n$ and deduce from (2.10) that, for some constants $C_1, C_2 > 0$ and for all $\lambda$ satisfying the condition

$$\lambda U^{(\alpha)} \left( \log \frac{U^{(\alpha)}}{\sigma} \right)^{1/\alpha} \le C_2, \tag{2.11}$$

we have

$$\mathbb{E}\mathrm{tr}(e^{\lambda Y_n}) \le \mathrm{tr} \exp\left\{\frac{C_1}{2}\lambda^2(\mathbb{E}X_1^2 + \cdots + \mathbb{E}X_n^2 + n\sigma^2 I_m)\right\},$$

which further implies that

$$\mathbb{E}\mathrm{tr}(e^{\lambda Y_n}) \le m \exp\{C_1\lambda^2 n\sigma^2)\}.$$

Combining this bound with (2.8) and (2.9), we get

$$\mathbb{P}\{\|Y_n\| \ge t\} \le 2m \exp\left\{-\lambda t + C_1\lambda^2 n\sigma^2\right\}.$$

The last bound can be now minimized with respect to all $\lambda$ satisfying (2.11), which yields that, for some constant $K > 0$,

$$\mathbb{P}\{\|Y_n\| \ge t\} \le 2m \exp\left\{-\frac{1}{K}\frac{t^2}{n\sigma^2 + tU^{(\alpha)}\log^{1/\alpha}(U^{(\alpha)}/\sigma)}\right\}.$$

This proves inequality (2.7).                                                    □

The next bounds immediately follow from (2.6) and (2.7): for all $t > 0$, with probability at least $1 - e^{-t}$

$$\left\|\frac{X_1 + \cdots + X_n}{n}\right\| \le 2\left(\sigma\sqrt{\frac{t + \log(2m)}{n}} \bigvee U\frac{t + \log(2m)}{n}\right) \qquad (2.12)$$

and, with some constant $C > 0$,

$$\left\|\frac{X_1 + \cdots + X_n}{n}\right\| \le C\left(\sigma\sqrt{\frac{t + \log(2m)}{n}} \bigvee\right.$$
$$\left. U^{(\alpha)}\left(\log\frac{U^{(\alpha)}}{\sigma}\right)^{1/\alpha}\frac{t + \log(2m)}{n}\right). \qquad (2.13)$$

Note that the size $m$ of the matrices has only logarithmic impact on the bounds.

It is easy to derive Bernstein type exponential inequalities for rectangular $m_1 \times m_2$ random matrices from the inequalities of Theorem 2.7 for Hermitian matrices. This is based on the following well known isomorphism trick (sometimes called *Paulsen dilation*). Denote by $\mathbb{M}_{m_1,m_2}(\mathbb{R})$ the space of all $m_1 \times m_2$ matrices with real entries and by $\mathbb{H}_m(\mathbb{C})$ the space of all Hermitian $m \times m$ matrices. Define the following linear mapping

$$J : \mathbb{M}_{m_1,m_2}(\mathbb{R}) \mapsto \mathbb{H}_{m_1+m_2}(\mathbb{C}), \quad \text{where } JS := \begin{pmatrix} O & S \\ S^* & O \end{pmatrix}.$$

Clearly,

$$(JS)^2 := \begin{pmatrix} SS^* & 0 \\ 0 & S^*S \end{pmatrix}.$$

Therefore,

$$\|JS\| = \|SS^*\|^{1/2} \vee \|S^*S\|^{1/2} = \|S\|$$

and, for independent random matrices $X_1, \ldots, X_n$ in $\mathbb{M}_{m_1,m_2}(\mathbb{R})$ with $\mathbb{E}X_j = 0$, we have

$$\sigma^2 := n^{-1}\left(\|\mathbb{E}(X_1 X_1^*) + \cdots + \mathbb{E}(X_n X_n^*)\| \vee \|\mathbb{E}(X_1^* X_1) + \cdots + \mathbb{E}(X_n^* X_n)\|\right)$$

$$= n^{-1}\|\mathbb{E}((JX_1)^2 + \cdots + (JX_n)^2)\|.$$

The following statement immediately follows from Theorem 2.7 by applying it to the Hermitian random matrices $JX_1, \ldots, JX_n$.

**Corollary 2.1.** *1. Let $m := m_1 + m_2$. Suppose that, for some $U > 0$ and for all $j = 1, \ldots, n$, $\|X_j\| \leq U$. Then*

$$\mathbb{P}\left\{\|X_1 + \cdots + X_n\| \geq t\right\} \leq 2m \exp\left\{-\frac{t^2}{2\sigma^2 n + 2Ut/3}\right\}. \tag{2.14}$$

*2. Let $\alpha \geq 1$ and suppose that for some $U^{(\alpha)} > 0$ and for all $j = 1, \ldots, n$,*

$$\left\|\,\|X_j\|\,\right\|_{\psi_\alpha} \vee 2\mathbb{E}^{1/2}\|X\|^2 \leq U^{(\alpha)}.$$

*Then, there exists a constant $K > 0$ such that*

$$\mathbb{P}\{\|X_1 + \cdots + X_n\| \geq t\} \leq 2m \exp\left\{-\frac{1}{K}\frac{t^2}{n\sigma^2 + tU^{(\alpha)}\log^{1/\alpha}(U^{(\alpha)}/\sigma_X)}\right\}. \tag{2.15}$$

## 2.5  Further Comments

Initially, the theory of empirical processes dealt with asymptotic problems: uniform versions of laws of large numbers, central limit theorem and laws of iterated logarithm. It started with the work by Vapnik and Chervonenkis (see [147] and references therein) on Glivenko-Cantelli problem and by Dudley [59] on the central limit theorem (extensions of Kolmogorov–Donsker theorems). Other early references include Koltchinskii [80], Pollard [122] and Giné and Zinn [69]. Since Talagrand [138, 139] developed his concentration inequalities, the focus of the theory has shifted to the development of bounds on sup-norms of empirical processes with applications to a variety of problems in statistics, learning theory, asymptotic geometric analysis, etc (see also [137]).

Symmetrization inequalities of Sect. 2.1 were introduced to the theory of empirical processes by Giné and Zinn [69] (an earlier form of Rademacher symmetrization was used by Koltchinskii [80] and Pollard [122]).

In Sect. 2.2, we follow the proof of Talagrand's comparison inequality for Rademacher sums given by Ledoux and Talagrand [101], Theorem 4.12.

Talagrand's concentration inequalities for product measures and empirical processes were proved in [138, 139]. Another approach to their proof, the entropy method based on logarithmic Sobolev inequalities, was introduced by Ledoux. It is discussed in detail in [100] and [107] (see also [30]). The bounded difference inequality based on the martingale method is well known and can be found in many books (e.g., [51, 107]).

Noncommutative Bernstein's inequality (2.6) was discovered by Ahlswede and Winter [4]. This inequality and its extensions proved to be very useful in the recent work on low rank matrix recovery (see Gross et al. [71], Gross [70], Recht [124], Koltchinskii [88]). Tropp [142] provides a detailed review of various inequalities of this type.

# Chapter 3
# Bounding Expected Sup-Norms of Empirical and Rademacher Processes

In what follows, we will use a number of bounds on expectation of suprema of empirical and Rademacher processes. Because of symmetrization inequalities, the problems of bounding expected suprema for these two stochastic processes are equivalent. The bounds are usually based on various complexity measures of function classes (such as linear dimension, VC-dimension, shattering numbers, uniform covering numbers, random covering numbers, bracketing numbers, generic chaining complexities, etc). It would be of interest to develop the bounds with precise dependence on such geometric parameters as the $L_2(P)$-diameter of the class. Combining the bounds on expected suprema with Talagrand's concentration inequalities yields exponential inequalities for the tail probabilities of sup-norms.

## 3.1  Gaussian and Subgaussian Processes, Metric Entropies and Generic Chaining Complexities

Recall that a random variable $Y$ is called *subgaussian* with parameter $\sigma^2$, or $Y \in SG(\sigma^2)$, iff for all $\lambda \in \mathbb{R}$

$$\mathbb{E}e^{\lambda Y} \leq e^{\lambda^2 \sigma^2 / 2}.$$

Normal random variable with mean 0 and variance $\sigma^2$ belongs to $SG(\sigma^2)$. If $\varepsilon$ is a Rademacher r.v., then $\varepsilon \in SG(1)$.

The next proposition gives two simple and important properties of subgaussian random variables (see, e.g., [148], Sect. 2.2.1 for the proof of property (ii)).

**Proposition 3.1.** *(i) If $Y_1, \ldots, Y_n$ are independent random variables and $Y_j \in SG(\sigma_j^2)$, then*

$$Y_1 + \cdots + Y_n \in SG(\sigma_1^2 + \cdots + \sigma_n^2).$$

*(ii) For arbitrary $Y_1, \ldots, Y_N$, $N \geq 2$ such that $Y_j \in SG(\sigma_j^2)$, $j = 1, \ldots, N$,*

V. Koltchinskii, *Oracle Inequalities in Empirical Risk Minimization and Sparse Recovery Problems*, Lecture Notes in Mathematics 2033, DOI 10.1007/978-3-642-22147-7_3, © Springer-Verlag Berlin Heidelberg 2011

$$\mathbb{E} \max_{1 \leq j \leq N} |Y_j| \leq C \max_{1 \leq j \leq N} \sigma_j \sqrt{\log N},$$

*where C is a numerical constant.*

Let $(T, d)$ be a pseudo-metric space and $Y(t), t \in T$ be a stochastic process. It is called *subgaussian* with respect to $d$ iff, for all $t, s \in T$, $Y(t) - Y(s) \in SG(d^2(t,s))$.

Denote $D(T) = D(T, d)$ the diameter of the space $T$. Let $N(T, d, \varepsilon)$ be the $\varepsilon$-covering number of $(T, d)$, that is, the minimal number of balls of radius $\varepsilon$ needed to cover $T$. Let $M(T, d, \varepsilon)$ be the $\varepsilon$-packing number of $(T, d)$, i.e., the largest number of points in $T$ separated from each other by at least a distance of $\varepsilon$. Obviously,

$$N(T, d, \varepsilon) \leq M(T, d, \varepsilon) \leq N(T, d, \varepsilon/2), \ \varepsilon \geq 0.$$

As always,
$$H(T, d, \varepsilon) = \log N(T, d, \varepsilon)$$

is called *the $\varepsilon$-entropy* of $(T, d)$.

**Theorem 3.1 (Dudley's entropy bounds).** *If $Y(t), t \in T$ is a subgaussian process with respect to $d$, then the following bounds hold with some numerical constant $C > 0$:*

$$\mathbb{E} \sup_{t \in T} Y(t) \leq C \int_0^{D(T)} H^{1/2}(T, d, \varepsilon) d\varepsilon$$

*and for all $t_0 \in T$*

$$\mathbb{E} \sup_{t \in T} |Y(t) - Y(t_0)| \leq C \int_0^{D(T)} H^{1/2}(T, d, \varepsilon) d\varepsilon.$$

The integral in the right hand side of the bound is often called *Dudley's entropy integral.*

For Gaussian processes, the following lower bound is also true (see [101], Sect. 3.3).

**Theorem 3.2 (Sudakov's entropy bound).** *If $Y(t), t \in T$ is a Gaussian process and*

$$d(t, s) := \mathbb{E}^{1/2}(X(t) - X(s))^2, \ t, s \in T,$$

*then the following bound holds with some numerical constant $C > 0$:*

$$\mathbb{E} \sup_{t \in T} Y(t) \geq C \sup_{\varepsilon > 0} \varepsilon H^{1/2}(T, d, \varepsilon).$$

Note that, if $Z$ is a standard normal vector in $\mathbb{R}^N$ and $T \subset \mathbb{R}^N$, then Sudakov's entropy bound immediately implies that, with some numerical constant $C' > 0$,

$$\sup_{\varepsilon > 0} \varepsilon H^{1/2}(T, \| \cdot \|_{\ell_2}, \varepsilon) \leq C' \mathbb{E} \sup_{t \in T} \langle Z, t \rangle. \tag{3.1}$$

We will also need another inequality of a similar flavor that is often called *dual Sudakov's inequality* (see Pajor and Tomczak-Jaegermann [121]). Namely, let $K \subset \mathbb{R}^N$ be a symmetric convex set (that is, $u \in K$ implies $-u \in K$). Denote

$$\|t\|_K := \sup_{u \in K} \langle u, t \rangle.$$

Finally, denote $B_2^N$ the unit ball in the space $l_2^N$ (that is, in $\mathbb{R}^N$ equipped with the $l_2$-norm). Then, the following bound holds with a numerical constant $C' > 0$:

$$\sup_{\varepsilon > 0} \varepsilon H^{1/2}(B_2^N, \|\cdot\|_K, \varepsilon) \le C' \mathbb{E} \sup_{t \in K} \langle Z, t \rangle. \tag{3.2}$$

Note that, for $T = K$, (3.1) provides an upper bound on the cardinality of minimal coverings of the symmetric convex set $K$ by the Euclidean balls of radius $\varepsilon$. On the other hand, (3.2) is a bound on the cardinality of minimal coverings of the Euclidean unit ball $B_2^N$ by the translations of the convex set $\varepsilon K^\circ$, $K^\circ$ being the polar set of $K$. In both cases, the bounds are dimension free.

The proof of Theorem 3.1 is based on the well known *chaining method* (see, e.g., [101], Sect. 11.1) that also leads to more refined *generic chaining bounds* (see Talagrand [140]). Talagrand's *generic chaining complexity* of a metric space $(T, d)$ is defined as follows. An admissible sequence $\{\Delta_n\}_{n \ge 0}$ is an increasing sequence of partitions of $T$ (that is, each next partition is a refinement of the previous one) such that $\mathrm{card}(\Delta_0) = 1$ and $\mathrm{card}(\Delta_n) \le 2^{2^n}$, $n \ge 1$. Given $t \in T$, let $\Delta_n(t)$ denote the unique subset from $\Delta_n$ that contains $t$. For a set $A \subset T$, let $D(A)$ denote its diameter. Define the generic chaining complexity $\gamma_2(T; d)$ as

$$\gamma_2(T; d) := \inf_{\{\Delta_n\}_{n \ge 0}} \sup_{t \in T} \sum_{n \ge 0} 2^{n/2} D(\Delta_n(t)),$$

where the inf is taken over all admissible sequences of partitions.

**Theorem 3.3 (Talagrand's generic chaining bounds).** *If $Y(t), t \in T$ is a centered Gaussian process with*

$$d(t, s) := \mathbb{E}^{1/2}(Y(t) - Y(s))^2, \ t, s \in T,$$

*then*

$$K^{-1} \gamma_2(T; d) \le \mathbb{E} \sup_{t \in T} Y(t) \le K \gamma_2(T; d),$$

*where $K > 0$ is a universal constant. The upper bound also holds for all subgaussian processes with respect to $d$.*

Of course, Talagrand's generic chaining complexity is upper bounded by Dudley's entropy integral. In special cases, other upper bounds are also available that might be sharper in specific applications. For instace, if $T \subset H$ is the unit ball in a Hilbert space $H$ and $d$ is the metric generated by an arbitrary norm in $H$, then, for some constant $C > 0$,

$$\gamma_2(T;d) \leq C \left( \int_0^\infty \varepsilon H(T;d;\varepsilon) d\varepsilon \right)^{1/2}. \tag{3.3}$$

This follows from a more general result by Talagrand [140] that applies also to Banach spaces with $p$-convex norms for $p \geq 2$.

In addition to Gaussian processes, Rademacher sums provide another important example of subgaussian processes. Given $T \subset \mathbb{R}^n$, define

$$Y(t) := \sum_{i=1}^n \varepsilon_i t_i, \; t = (t_1, \ldots, t_n) \in T,$$

where $\{\varepsilon_i\}$ are i.i.d. Rademacher random variables. The stochastic process $Y(t)$, $t \in T$ is called *the Rademacher sum* indexed by $T$. It is a subgaussian process with respect to the Euclidean distance in $\mathbb{R}^n$ :

$$d(t,s) = \left( \sum_{i=1}^n (t_i - s_i)^2 \right)^{1/2}.$$

The following result by Talagrand is a version of Sudakov's type lower bound for Rademacher sums (see [101], Sect. 4.5).

Denote

$$R(T) := \mathbb{E}_\varepsilon \sup_{t \in T} \left| \sum_{i=1}^n \varepsilon_i t_i \right|.$$

**Theorem 3.4 (Talagrand).** *There exists a universal constant $L$ such that*

$$R(T) \geq \frac{1}{L} \delta H^{1/2}(T, d, \delta) \tag{3.4}$$

*whenever*

$$R(T) \sup_{t \in T} \|t\|_{\ell_\infty} \leq \frac{\delta^2}{L}. \tag{3.5}$$

## 3.2    Finite Classes of Functions

Suppose $\mathscr{F}$ is a finite class of measurable functions uniformly bounded by a constant $U > 0$. Let $N := \mathrm{card}(\mathscr{F}) \geq 2$. Denote $\sigma^2 := \sup_{f \in \mathscr{F}} Pf^2$.

**Theorem 3.5.** *There exist universal constants $K_1, K_2$ such that*

$$\mathbb{E} \|R_n\|_{\mathscr{F}} \leq K_1 U \sqrt{\frac{\log N}{n}}$$

*and*

$$\mathbb{E}\|R_n\|_{\mathscr{F}} \le K_2\left[\sigma\sqrt{\frac{\log N}{n}} \bigvee U\frac{\log N}{n}\right].$$

*Proof.* Conditionally on $X_1, \ldots, X_n$, the random variable

$$\sqrt{n}R_n(f) = \frac{1}{\sqrt{n}}\sum_{j=1}^{n}\varepsilon_j f(X_j), \ f \in \mathscr{F}$$

is subgaussian with parameter $\|f\|_{L_2(P_n)}$. Therefore, it follows from Proposition 3.1, (ii) that

$$\mathbb{E}_\varepsilon\|R_n\|_{\mathscr{F}} \le K \sup_{f\in\mathscr{F}} \|f\|_{L_2(P_n)}\sqrt{\frac{\log N}{n}}.$$

The first bound now follows since $\sup_{f\in\mathscr{F}}\|f\|_{L_2(P_n)} \le U$. To prove the second bound, denote $\mathscr{F}^2 := \{f^2 : f \in \mathscr{F}\}$ and observe that

$$\sup_{f\in\mathscr{F}}\|f\|_{L_2(P_n)} \le \sup_{f\in\mathscr{F}}\|f\|_{L_2(P)} + \sqrt{\|P_n - P\|_{\mathscr{F}^2}},$$

which implies

$$\mathbb{E}\sup_{f\in\mathscr{F}}\|f\|_{L_2(P_n)} \le \sigma + \sqrt{\mathbb{E}\|P_n - P\|_{\mathscr{F}^2}}.$$

Using symmetrization and contraction inequalities, we get

$$\mathbb{E}\|P_n - P\|_{\mathscr{F}^2} \le 2\mathbb{E}\|R_n\|_{\mathscr{F}^2} \le 8U\mathbb{E}\|R_n\|_{\mathscr{F}}.$$

Hence,

$$\mathbb{E}\|R_n\|_{\mathscr{F}} \le K\mathbb{E}\sup_{f\in\mathscr{F}}\|f\|_{L_2(P_n)}\sqrt{\frac{\log N}{n}} \le K\left(\sigma + \sqrt{8U\mathbb{E}\|R_n\|_{\mathscr{F}}}\right)\sqrt{\frac{\log N}{n}}.$$

The result now follows by bounding the solution with respect to $\mathbb{E}\|R_n\|_{\mathscr{F}}$ of the above inequality.                                                              □

The same result can be also deduced from the following theorem (it is enough to take $q = \log N$).

**Theorem 3.6.** *There exists a universal constants $K$ such that for all $q \ge 2$*

$$\mathbb{E}^{1/q}\|R_n\|_{\mathscr{F}}^q \le \mathbb{E}^{1/q}\|R_n\|_{\ell_q(\mathscr{F})}^q := \mathbb{E}^{1/q}\sum_{f\in\mathscr{F}}|R_n(f)|^q$$

$$\le K\left[\sigma\frac{(q-1)^{1/2}N^{1/q}}{n^{1/2}} \bigvee U\frac{(q-1)N^{2/q}}{n}\right].$$

*Proof.* We will need the following simple property of Rademacher sums: for all $q \geq 2$,

$$\mathbb{E}^{1/q} \left| \sum_{i=1}^{n} \alpha_i \varepsilon_i \right|^q \leq (q-1)^{1/2} \left( \sum_{i=1}^{n} \alpha_i^2 \right)^{1/2}$$

(see, e.g., de la Pena and Giné [50], p. 21). Using this inequality, we get

$$\mathbb{E}_\varepsilon \| R_n \|_{\mathscr{F}}^q \leq \sum_{f \in \mathscr{F}} \mathbb{E}_\varepsilon |R_n(f)|^q \leq (q-1)^{q/2} n^{-q/2} \sum_{f \in \mathscr{F}} \| f \|_{L_2(P_n)}^q$$

$$\leq (q-1)^{q/2} n^{-q/2} N \left( \sup_{f \in \mathscr{F}} P_n f^2 \right)^{q/2}$$

$$\leq (q-1)^{q/2} n^{-q/2} N \left( \sigma^2 + \| P_n - P \|_{\mathscr{F}^2} \right)^{q/2}.$$

This easily implies

$$\mathbb{E}^{1/q} \| R_n \|_{\mathscr{F}}^q \leq \mathbb{E}^{1/q} \sum_{f \in \mathscr{F}} |R_n(f)|^q \tag{3.6}$$

$$\leq (q-1)^{1/2} n^{-1/2} N^{1/q} 2^{1/2-1/q} \left( \sigma + \mathbb{E}^{1/q} \| P_n - P \|_{\mathscr{F}^2}^{q/2} \right).$$

It remains to use symmetrization and contraction inequalities to get

$$\mathbb{E}^{1/q} \| P_n - P \|_{\mathscr{F}^2}^{q/2} \leq 2U^{1/2} \mathbb{E}^{1/q} \| R_n \|_{\mathscr{F}}^{q/2} \leq 2U^{1/2} \sqrt{\mathbb{E}^{1/q} \| R_n \|_{\mathscr{F}}^q},$$

to substitute this bound into (3.6) and to solve the resulting inequality for $\mathbb{E}^{1/q} \| R_n \|_{\mathscr{F}}^q$ to complete the proof.    □

## 3.3  Shattering Numbers and VC-classes of Sets

Let $\mathscr{C}$ be a class of subsets of $S$. Given a finite set $F \subset S$, denote

$$\Delta^{\mathscr{C}}(F) := \mathrm{card}\{\mathscr{C} \cap F\},$$

where $\mathscr{C} \cap F := \left\{ C \cap F : C \in \mathscr{C} \right\}$. Clearly,

$$\Delta^{\mathscr{C}}(F) \leq 2^{\mathrm{card}(F)}.$$

If $\Delta^{\mathscr{C}}(F) = 2^{\mathrm{card}(F)}$, it is said that $F$ is shattered by $\mathscr{C}$. The numbers $\Delta^{\mathscr{C}}(F)$ are called *the shattering numbers* of the class $\mathscr{C}$.

Define

$$m^{\mathscr{C}}(n) := \sup\left\{\Delta^{\mathscr{C}}(F) : F \subset S, \mathrm{card}(F) \leq n\right\}.$$

Clearly, $m^{\mathscr{C}}(n) \leq 2^n$, $n \geq 1$, and if, for some $n$, $m^{\mathscr{C}}(n) < 2^n$, then $m^{\mathscr{C}}(k) < 2^k$ for all $k \geq n$.

Let

$$V(\mathscr{C}) := \min\{n \geq 1 : m^{\mathscr{C}}(n) < 2^n\}.$$

If $m^{\mathscr{C}}(n) = 2^n$ for all $n \geq 1$, set $V(\mathscr{C}) = \infty$. The number $V(\mathscr{C})$ is called the *Vapnik–Chervonenkis dimension (or the VC-dimension)* of class $\mathscr{C}$. If $V(\mathscr{C}) < +\infty$, then $\mathscr{C}$ is called the Vapnik–Chervonenkis class (or VC-class). It means that no set $F$ of cardinality $n \geq V(\mathscr{C})$ is shattered by $\mathscr{C}$.

Denote

$$\binom{n}{\leq k} := \binom{n}{0} + \cdots + \binom{n}{k}.$$

The following lemma (proved independently in somewhat different forms by Sauer, Shelah, and also by Vapnik and Chervonenkis) is one of the main combinatorial facts related to VC-classes.

**Theorem 3.7 (Sauer's Lemma).** *Let $F \subset S$, $\mathrm{card}(F) = n$. If*

$$\Delta^{\mathscr{C}}(F) > \binom{n}{\leq k - 1},$$

*then there exists a subset $F' \subset F$, $\mathrm{card}(F') = k$ such that $F'$ is shattered by $\mathscr{C}$.*

The Sauer's Lemma immediately implies that, for a VC-class $\mathscr{C}$,

$$m^{\mathscr{C}}(n) \leq \binom{n}{\leq V(\mathscr{C}) - 1},$$

which can be further bounded by $\left(\frac{ne}{V(\mathscr{C})-1}\right)^{V(\mathscr{C})-1}$.

We will view $P$ and $P_n$ as functions defined on a class $\mathscr{C}$ of measurable sets $C \mapsto P(C), C \mapsto P_n(C)$ and the Rademacher process will be also indexed by sets:

$$R_n(C) := n^{-1} \sum_{j=1}^{n} \varepsilon_j I_C(X_j).$$

For $Y : \mathscr{C} \mapsto \mathbb{R}$, we still write $\|Y\|_{\mathscr{C}} := \sup_{C \in \mathscr{C}} |Y(C)|$. Denote $\mathscr{F} := \{I_C : C \in \mathscr{C}\}$.

**Theorem 3.8.** *There exists a numerical constant $K > 0$ such that*

$$\mathbb{E}\|P_n - P\|_{\mathscr{C}} \leq K\mathbb{E}\sqrt{\frac{\log \Delta^{\mathscr{C}}(X_1, \ldots, X_n)}{n}} \leq K\sqrt{\frac{\mathbb{E}\log \Delta^{\mathscr{C}}(X_1, \ldots, X_n)}{n}}.$$

The drawback of this result is that it does not take into account the "size" of the sets in the class $\mathscr{C}$. A better bound is possible in the case when, for all $C \in \mathscr{C}$, $P(C)$ is small. We will derive such an inequality in which the size of $\mathbb{E}\|P_n - P\|_{\mathscr{C}}$ is controlled in terms of random shattering numbers $\Delta^{\mathscr{C}}(X_1, \ldots, X_n)$ and of

$$\|P\|_{\mathscr{C}} = \sup_{C \in \mathscr{C}} P(C)$$

(and which implies the inequality of Theorem 3.8).

**Theorem 3.9.** *There exists a numerical constant $K > 0$ such that*

$$\mathbb{E}\|P_n - P\|_{\mathscr{C}} \leq K\|P\|_{\mathscr{C}}^{1/2}\mathbb{E}\sqrt{\frac{\log \Delta^{\mathscr{C}}(X_1, \ldots, X_n)}{n}} \bigvee K\frac{\mathbb{E}\log \Delta^{\mathscr{C}}(X_1, \ldots, X_n)}{n}$$

$$\leq K\|P\|_{\mathscr{C}}^{1/2}\sqrt{\frac{\mathbb{E}\log \Delta^{\mathscr{C}}(X_1, \ldots, X_n)}{n}} \bigvee K\frac{\mathbb{E}\log \Delta^{\mathscr{C}}(X_1, \ldots, X_n)}{n}.$$

*Proof.* Let

$$T := \left\{(I_C(X_1), \ldots, I_C(X_n)) : C \in \mathscr{C}\right\}.$$

Clearly, $\text{card}(T) = \Delta^{\mathscr{C}}(X_1, \ldots, X_n)$ and

$$\mathbb{E}_{\varepsilon}\|R_n\|_{\mathscr{C}} = \mathbb{E}_{\varepsilon}\sup_{t \in T}\left|n^{-1}\sum_{i=1}^{n}\varepsilon_i t_i\right|.$$

For all $t \in T$, $n^{-1}\sum_{i=1}^{n}\varepsilon_i t_i$ is a subgaussian random variable with parameter $n^{-1}\|t\|_{\ell_2}$. Therefore, by Proposition 3.1,

$$\mathbb{E}_{\varepsilon}\sup_{t \in T}\left|n^{-1}\sum_{i=1}^{n}\varepsilon_i t_i\right| \leq Kn^{-1}\sup_{t \in T}\|t\|_{\ell_2}\sqrt{\log \Delta^{\mathscr{C}}(X_1, \ldots, X_n)}.$$

Note that

$$n^{-1}\sup_{t \in T}\|t\|_{\ell_2} = n^{-1/2}(\sup_{C \in \mathscr{C}} P_n(C))^{1/2}.$$

Hence,

$$\mathbb{E}_\varepsilon \|R_n\|_\mathscr{C} \leq K n^{-1/2} \mathbb{E}\|P_n\|_\mathscr{C}^{1/2} \sqrt{\log \Delta^\mathscr{C}(X_1, \ldots, X_n)}$$

$$\leq K n^{-1/2} \mathbb{E}\sqrt{\|P_n - P\|_\mathscr{C} + \|P\|_\mathscr{C}} \sqrt{\log \Delta^\mathscr{C}(X_1, \ldots, X_n)}$$

$$\leq K n^{-1/2} \mathbb{E}\sqrt{\|P_n - P\|_\mathscr{C}} \sqrt{\log \Delta^\mathscr{C}(X_1, \ldots, X_n)}$$

$$+ K n^{-1/2} \sqrt{\|P\|_\mathscr{C}} \mathbb{E}\sqrt{\log \Delta^\mathscr{C}(X_1, \ldots, X_n)}.$$

By symmetrization inequality,

$$\mathbb{E}\|P_n - P\|_\mathscr{C} \leq 2K n^{-1/2} \mathbb{E}\sqrt{\|P_n - P\|_\mathscr{C}} \sqrt{\log \Delta^\mathscr{C}(X_1, \ldots, X_n)}$$

$$+ 2K n^{-1/2} \sqrt{\|P\|_\mathscr{C}} \mathbb{E}\sqrt{\log \Delta^\mathscr{C}(X_1, \ldots, X_n)}$$

$$\leq 2K n^{-1/2} \sqrt{\mathbb{E}\|P_n - P\|_\mathscr{C}} \sqrt{\mathbb{E}\log \Delta^\mathscr{C}(X_1, \ldots, X_n)}$$

$$+ 2K n^{-1/2} \sqrt{\|P\|_\mathscr{C}} \mathbb{E}\sqrt{\log \Delta^\mathscr{C}(X_1, \ldots, X_n)},$$

where we also used Cauchy–Schwarz inequality. It remains to solve the resulting inequality with respect to $\mathbb{E}\|P_n - P\|_\mathscr{C}$ (or just to upper bound its solution) to get the result. □

In the case of VC-classes,

$$\log \Delta^\mathscr{C}(X_1, \ldots, X_n) \leq \log m^\mathscr{C}(n) \leq K V(\mathscr{C}) \log n$$

with some numerical constant $K > 0$. Thus, Theorem 3.9 yields the bound

$$\mathbb{E}\|P_n - P\|_\mathscr{C} \leq K \left( \|P\|_\mathscr{C}^{1/2} \sqrt{\frac{V(\mathscr{C}) \log n}{n}} \bigvee \frac{V(\mathscr{C}) \log n}{n} \right).$$

However, this bound is not sharp: the logarithmic factor involved in it can be eliminated. To this end, the following bound on the covering numbers of a VC-class $\mathscr{C}$ is needed. For an arbitrary probability measure $Q$ on $(S, \mathscr{A})$, define the distance

$$d_Q(C_1, C_2) = Q(C_1 \triangle C_2), \quad C_1, C_2 \in \mathscr{C}.$$

**Theorem 3.10.** *There exists a universal constant $K > 0$ such that for any VC-class $\mathscr{C} \subset \mathscr{A}$ and for all probability measures $Q$ on $(S, \mathscr{A})$*

$$N(\mathscr{C}; d_Q; \varepsilon) \leq K V(\mathscr{C})(4e)^{V(\mathscr{C})} \left(\frac{1}{\varepsilon}\right)^{V(\mathscr{C})-1}, \quad \varepsilon \in (0, 1).$$

This result is due to Haussler and it is an improvement of an earlier bound by Dudley (the proof and precise references can be found, e.g., in van der Vaart and Wellner [148]).

By Theorem 3.10, we get

$$N(\mathscr{C}; d_{P_n}; \varepsilon) \le K V(\mathscr{C})(4e)^{V(\mathscr{C})} \left(\frac{1}{\varepsilon}\right)^{V(\mathscr{C})-1}, \quad \varepsilon \in (0,1).$$

Using this fact one can prove the following inequality:

$$\mathbb{E}\|P_n - P\|_{\mathscr{C}} \le K\left(\|P\|_{\mathscr{C}}^{1/2} \sqrt{\log \frac{K}{\|P\|_{\mathscr{C}}}} \sqrt{\frac{V(\mathscr{C})}{n}} \bigvee \frac{V(\mathscr{C}) \log \frac{K}{\|P\|_{\mathscr{C}}}}{n}\right).$$

We are not giving its proof here. However, in the next section, we establish more general results for VC-type classes of functions (see (3.17)) that do imply the above bound.

## 3.4  Upper Entropy Bounds

Let $N(\mathscr{F}; L_2(P_n); \varepsilon)$ denote the minimal number of $L_2(P_n)$-balls of radius $\varepsilon$ covering $\mathscr{F}$ and let

$$\sigma_n^2 := \sup_{f \in \mathscr{F}} P_n f^2.$$

Also denote by $\gamma_2(\mathscr{F}; L_2(P_n))$ Talagrand's generic chaining complexity of $\mathscr{F}$ with respect to the $L_2(P_n)$-distance.

**Theorem 3.11.** *The following bound holds with a numerical constant $C > 0$ :*

$$\mathbb{E}\|R_n\|_{\mathscr{F}} \le \frac{C}{\sqrt{n}} \mathbb{E}\gamma_2(\mathscr{F}; L_2(P_n)).$$

*As a consequence,*

$$\mathbb{E}\|R_n\|_{\mathscr{F}} \le \frac{C}{\sqrt{n}} \mathbb{E} \int_0^{2\sigma_n} \sqrt{\log N(\mathscr{F}; L_2(P_n); \varepsilon)} d\varepsilon$$

*with some constant $C > 0$.*

*Proof.* Conditionally on $X_1, \ldots, X_n$, the process

$$\sqrt{n}R_n(f) = \frac{1}{\sqrt{n}} \sum_{j=1}^n \varepsilon_j f(X_j), \ f \in \mathscr{F}$$

is subgaussian with respect to the distance of the space $L_2(P_n)$. Hence, it follows from Theorem 3.3 that

$$\mathbb{E}_\varepsilon \|R_n\|_{\mathscr{F}} \le C n^{-1/2} \gamma_2(\mathscr{F}; L_2(P_n)). \tag{3.7}$$

Taking expectation of both sides, yields the first inequality. The second inequality follows by bounding Talagrand's generic chaining complexity from above by Dudley's entropy integral.                                                                    □

Following Giné and Koltchinskii [66], we will derive from Theorem 3.11 several bounds under more special conditions on the random entropy. Assume that the functions in $\mathscr{F}$ are uniformly bounded by a constant $U > 0$ and let $F \le U$ denote a measurable envelope of $\mathscr{F}$, that is,

$$|f(x)| \le F(x), x \in S, f \in \mathscr{F}.$$

We will assume that $\sigma^2$ is a number such that

$$\sup_{f \in \mathscr{F}} Pf^2 \le \sigma^2 \le \|F\|^2_{L_2(P)}$$

Most often, we will use $\sigma^2 = \sup_{f \in \mathscr{F}} Pf^2$.

Let $H : [0, \infty) \mapsto [0, \infty)$ be a regularly varying function of exponent $0 \le \alpha < 2$, strictly increasing for $u \ge 1/2$ and such that $H(u) = 0$ for $0 \le u < 1/2$.

**Theorem 3.12.** *If, for all $\varepsilon > 0$ and $n \ge 1$,*

$$\log N(\mathscr{F}, L_2(P_n), \varepsilon) \le H\left(\frac{\|F\|_{L_2(P_n)}}{\varepsilon}\right), \tag{3.8}$$

*then there exists a constant $C > 0$ that depends only on $H$ and such that*

$$\mathbb{E}\|R_n\|_{\mathscr{F}} \le C\left[\frac{\sigma}{\sqrt{n}}\sqrt{H\left(\frac{\|F\|_{L_2(P)}}{\sigma}\right)} \vee \frac{U}{n} H\left(\frac{\|F\|_{L_2(P)}}{\sigma}\right)\right]. \tag{3.9}$$

*In particular, if, for some $C_1 > 0$,*

$$n\sigma^2 \ge C_1 U^2 H\left(\frac{\|F\|_{L_2(P)}}{\sigma}\right),$$

*then*

$$\mathbb{E}\|R_n\|_{\mathscr{F}} \le \frac{C\sigma}{\sqrt{n}}\sqrt{H\left(\frac{\|F\|_{L_2(P)}}{\sigma}\right)} \tag{3.10}$$

*with a constant $C > 0$ that depends only on $H$ and $C_1$.*

*Proof.* Without loss of generality, assume that $U = 1$ (otherwise the result follows by a simple rescaling of the class $\mathscr{F}$). Given function $H$, we will use constants $C_H > 0$ and $D_H > 0$ for which

$$\sup_{v \geq 1} \frac{\int_v^\infty u^{-2}\sqrt{H(u)}\,du}{v^{-1}\sqrt{H(v)}} \bigvee 1 \leq C_H, \quad \int_0^2 \sqrt{H(1/u)}\,du = \int_{1/2}^\infty u^{-2}\sqrt{H(u)}\,du \leq D_H.$$

The bound of Theorem 3.11 implies that with some numerical constant $C > 0$[1]

$$\mathbb{E}\|R_n\|_{\mathscr{F}} \leq Cn^{-1/2}\mathbb{E}\int_0^{2\sigma_n} \sqrt{\log N(\mathscr{F}, L_2(P_n), \varepsilon)}\,d\varepsilon$$

$$\leq Cn^{-1/2}\mathbb{E}\int_0^{2\sigma_n} \sqrt{H\left(\frac{\|F\|_{L_2(P_n)}}{\varepsilon}\right)}\,d\varepsilon$$

$$\leq Cn^{-1/2}\mathbb{E}\int_0^{2\sigma_n} \sqrt{H\left(\frac{2\|F\|_{L_2(P)}}{\varepsilon}\right)}\,d\varepsilon\, I\left(\|F\|_{L_2(P_n)} \leq 2\|F\|_{L_2(P)}\right) \quad (3.11)$$

$$+ Cn^{-1/2}\mathbb{E}\int_0^{2\sigma_n} \sqrt{H\left(\frac{\|F\|_{L_2(P_n)}}{\varepsilon}\right)}\,d\varepsilon\, I\left(\|F\|_{L_2(P_n)} > 2\|F\|_{L_2(P)}\right).$$

It is very easy to bound the second term in the sum. First note that

$$\int_0^{2\sigma_n} \sqrt{H\left(\frac{\|F\|_{L_2(P_n)}}{\varepsilon}\right)}\,d\varepsilon \leq \|F\|_{L_2(P_n)}\int_0^2 \sqrt{H(1/u)}\,du \leq D_H\|F\|_{L_2(P_n)}.$$

Then use Hölder's inequality and Bernstein's inequality to get

$$n^{-1/2}\mathbb{E}\left[\int_0^{2\sigma_n} \sqrt{H\left(\frac{\|F\|_{L_2(P_n)}}{\varepsilon}\right)}\,d\varepsilon I\left(\|F\|_{L_2(P_n)} > 2\|F\|_{L_2(P)}\right)\right] \quad (3.12)$$

$$\leq D_H n^{-1/2}\|F\|_{L_2(P)} \exp\left\{-\frac{9}{8}n\|F\|_{L_2(P)}^2\right\} \leq \frac{D_H}{2n}.$$

Bounding the first term is slightly more complicated. Recall the notation

$$\mathscr{F}^2 := \{f^2 : f \in \mathscr{F}\}.$$

Using symmetrization and contraction inequalities, we get

$$\mathbb{E}\sigma_n^2 \leq \sigma^2 + \mathbb{E}\|P_n - P\|_{\mathscr{F}^2} \leq \sigma^2 + 2\mathbb{E}\|R_n\|_{\mathscr{F}^2} \leq \sigma^2 + 8\mathbb{E}\|R_n\|_{\mathscr{F}} =: B^2. \quad (3.13)$$

---

[1]The value of $C$ might change from place to place.

Since, for nonincreasing $h$, the function

$$u \mapsto \int_0^u h(t)dt$$

is concave, we have, by the properties of $H$, that

$$n^{-1/2}\mathbb{E}\int_0^{2\sigma_n}\sqrt{H\left(\frac{2\|F\|_{L_2(P)}}{\varepsilon}\right)}d\varepsilon\, I(\|F\|_{L_2(P_n)} \leq 2\|F\|_{L_2(P)})$$

$$\leq n^{-1/2}\mathbb{E}\int_0^{2\sigma_n}\sqrt{H\left(\frac{2\|F\|_{L_2(P)}}{\varepsilon}\right)}d\varepsilon$$

$$\leq n^{-1/2}\int_0^{2(\mathbb{E}\sigma_n^2)^{1/2}}\sqrt{H\left(\frac{2\|F\|_{L_2(P)}}{\varepsilon}\right)}d\varepsilon$$

$$\leq n^{-1/2}\int_0^{2B}\sqrt{H\left(\frac{2\|F\|_{L_2(P)}}{\varepsilon}\right)}d\varepsilon$$

$$= 2\|F\|_{L_2(P)}n^{-1/2}\int_0^{B/\|F\|_{L_2(P)}}\sqrt{H\left(\frac{1}{\varepsilon}\right)}d\varepsilon$$

$$= 2n^{-1/2}\|F\|_{L_2(P)}\int_{\|F\|_{L_2(P)}/B}^{+\infty}u^{-2}\sqrt{H(u)}du. \qquad (3.14)$$

In the case when $B \leq \|F\|_{L_2(P)}$, this yields the bound

$$n^{-1/2}\mathbb{E}\int_0^{2\sigma_n}\sqrt{H\left(\frac{2\|F\|_{L_2(P)}}{\varepsilon}\right)}d\varepsilon\, I(\|F\|_{L_2(P_n)} \leq 2\|F\|_{L_2(P)})$$

$$\leq 2C_H n^{-1/2}B\sqrt{H\left(\frac{\|F\|_{L_2(P)}}{B}\right)} \leq 2C_H n^{-1/2}B\sqrt{H\left(\frac{\|F\|_{L_2(P)}}{\sigma}\right)}.$$

In the case when $B > \|F\|_{L_2(P)}$, the bound becomes

$$n^{-1/2}\mathbb{E}\int_0^{2\sigma_n}\sqrt{H\left(\frac{2\|F\|_{L_2(P)}}{\varepsilon}\right)}d\varepsilon\, I(\|F\|_{L_2(P_n)} \leq 2\|F\|_{L_2(P)})$$

$$\leq 2n^{-1/2}\|F\|_{L_2(P)}\int_{1/2}^{+\infty}u^{-2}\sqrt{H(u)}du$$

$$\leq 2\frac{D_H}{\sqrt{H(1)}}n^{-1/2}\|F\|_{L_2(P)}\sqrt{H(1)} \leq 2\frac{D_H}{\sqrt{H(1)}}n^{-1/2}B\sqrt{H\left(\frac{\|F\|_{L_2(P)}}{\sigma}\right)},$$

where we also used the assumption that

$$\sup_{f \in \mathscr{F}} Pf^2 \leq \sigma^2 \leq \|F\|_{L_2(P)}^2.$$

Thus, in both cases we have

$$n^{-1/2}\mathbb{E}\int_0^{2\sigma_n} \sqrt{H\left(\frac{2\|F\|_{L_2(P)}}{\varepsilon}\right)}d\varepsilon \, I(\|F\|_{L_2(P_n)} \leq 2\|F\|_{L_2(P)})$$

$$\leq Cn^{-1/2}B\sqrt{H\left(\frac{\|F\|_{L_2(P)}}{\sigma}\right)} \tag{3.15}$$

with a constant $C$ depending only on $H$.

Now, we deduce from inequality (3.15) that

$$n^{-1/2}\mathbb{E}\left[\int_0^{2\sigma_n} \sqrt{H\left(\frac{\|F\|_{L_2(P_n)}}{\varepsilon}\right)}d\varepsilon \, I\left(\|F\|_{L_2(P_n)} \leq 2\|F\|_{L_2(P)}\right)\right]$$

$$\leq Cn^{-1/2}\sigma\sqrt{H\left(\frac{\|F\|_{L_2(P)}}{\sigma}\right)} + \sqrt{8}Cn^{-1/2}\sqrt{\mathbb{E}\|R_n\|_{\mathscr{F}}}\sqrt{H\left(\frac{\|F\|_{L_2(P)}}{\sigma}\right)}.$$

We will use the last bound together with inequalities (3.11) and (3.12). Denote

$$E := \mathbb{E}\|R_n\|_{\mathscr{F}}.$$

Then, we end up with the following inequality

$$E \leq CD_H n^{-1} + Cn^{-1/2}\sigma\sqrt{H\left(\frac{\|F\|_{L_2(P)}}{\sigma}\right)} + \sqrt{8}Cn^{-1/2}\sqrt{E}\sqrt{H\left(\frac{\|F\|_{L_2(P)}}{\sigma}\right)}.$$

Solving it with respect to $E$ completes the proof.          □

The next bounds follow from Theorem 3.12 with $\sigma^2 := \sup_{f \in \mathscr{F}} Pf^2$. If for some $A > 0, V > 0$ and for all $\varepsilon > 0$,

$$N(\mathscr{F}; L_2(P_n); \varepsilon) \leq \left(\frac{A\|F\|_{L_2(P_n)}}{\varepsilon}\right)^V, \tag{3.16}$$

then with some universal constant $C > 0$ (for $\sigma^2 \geq \text{const } n^{-1}$)

$$\mathbb{E}\|R_n\|_{\mathscr{F}} \leq C\left[\sqrt{\frac{V}{n}}\sigma\sqrt{\log\frac{A\|F\|_{L_2(P)}}{\sigma}} \vee \frac{VU}{n}\log\frac{A\|F\|_{L_2(P)}}{\sigma}\right]. \tag{3.17}$$

If for some $A > 0, \rho \in (0, 1)$ and for all $\varepsilon > 0$,

$$\log N(\mathscr{F}; L_2(P_n); \varepsilon) \leq \left( \frac{A \|F\|_{L_2(P_n)}}{\varepsilon} \right)^{2\rho}, \tag{3.18}$$

then

$$\mathbb{E} \|R_n\|_{\mathscr{F}} \leq C \left[ \frac{A^\rho \|F\|_{L_2(P)}^\rho}{\sqrt{n}} \sigma^{1-\rho} \bigvee \frac{A^{2\rho/(\rho+1)} \|F\|_{L_2(P)}^{2\rho/(\rho+1)} U^{(1-\rho)/(1+\rho)}}{n^{1/(1+\rho)}} \right]. \tag{3.19}$$

A function class $\mathscr{F}$ is called VC-subgraph iff

$$\left\{ \{(x, t) : 0 \leq f(x) \leq t\} \cup \{(x, t) : 0 \geq f(x) \geq t\} : f \in \mathscr{F} \right\}$$

is a VC-class. For a VC-subgraph class $\mathscr{F}$, the following bound holds with some constants $A, V > 0$ and for all probability measures $Q$ on $(S, \mathscr{A})$ :

$$N(\mathscr{F}; L_2(Q); \varepsilon) \leq \left( \frac{A \|F\|_{L_2(Q)}}{\varepsilon} \right)^V, \varepsilon > 0 \tag{3.20}$$

(see, e.g., van der Vaart and Wellner [148], Theorem 2.6.7). Of course, this *uniform covering numbers* condition does imply (3.16) and, as a consequence, (3.17).

We will call the function classes satisfying (3.16) *VC-type classes*.

If $\mathscr{H}$ is VC-type, then its convex hull conv($\mathscr{H}$) satisfies (3.18) with $\rho := \frac{V}{V+2}$ (see van der Vaart and Wellner [148], Theorem 2.6.9). More precisely, the following result holds.

**Theorem 3.13.** *Let $\mathscr{H}$ be a class of measurable functions on $(S, \mathscr{A})$ with a measurable envelope $F$ and let $Q$ be a probability measure on $(S, \mathscr{A})$. Suppose that $F \in L_2(Q)$ and*

$$N(\mathscr{H}; L_2(Q); \varepsilon) \leq \left( \frac{A \|F\|_{L_2(Q)}}{\varepsilon} \right)^V, \quad \varepsilon \leq \|F\|_{L_2(Q)}.$$

*Then*

$$\log N(\text{conv}(\mathscr{H}); L_2(Q); \varepsilon) \leq \left( \frac{B \|F\|_{L_2(Q)}}{\varepsilon} \right)^{2V/(V+2)}, \quad \varepsilon \leq \|F\|_{L_2(Q)}$$

*for some constant $B$ that depends on $A$ and $V$.*

So, one can use the bound (3.19) for $\mathscr{F} \subset \text{conv}(\mathscr{H})$. Note that in this bound the envelope $F$ of the class $\mathscr{H}$ itself should be used rather than an envelope of a subset $\mathscr{F}$ of its convex hull (which might be smaller than $F$).

## 3.5   Lower Entropy Bounds

In this section, lower bounds on $\mathbb{E}\|R_n\|_{\mathscr{F}}$ expressed in terms of entropy of the class $\mathscr{F}$ will be proved. Again, we follow the paper by Giné and Koltchinskii [66]. In what follows, the function $H$ satisfies the conditions of Theorem 3.12. Denote $\sigma^2 = \sup_{f \in \mathscr{F}} Pf^2$.

Under the notations of Sect. 3.4, we introduce the following condition: with some constant $c > 0$

$$\log N(\mathscr{F}, L_2(P), \sigma/2) \geq cH\left(\frac{\|F\|_{L_2(P)}}{\sigma}\right). \tag{3.21}$$

**Theorem 3.14.** *Suppose that $\mathscr{F}$ satisfies condition (3.8). There exist a universal constant $B > 0$ and a constant $C_1$ that depends only on $H$ such that*

$$\mathbb{E}\|R_n\|_{\mathscr{F}} \geq B\frac{\sigma}{\sqrt{n}}\sqrt{\log N(\mathscr{F}, L_2(P), \sigma/2)} \tag{3.22}$$

*provided that*

$$n\sigma^2 \geq C_1 U^2 H\left(\frac{6\|F\|_{L_2(P)}}{\sigma}\right). \tag{3.23}$$

*Moreover, if in addition (3.21) holds, then, for some constants $C_2$ depending only on $c$, constant $C_3$ depending only on $H$, and for all $n$ such that (3.23) holds,*

$$C_2\frac{\sigma}{\sqrt{n}}\sqrt{H\left(\frac{\|F\|_{L_2(P)}}{\sigma}\right)} \leq \mathbb{E}\|R_n\|_{\mathscr{F}} \leq C_3\frac{\sigma}{\sqrt{n}}\sqrt{H\left(\frac{\|F\|_{L_2(P)}}{\sigma}\right)}. \tag{3.24}$$

*Proof.* Without loss of generality, we can assume that $U = 1$, so, the functions in the class $\mathscr{F}$ are bounded by 1. The general case would follow by a simple rescaling. First note that, under the assumptions of the theorem, inequality (3.10) holds, so, we have with some constant $C$ depending only on $H$

$$\mathbb{E}\|R_n\|_{\mathscr{F}} \leq C\frac{\sigma}{\sqrt{n}}\sqrt{H\left(\frac{\|F\|_{L_2(P)}}{\sigma}\right)}.$$

This already proves the right hand side of inequality (3.24).

It follows from Theorem 3.4 that

$$\mathbb{E}_\varepsilon\|R_n\|_{\mathscr{F}} \geq \frac{1}{8L}\frac{\sigma}{\sqrt{n}}\sqrt{\log N(\mathscr{F}, L_2(P_n), \sigma/8)}, \tag{3.25}$$

as soon as

$$\mathbb{E}_\varepsilon\|R_n\|_{\mathscr{F}} \leq \frac{\sigma^2}{64L}. \tag{3.26}$$

To use this result, we will derive a lower bound on the right hand side of (3.25) and an upper bound on the left hand side of (3.26) that hold with a high probability. Let us bound first the right hand side of (3.25).

Let

$$M := M(\mathscr{F}, L_2(P), \sigma/2)$$

(recall that $M(\mathscr{F}, L_2(P), \sigma/2)$ denotes the $\sigma/2$-packing number of the class $\mathscr{F} \subset L_2(P)$). We apply the law of large numbers to $M$ functions in a maximal $\sigma/2$-separated subset of $\mathscr{F}$ and also to the envelope $F$. It implies that, for all $\varepsilon > 0$, there exists $n$ and $\omega$ such that

$$M(\mathscr{F}, L_2(P), \sigma/2) \le M(\mathscr{F}, L_2(P_n(\omega)), (1-\varepsilon)\sigma/2)$$
$$\le N(\mathscr{F}, L_2(P_n(\omega)), (1-\varepsilon)\sigma/4)$$

and

$$\|F\|_{L_2(P_n(\omega))} \le (1+\varepsilon)\|F\|_{L_2(P)}.$$

Take $\varepsilon = 1/5$. Then, by (3.8),

$$M(\mathscr{F}, L_2(P), \sigma/2) \le \exp\left\{ H\left( \frac{6\|F\|_{L_2(P)}}{\sigma} \right) \right\}. \tag{3.27}$$

Let $f_1, \ldots, f_M$ be a maximal subset of $\mathscr{F}$ such that

$$P(f_i - f_j)^2 \ge \sigma^2/4 \text{ for all } 1 \le i \ne j \le M.$$

In addition, we have

$$P(f_i - f_j)^4 \le 4P(f_i - f_j)^2 \le 16\sigma^2.$$

Bernstein's inequality implies that

$$\mathbb{P}\left\{ \max_{1 \le i \ne j \le M} \left( nP(f_i - f_j)^2 - \sum_{k=1}^{n}(f_i - f_j)^2(X_k) \right) > \frac{8}{3}t + \sqrt{32tn\sigma^2} \right\} \le M^2 e^{-t}.$$

Let $t = \delta n\sigma^2$. Since $P(f_i - f_j)^2 \ge \sigma^2/4$ and (3.27) holds, we get

$$\mathbb{P}\left\{ \min_{1 \le i \ne j \le M} \frac{1}{n}\sum_{k=1}^{n}(f_i - f_j)^2(X_k) \le \sigma^2\left( 1/4 - 8\delta/3 - \sqrt{32\delta} \right) \right\}$$

$$\le \exp\left\{ 6H\left( \frac{3\|F\|_{L_2(P)}}{\sigma} \right) - \delta n\sigma^2 \right\}.$$

For $\delta = 1/(32 \cdot 8^3)$, this yields

$$\mathbb{P}\left\{\min_{1\leq i\neq j\leq M} P_n(f_i - f_j)^2 \leq \frac{\sigma^2}{16}\right\} \leq \exp\left\{H\left(\frac{6\|F\|_{L_2(P)}}{\sigma}\right) - \frac{n\sigma^2}{32\cdot 8^3}\right\}. \quad (3.28)$$

Denote

$$E_1 := \left\{\omega : M(\mathscr{F}, L_2(P_n), \sigma/4) \geq M\right\}.$$

On this event,

$$N(\mathscr{F}, L_2(P_n), \sigma/8) \geq M(\mathscr{F}, L_2(P_n), \sigma/4) \geq$$
$$M = M(\mathscr{F}, L_2(P), \sigma/2) \geq N(\mathscr{F}, L_2(P), \sigma/2)$$

and

$$\mathbb{P}(E_1) \geq 1 - \exp\left\{H\left(\frac{6\|F\|_{L_2(P)}}{\sigma}\right) - \frac{n\sigma^2}{32\cdot 8^3}\right\}. \quad (3.29)$$

Using symmetrization and contraction inequalities and condition (3.23), we have

$$\mathbb{E}\|P_n - P\|_{\mathscr{F}^2} \leq 2\mathbb{E}\|R_n\|_{\mathscr{F}^2} \leq 8\mathbb{E}\|R_n\|_{\mathscr{F}} \leq C\frac{\sigma}{\sqrt{n}}\sqrt{H\left(\frac{\|F\|_{L_2(P)}}{\sigma}\right)} \leq 6\sigma^2 \quad (3.30)$$

(with a proper choice of constant $C_1$ in (3.23)). Next, Bousquet's version of Talagrand's inequality (see Sect. 2.3) yields the bound

$$\mathbb{P}\left\{\|P_n - P\|_{\mathscr{F}^2} \geq 6\sigma^2 + \sigma\sqrt{\frac{26t}{n}} + \frac{t}{3n}\right\} \leq e^{-t}.$$

We take $t = 26n\sigma^2$. Then

$$\mathbb{P}\left\{\|P_n - P\|_{\mathscr{F}^2} \geq 41\sigma^2\right\} \leq \exp\{-26n\sigma^2\}.$$

Denote

$$E_2 := \left\{\omega : \sigma_n^2 = \sup_{f\in\mathscr{F}} P_n f^2 < 42\sigma^2\right\}. \quad (3.31)$$

Then

$$\mathbb{P}(E_2) > 1 - \exp\{-26n\sigma^2\}. \quad (3.32)$$

Also, by Bernstein's inequality, the event

$$E_3 = \{\omega : \|F\|_{L_2(P_n)} \leq 2\|F\|_{L_2(P)}\} \quad (3.33)$$

has probability

$$\mathbb{P}(E_3) \geq 1 - \exp\left\{-\frac{9}{4}n\|F\|_{L_2(P)}^2\right\}. \quad (3.34)$$

On the event $E_2 \cap E_3$, (3.7) and (3.23) yield that, with some constant $C$ depending only on $H$, the following bounds hold[2]:

$$
\mathbb{E}_\varepsilon \|R_n\|_{\mathscr{F}} \leq \frac{C}{\sqrt{n}} \int_0^{2\sigma_n} \sqrt{H\left(\frac{\|F\|_{L_2(P_n)}}{\varepsilon}\right)} d\varepsilon
$$

$$
\leq \frac{C}{\sqrt{n}} \int_0^{2\sqrt{42}\sigma} \sqrt{H\left(\frac{2\|F\|_{L_2(P)}}{\varepsilon}\right)} d\varepsilon
$$

$$
\leq \frac{2C}{\sqrt{n}} \|F\|_{L_2(P)} \int_0^{2\sqrt{42}\sigma/\|F\|_{L_2(P)}} \sqrt{H\left(\frac{1}{\varepsilon}\right)} d\varepsilon
$$

$$
= \frac{2C}{\sqrt{n}} \|F\|_{L_2(P)} \int_{(2\sqrt{42})^{-1}\|F\|_{L_2(P)}/\sigma}^{+\infty} u^{-2} \sqrt{H(u)} du.
$$

Arguing as in the derivation of (3.15), the integral in the right hand side can be bounded from above by

$$
C \frac{\sigma}{\sqrt{n}} \sqrt{H\left(\frac{\|F\|_{L_2(P)}}{\sigma}\right)}
$$

with a constant $C$ depending only on $H$. This leads to the following bound

$$
\mathbb{E}_\varepsilon \|R_n\|_{\mathscr{F}} \leq C \frac{\sigma}{\sqrt{n}} \sqrt{H\left(\frac{\|F\|_{L_2(P)}}{\sigma}\right)} < \frac{\sigma^2}{64L} \tag{3.35}
$$

(which again holds with a proper choice of constant $C_1$ in (3.23)). It follows from (3.25) to (3.35) that

$$
\mathbb{E}\|R_n\|_{\mathscr{F}} \geq \frac{1}{8L} \frac{\sigma}{\sqrt{n}} \sqrt{\log N(\mathscr{F}, L_2(P), \sigma/2)} \mathbb{P}(E_1 \cap E_2 \cap E_3) \tag{3.36}
$$

and that

$$
\mathbb{P}(E_1 \cap E_2 \cap E_3) \geq
$$

$$
1 - \exp\left\{ H\left(\frac{6\|F\|_{L_2(P)}}{\sigma}\right) - \frac{n\sigma^2}{32 \cdot 8^3}\right\} - \exp\{-26n\sigma^2\} - \exp\{-9n\sigma^2/4\}.
$$

This last probability is larger than $1/2$ by condition (3.23) with a proper value of $C_1$. Thus, (3.36) implies inequality (3.22). The left hand side of inequality (3.24) now follows from (3.22) and (3.21), completing the proof.                                  □

---

[2]Note that $C$ might change its value from place to place.

## 3.6    Generic Chaining Complexities and Bounding Empirical Processes Indexed by $\mathscr{F}^2$

Generic chaining complexities can be used to control the size of empirical processes indexed by a function class $\mathscr{F}$ (see [140]). For instance, one can define the complexity $\gamma_2(\mathscr{F}; L_2(P))$, that is, $\gamma_2(\mathscr{F}; d)$, where $d$ is the $L_2(P)$-distance. Another useful distance is based on the $\psi_2$-norm for random variables on the probability space $(S, \mathscr{A}, P)$ (see Sect. A.1). The generic chaining complexity that corresponds to the $\psi_2$-distance will be denoted by $\gamma_2(\mathscr{F}; \psi_2)$. In particular, these complexities were used to bound the sup-norm of the empirical process indexed by the class $\mathscr{F}^2 := \{f^2 : f \in \mathscr{F}\}$. This is of importance in a variety of applications including sparse recovery problems. The goal is to control this empirical process in terms of complexity measures of the class $\mathscr{F}$ rather than the class $\mathscr{F}^2$. A standard approach to this problem is to use the symmetrization inequality (to replace the empirical process by the Rademacher process) followed by the comparison inequality for Rademacher sums. However, for this approach, one has to deal with the uniformly bounded class $\mathscr{F}$ (the Lipschitz constant in the comparison inequality would be in this case $2 \sup_{f \in \mathscr{F}} \|f\|_\infty$). In many interesting applications (for instance, in sparse recovery) the quantity $\sup_{f \in \mathscr{F}} \|f\|_\infty$ might be infinite, or very large. To overcome this difficulty Klartag and Mendelson [76]) started developing another approach based on generic chaining bounds for empirical processes. Quite recently, following this path, Mendelson [115] proved the following deep result.

**Theorem 3.15.** *Suppose that $\mathscr{F}$ is a symmetric class, that is, $f \in \mathscr{F}$ implies $-f \in \mathscr{F}$, and $Pf = 0, f \in \mathscr{F}$. Then, for some universal constant $K > 0$,*

$$\mathbb{E}\|P_n - P\|_{\mathscr{F}^2} \leq K\left[\sup_{f \in \mathscr{F}} \|f\|_{\psi_1} \frac{\gamma_2(\mathscr{F}; \psi_2)}{\sqrt{n}} \bigvee \frac{\gamma_2^2(\mathscr{F}; \psi_2)}{n}\right].$$

We will discuss one more result in the same direction which provides a bound on $\mathbb{E}\|P_n - P\|_{\mathscr{F}^2}$ in terms of $L_\infty(P_n)$ generic chaining complexity $\gamma_2(\mathscr{F}; L_\infty(P_n))$ of class $\mathscr{F}$. Denote $\sigma^2 := \sup_{f \in \mathscr{F}} Pf^2$ and

$$\Gamma_{n,\infty}(\mathscr{F}) := \mathbb{E}\gamma_2^2(\mathscr{F}; L_\infty(P_n)).$$

**Theorem 3.16.** *There exists a universal constant $K > 0$ such that*

$$\mathbb{E} \sup_{f \in \mathscr{F}} |P_n f^2 - Pf^2| \leq K\left[\sigma\sqrt{\frac{\Gamma_{n,\infty}(\mathscr{F})}{n}} \bigvee \frac{\Gamma_{n,\infty}(\mathscr{F})}{n}\right]. \tag{3.37}$$

*Proof.* We start with the first bound of Theorem 3.11 and apply it together with symmetrization inequality to class $\mathscr{F}^2$ to get that with some constant $C > 0$

$$\mathbb{E} \sup_{f \in \mathscr{F}} |P_n f^2 - P f^2| \leq \frac{C}{\sqrt{n}} \mathbb{E} \gamma_2(\mathscr{F}^2; L_2(P_n)). \tag{3.38}$$

Next we have

$$\|f^2 - g^2\|^2_{L_2(P_n)} = n^{-1} \sum_{j=1}^{n} |f^2(X_j) - g^2(X_j)|^2$$

$$= n^{-1} \sum_{j=1}^{n} (f(X_j) - g(X_j))^2 (f(X_j) + g(X_j))^2 \leq 4 \sup_{f \in \mathscr{F}} P_n f^2 \|f - g\|^2_{L_\infty(P_n)},$$

which implies

$$\|f^2 - g^2\|_{L_2(P_n)} \leq 2\sigma_n \|f - g\|_{L_\infty(P_n)}, \tag{3.39}$$

where $\sigma_n^2 = \sup_{f \in \mathscr{F}} P_n f^2$. It follows from (3.39) that

$$\gamma_2(\mathscr{F}^2; L_2(P_n)) \leq 2\sigma_n \gamma_2(\mathscr{F}; L_\infty(P_n)).$$

and (3.38) implies that

$$E := \mathbb{E} \sup_{f \in \mathscr{F}} |P_n f^2 - P f^2| \leq \frac{2C}{\sqrt{n}} \mathbb{E} \sigma_n \gamma_2(\mathscr{F}; L_\infty(P_n))$$

$$\leq \frac{2C}{\sqrt{n}} \mathbb{E}^{1/2} \sigma_n^2 \sqrt{\Gamma_{n,\infty}(\mathscr{F})}.$$

Note also that

$$\mathbb{E}^{1/2} \sigma_n^2 = \mathbb{E}^{1/2} \sup_{f \in \mathscr{F}} P_n f^2 \leq \mathbb{E}^{1/2} \left( \sup_{f \in \mathscr{F}} |P_n f^2 - P f^2| + \sigma^2 \right) \leq \sqrt{E} + \sigma.$$

Therefore, (3.40) implies that with some constant $C > 0$

$$E \leq \frac{2C}{\sqrt{n}} (\sqrt{E} + \sigma) \sqrt{\Gamma_{n,\infty}(\mathscr{F})},$$

and bound (3.37) easily follows by solving the last inequality for $E$.     □

## 3.7   Function Classes in Hilbert Spaces

Suppose that $L$ is a finite dimensional subspace of $L_2(P)$ with $\dim(L) = d$. Denote

$$\psi_L(x) := \frac{1}{\sqrt{d}} \sup_{f \in L, \|f\|_{L_2(P)} \leq 1} |f(x)|.$$

We will use the following $L_p$-version of Hoffmann–Jørgensen inequality: for all independent mean zero random variables $Y_j$, $j = 1, \ldots, n$ with values in a Banach space $B$ and with $\mathbb{E}\|Y_j\|^p < +\infty$, for some $p \geq 1$,

$$\mathbb{E}^{1/p} \left\| \sum_{j=1}^{n} Y_j \right\|^p \leq K_p \left( \mathbb{E} \left\| \sum_{j=1}^{n} Y_j \right\| + \mathbb{E}^{1/p} \left( \max_{1 \leq i \leq n} \|Y_i\| \right)^p \right), \tag{3.40}$$

where $K_p$ is a constant depending only on $p$ (see Ledoux and Talagrand [101], Theorem 6.20).

**Proposition 3.2.** *Let* $\mathscr{F} := \{f \in L : \|f\|_{L_2(P)} \leq r\}$. *Then*

$$\mathbb{E}\|R_n\|_{\mathscr{F}} \leq \mathbb{E}^{1/2}\|R_n\|_{\mathscr{F}}^2 = r\sqrt{\frac{d}{n}}.$$

*Moreover, there exists a universal constant $K$ such that whenever*

$$\mathbb{E} \max_{1 \leq i \leq n} \psi_L^2(X_i) \leq \frac{n}{K^2},$$

*we have*

$$\mathbb{E}\|R_n\|_{\mathscr{F}} \geq \frac{1}{K} r \sqrt{\frac{d}{n}}.$$

*Proof.* Let $\phi_1, \ldots, \phi_d$ be an orthonormal basis of $L$. Then

$$\|R_n\|_{\mathscr{F}} := \sup_{f \in L, \|f\|_{L_2(P)} \leq r} |R_n(f)| = \sup \left\{ \left| R_n \left( \sum_{j=1}^{d} \alpha_j \phi_j \right) \right| : \sum_{j=1}^{d} \alpha_j^2 \leq r^2 \right\}$$

$$= \sup \left\{ \left| \sum_{j=1}^{d} \alpha_j R_n(\phi_j) \right| : \sum_{j=1}^{d} \alpha_j^2 \leq r^2 \right\} = r \left( \sum_{j=1}^{d} R_n^2(\phi_j) \right)^{1/2}.$$

Therefore,

$$\mathbb{E}\|R_n\|_{\mathscr{F}}^2 = r^2 \sum_{j=1}^{d} \mathbb{E} R_n^2(\phi_j),$$

and the first statement follows since

$$\mathbb{E} R_n^2(\phi_j) = \frac{P\phi_j^2}{n}, \quad j = 1, \ldots, n.$$

The proof of the second statement follows from the first statement and inequality (3.40), which immediately yields

$$r\sqrt{\frac{d}{n}} = \mathbb{E}^{1/2} \|R_n\|_{\mathscr{F}}^2 \leq K_2 \left( \mathbb{E} \|R_n\|_{\mathscr{F}} + r\sqrt{\frac{d}{n}} \frac{1}{\sqrt{n}} \mathbb{E}^{1/2} \max_{1 \leq i \leq n} \psi_L^2(X_i) \right),$$

and the result follows by assuming that $K = 2K_2$. $\qquad\square$

Let $K$ be a symmetric nonnegatively definite square integrable kernel on $S \times S$ and let $\mathscr{H}_K$ be the corresponding *reproducing kernel Hilbert space (RKHS)*, i.e., $\mathscr{H}_K$ is the completion of the linear span of functions $\{K(x, \cdot) : x \in S\}$ with respect to the following inner product:

$$\left\langle \sum_i \alpha_i K(x_i, \cdot), \sum_j \beta_j K(y_i, \cdot) \right\rangle_K = \sum_{i,j} \alpha_i \beta_j K(x_i, y_j).$$

The corresponding norm will be denoted by $\| \cdot \|_K$. Let

$$\mathscr{F} := \{ f \in \mathscr{H}_K : \|f\|_K \leq 1 \text{ and } \|f\|_{L_2(P)} \leq r \}$$

Finally, let $A_K$ denote the linear integral operator from $L_2(P)$ into $L_2(P)$ with kernel $K$,

$$A_K f(x) = \int_S K(x, y) f(y) P(dy),$$

let $\{\lambda_i\}$ denote its eigenvalues arranged in decreasing order and $\{\phi_i\}$ denote the corresponding $L_2(P)$-orthonormal eigenfunctions.

The following result is due to Mendelson [113].

**Proposition 3.3.** *There exist universal constants* $C_1, C_2 > 0$ *such that*

$$C_1 \left( n^{-1} \sum_{j=1}^{\infty} (\lambda_j \wedge r^2) \right)^{1/2} \leq \mathbb{E}^{1/2} \|R_n\|_{\mathscr{F}}^2 \leq C_2 \left( n^{-1} \sum_{j=1}^{\infty} (\lambda_j \wedge r^2) \right)^{1/2}.$$

*In addition, there exists a universal constant* $C$ *such that*

$$\mathbb{E} \|R_n\|_{\mathscr{F}} \geq \frac{1}{C} \left( n^{-1} \sum_{j=1}^{\infty} (\lambda_j \wedge r^2) \right)^{1/2} - \frac{\sqrt{\sup_{x \in S} K(x, x)}}{n}.$$

*Proof.* By the well known properties of RKHS,

$$\mathscr{F} = \left\{ \sum_{k=1}^{\infty} c_k \phi_k : c = (c_1, c_2, \dots) \in \mathscr{E}_1 \cap \mathscr{E}_2 \right\},$$

where

$$\mathscr{E}_1 := \left\{ c : \sum_{k=1}^{\infty} \frac{c_k^2}{\lambda_k} \leq 1 \right\} \text{ and } \mathscr{E}_2 := \left\{ c : \sum_{k=1}^{\infty} \frac{c_k^2}{r^2} \leq 1 \right\}.$$

In other words, the set $\mathscr{E}_1$ is the ellipsoid in $\ell_2$ (with the center at the origin) with "half-axes" $\sqrt{\lambda_k}$ and $\mathscr{E}_2$ is the ellipsoid with "half-axes" $r$ (a ball of radius $r$). Let

$$\mathscr{E} := \left\{ c : \sum_{k=1}^{\infty} \frac{c_k^2}{\lambda_k \wedge r^2} \leq 1 \right\}$$

denote the ellipsoid with "half-axes" $\sqrt{\lambda_k} \wedge r$. A straightforward argument shows that $\mathscr{E} \subset \mathscr{E}_1 \cap \mathscr{E}_2 \subset \sqrt{2}\mathscr{E}$. Hence,

$$\sup_{c \in \mathscr{E}} \left| R_n \left( \sum_{k=1}^{\infty} c_k \phi_k \right) \right| \leq \| R_n \|_{\mathscr{F}} \leq \sqrt{2} \sup_{c \in \mathscr{E}} \left| R_n \left( \sum_{k=1}^{\infty} c_k \phi_k \right) \right|.$$

Also, we have

$$\sup_{c \in \mathscr{E}} \left| R_n \left( \sum_{k=1}^{\infty} c_k \phi_k \right) \right|^2 = \sup_{c \in \mathscr{E}} \left| \sum_{k=1}^{\infty} \frac{c_k}{\sqrt{\lambda_k} \wedge r} \left( \sqrt{\lambda_k} \wedge r \right) R_n(\phi_k) \right|^2$$

$$= \sum_{k=1}^{\infty} \left( \lambda_k \wedge r^2 \right) R_n^2(\phi_k).$$

Hence,

$$\mathbb{E} \sup_{c \in \mathscr{E}} \left| R_n \left( \sum_{k=1}^{\infty} c_k \phi_k \right) \right|^2 = \sum_{k=1}^{\infty} \left( \lambda_k \wedge r^2 \right) \mathbb{E} R_n^2(\phi_k).$$

Since $P\phi_k^2 = 1$, $\mathbb{E} R_n^2(\phi_k) = \frac{1}{n}$, we get

$$\mathbb{E} \sup_{c \in \mathscr{E}} \left| R_n \left( \sum_{k=1}^{\infty} c_k \phi_k \right) \right|^2 = n^{-1} \sum_{k=1}^{\infty} (\lambda_k \wedge r^2),$$

and the first bound follows.

The proof of the second bound is based on the observation that

$$\sup_{f \in \mathscr{F}} |f(x)| \leq \sqrt{\sup_{x \in S} K(x, x)}$$

and on the same application of Hoffmann–Jørgensen inequality as in the previous proposition. □

A similar result with the identical proof holds for data-dependent Rademacher complexity $\mathbb{E}_\varepsilon \|R_n\|_{\mathscr{F}}$. In this case, let $\{\lambda_i^{(n)}\}$ be the eigenvalues (arranged in decreasing order) of the random matrix $\left(n^{-1} K(X_i, X_j)\right)_{i,j=1}^n$ (equivalently, of the integral operator from $L_2(P_n)$ into $L_2(P_n)$ with kernel $K$).

**Proposition 3.4.** *There exist universal constants $C_1, C_2 > 0$ such that*

$$C_1 \left( n^{-1} \sum_{j=1}^n (\lambda_j^{(n)} \wedge r^2) \right)^{1/2} \leq \mathbb{E}_\varepsilon^{1/2} \|R_n\|_{\mathscr{F}}^2 \leq C_2 \left( n^{-1} \sum_{j=1}^n (\lambda_j^{(n)} \wedge r^2) \right)^{1/2}.$$

*In addition, there exists a universal constant $C$ such that*

$$\mathbb{E}_\varepsilon \|R_n\|_{\mathscr{F}} \geq \frac{1}{C} \left( n^{-1} \sum_{j=1}^n (\lambda_j^{(n)} \wedge r^2) \right)^{1/2} - \frac{\sqrt{\sup_{x \in S} K(x, x)}}{n}.$$

## 3.8 Further Comments

The main reference to the generic chaining method is the book by Talagrand [140]. Shattering numbers and Vapnik–Chervonenkis classes have been discussed in many books [51, 59, 148].

Special cases of the inequalities discussed in Sect. 3.4 can be found in Talagrand [137], Einmahl and Mason [60], Giné and Guillou [64], Mendelson [112], Giné et al. [65]. Theorem 3.12 is given in Giné and Koltchinskii [66] (in a slightly more precise form). Lower bounds proved in Sect. 3.5 are due to Giné and Koltchinskii [66].

A number of other entropy bounds on suprema of empirical and Rademacher processes (in particular, in terms of so called bracketing numbers) can be found in Dudley [59] and van der Vaart and Wellner [148]. Recently, van der Vaart and Wellner [149] proved new versions of bounds under uniform entropy conditions (both for bounded and for unbounded function classes).

Generic chaining complexities were used by Klartag and Mendelson [76] to bound empirical processes indexed by the squares of functions. This method was further developed in [117] and, especially, in [115]. Another approach is based on $L_\infty(P_n)$-covering numbers and generic chaining complexities (see Theorem 3.16). It goes back to Rudelson [128] and it was used in learning theory and sparse recovery problems in [18, 116]. Similar idea was also used by Giné and Mason [68].

# Chapter 4
# Excess Risk Bounds

In this chapter, we develop distribution dependent and data dependent upper bounds on the excess risk $\mathcal{E}_P(\hat{f}_n)$ of an empirical risk minimizer

$$\hat{f}_n := \operatorname{argmin}_{f \in \mathscr{F}} P_n f. \tag{4.1}$$

We will assume that such a minimizer exists (a simple modification of the results is possible if $\hat{f}_n$ is an approximate solution of (4.1)). Our approach to this problem has been already outlined in Chap. 1 and it is closely related to the recent work of Massart [106], Koltchinskii and Panchenko [92], Bartlett et al. [15], Bousquet et al. [34], Koltchinskii [83], Bartlett and Mendelson [17].

## 4.1 Distribution Dependent Bounds and Ratio Bounds for Excess Risk

To simplify the matter, assume that the functions in $\mathscr{F}$ take their values in $[0, 1]$. Recall that the set

$$\mathscr{F}_P(\delta) := \left\{ f \in \mathscr{F} : \mathcal{E}_P(f) \leq \delta \right\}$$

is called the $\delta$-minimal set of the risk $P$. In particular, $\mathscr{F}_P(0)$ is its minimal set. Define $\rho_P : L_2(P) \times L_2(P) \mapsto [0, +\infty)$ such that

$$\rho_P^2(f, g) \geq P(f - g)^2 - (P(f - g))^2, \quad f, g \in L_2(P).$$

Usually, $\rho_P$ is also a (pseudo)metric, such as

$$\rho_P^2(f, g) = P(f - g)^2 \text{ or } \rho_P^2(f, g) = P(f - g)^2 - (P(f - g))^2.$$

V. Koltchinskii, *Oracle Inequalities in Empirical Risk Minimization and Sparse Recovery Problems*, Lecture Notes in Mathematics 2033, DOI 10.1007/978-3-642-22147-7_4,

Under the notations of Sect. 1.2,

$$D(\delta) := D_P(\mathcal{F}; \delta) := \sup_{f,g \in \mathcal{F}(\delta)} \rho_P(f, g)$$

is the $\rho_P$-diameter of the $\delta$-minimal set. Also, denote

$$\mathcal{F}'(\delta) := \left\{ f - g : f, g \in \mathcal{F}(\delta) \right\}$$

and

$$\phi_n(\delta) := \phi_n(\mathcal{F}; P; \delta) := \mathbb{E} \| P_n - P \|_{\mathcal{F}'(\delta)}.$$

Let $\{\delta_j\}_{j \geq 0}$ be a decreasing sequence of positive numbers with $\delta_0 = 1$ and let $\{t_j\}_{j \geq 0}$ be a sequence of positive numbers. For $\delta \in (\delta_{j+1}, \delta_j]$, define

$$U_n(\delta) := \phi_n(\delta_j) + \sqrt{2 \frac{t_j}{n}(D^2(\delta_j) + 2\phi_n(\delta_j))} + \frac{t_j}{2n}. \tag{4.2}$$

Finally, denote

$$\delta_n(\mathcal{F}; P) := \sup\{\delta \in (0, 1] : \delta \leq U_n(\delta)\}.$$

It is easy to check that

$$\delta_n(\mathcal{F}, P) \leq U_n(\delta_n(\mathcal{F}, P)).$$

Obviously, the definitions of $U_n$ and $\delta_n(\mathcal{F}, P)$ depend on the choice of $\{\delta_j\}$ and $\{t_j\}$.

We start with the following simple inequality that provides a distribution dependent upper bound on the excess risk $\mathcal{E}_P(\hat{f}_n)$.

**Theorem 4.1.** *For all $\delta \geq \delta_n(\mathcal{F}; P)$,*

$$\mathbb{P}\{\mathcal{E}(\hat{f}_n) > \delta\} \leq \sum_{\delta_j \geq \delta} e^{-t_j}.$$

*Proof.* It is enough to assume that $\delta > \delta_n(\mathcal{F}; P)$ (otherwise, the result follows by continuity). Denote $\hat{\delta} := \mathcal{E}(\hat{f}_n)$. If $\hat{\delta} \geq \delta \geq \varepsilon > 0$ and $g \in \mathcal{F}(\varepsilon)$, we have

$$\hat{\delta} = P\hat{f}_n - \inf_{g \in \mathcal{F}} Pg \leq P(\hat{f}_n - g) + \varepsilon$$

$$\leq P_n(\hat{f}_n - g) + (P - P_n)(f - g) + \varepsilon \leq \| P_n - P \|_{\mathcal{F}'(\hat{\delta})} + \varepsilon.$$

By letting $\varepsilon \to 0$, this gives $\hat{\delta} \leq \| P_n - P \|_{\mathcal{F}'(\hat{\delta})}$. Denote

$$E_{n,j} := \left\{ \| P_n - P \|_{\mathcal{F}'(\delta_j)} \leq U_n(\delta_j) \right\}.$$

It follows from Bousquet's version of Talagrand's inequality (see Sect. 2.3) that $\mathbb{P}(E_{n,j}) \geq 1 - e^{-t_j}$. Let

$$E_n := \bigcap_{\delta_j \geq \delta} E_{n,j}.$$

Then

$$\mathbb{P}(E_n) \geq 1 - \sum_{\delta_j \geq \delta} e^{-t_j}.$$

On the event $E_n$, for all $\sigma \geq \delta$, $\|P_n - P\|_{\mathscr{F}'(\sigma)} \leq U_n(\sigma)$, which holds by the definition of $U_n(\delta)$ and monotonicity of the function $\delta \mapsto \|P_n - P\|_{\mathscr{F}'(\delta)}$. Thus, on the event $\{\hat{\delta} \geq \delta\} \cap E_n$, we have

$$\delta \leq \|P_n - P\|_{\mathscr{F}'(\hat{\delta})} \leq U_n(\hat{\delta}),$$

which implies that $\delta \leq \hat{\delta} \leq \delta_n(\mathscr{F}; P)$, contradicting the assumption that $\delta > \delta_n(\mathscr{F}; P)$. Therefore, we must have $\{\hat{\delta} \geq \delta\} \subset E_n^c$, and the result follows.    □

We now turn to uniform bounds on the ratios of the excess empirical risk of a function $f \in \mathscr{F}$ to its true excess risk. The excess empirical risk is defined as

$$\hat{\mathscr{E}}_n(f) := \mathscr{E}_{P_n}(f).$$

Given $\psi : \mathbb{R}_+ \mapsto \mathbb{R}_+$, denote

$$\psi^\flat(\delta) := \sup_{\sigma \geq \delta} \frac{\psi(\sigma)}{\sigma}$$

and

$$\psi^\sharp(\varepsilon) := \inf\left\{\delta > 0 : \psi^\flat(\delta) \leq \varepsilon\right\}.$$

These transformations will be called the $\flat$-transform and the $\sharp$-transform of $\psi$, respectively. Some of their simple properties are summarized in Sect. A.3.

It happens that, with a high probability, the quantity

$$\sup_{f \in \mathscr{F}, \mathscr{E}(f) \geq \delta} \left| \frac{\hat{\mathscr{E}}_n(f)}{\mathscr{E}(f)} - 1 \right|$$

can be bounded from above by the function $\delta \mapsto V_n(\delta) := U_n^\flat(\delta)$.

**Theorem 4.2.** *For all $\delta \geq \delta_n(\mathscr{F}; P)$,*

$$\mathbb{P}\left\{ \sup_{f \in \mathscr{F}, \mathscr{E}(f) \geq \delta} \left| \frac{\hat{\mathscr{E}}_n(f)}{\mathscr{E}(f)} - 1 \right| > V_n(\delta) \right\} \leq \sum_{\delta_j \geq \delta} e^{-t_j}.$$

*Proof.* Consider the event $E_n$ defined in the proof of Theorem 4.1. For this event

$$\mathbb{P}(E_n) \geq 1 - \sum_{\delta_j \geq \delta} e^{-t_j},$$

so, it is enough to prove that the inequality

$$\sup_{f \in \mathscr{F}, \mathscr{E}(f) \geq \delta} \left| \frac{\hat{\mathscr{E}}_n(f)}{\mathscr{E}(f)} - 1 \right| \leq V_n(\delta)$$

holds on the event $E_n$. To this end, note that on $E_n$, by the proof of Theorem 4.1, $\hat{f}_n \in \mathscr{F}(\delta)$. For all $f \in \mathscr{F}$ such that $\sigma := \mathscr{E}(f) \geq \delta$, for arbitrary $\varepsilon \in (0, \delta)$ and $g \in \mathscr{F}(\varepsilon)$, the following bounds hold:

$$\sigma = \mathscr{E}(f) \leq Pf - Pg + \varepsilon \leq P_n f - P_n g + (P - P_n)(f - g) + \varepsilon$$
$$\leq \hat{\mathscr{E}}_n(f) + \|P_n - P\|_{\mathscr{F}'(\sigma)} + \varepsilon \leq \hat{\mathscr{E}}_n(f) + U_n(\sigma) + \varepsilon \leq \hat{\mathscr{E}}_n(f) + V_n(\delta)\sigma + \varepsilon,$$

which means that on the event $E_n$ the condition $\mathscr{E}(f) \geq \delta$ implies that

$$\hat{\mathscr{E}}_n(f) \geq \left(1 - V_n(\delta)\right)\mathscr{E}(f).$$

Similarly, on $E_n$, the condition $\sigma := \mathscr{E}(f) \geq \delta$ implies that

$$\hat{\mathscr{E}}_n(f) = P_n f - P_n \hat{f}_n \leq Pf - P\hat{f}_n + (P_n - P)(f - \hat{f}_n) \leq$$

$$\leq \mathscr{E}(f) + U_n(\sigma) \leq \mathscr{E}(f) + V_n(\delta)\sigma = \left(1 + V_n(\delta)\right)\mathscr{E}(f),$$

and the result follows.                                                                      □

A convenient choice of sequence $\{\delta_j\}$ is $\delta_j := q^{-j}$, $j \geq 0$ with some fixed $q > 1$. If $t_j = t > 0$, $j \geq 0$, the corresponding functions $U_n(\delta)$ and $V_n(\delta)$ will be denoted by $U_n(\delta; t)$ and $V_n(\delta; t)$, and $\delta_n(\mathscr{F}; P)$ will be denoted by $\delta_n(t)$.

The following corollary is obvious.

**Corollary 4.1.** *For all $t > 0$ and for all $\delta \geq \delta_n(t)$,*

$$\mathbb{P}\{\mathscr{E}(\hat{f}_n) \geq \delta\} \leq \left(\log_q \frac{q}{\delta}\right) e^{-t}$$

*and*

$$\mathbb{P}\left\{ \sup_{f \in \mathscr{F}, \mathscr{E}(f) \geq \delta} \left| \frac{\hat{\mathscr{E}}_n(f)}{\mathscr{E}(f)} - 1 \right| > V_n(\delta; t) \right\} \leq \left(\log_q \frac{q}{\delta}\right) e^{-t}.$$

It follows from the definition of $\delta_n(t)$ that $\delta_n(t) \geq \frac{t}{n}$. Because of this, the probabilities in Corollary 4.1 can be bounded from above by $\log_q \frac{n}{t} \exp\{-t\}$ (which depends neither on the class $\mathscr{F}$, nor on $P$). Most often, the logarithmic factor in front of the exponent does not create a problem: in typical applications, $\delta_n(t)$ is upper bounded by $\delta_n + \frac{t}{n}$, where $\delta_n$ is larger than $\frac{\log \log n}{n}$. Adding $\log \log n$ to $t$ is enough to eliminate the impact of the logarithm. However, if $\delta_n = O(n^{-1})$, the presence of the logarithmic factor would result in a suboptimal bound. To tackle this difficulty, we will use a slightly different choice of $\{\delta_j\}$, $\{t_j\}$.

For $q > 1$ and $t > 0$, denote

$$V_n^t(\sigma) := 2q\left[\phi_n^\flat(\sigma) + \sqrt{(D^2)^\flat(\sigma)}\sqrt{\frac{t}{n\sigma}} + \frac{t}{n\sigma}\right], \quad \sigma > 0.$$

Let

$$\sigma_n^t := \sigma_n^t(\mathscr{F}; P) := \inf\{\sigma : V_n^t(\sigma) \leq 1\}.$$

**Theorem 4.3.** *For all $t > 0$*

$$\mathbb{P}\{\mathscr{E}(\hat{f}_n) > \sigma_n^t\} \leq C_q e^{-t}$$

*and for all $\sigma \geq \sigma_n^t$*

$$\mathbb{P}\left\{\sup_{f \in \mathscr{F}, \mathscr{E}(f) \geq \sigma} \left|\frac{\hat{\mathscr{E}}_n(f)}{\mathscr{E}(f)} - 1\right| > V_n^t(\sigma)\right\} \leq C_q e^{-t},$$

*where*

$$C_q := \frac{q}{q-1} \vee e.$$

*Proof.* Let $\sigma > \sigma_n^t$. Take $\delta_j = q^{-j}$, $j \geq 0$ and $t_j := t\frac{\delta_j}{\sigma}$ for some $t > 0, \sigma > 0$. The function $U_n(\delta)$, the quantity $\delta_n(\mathscr{F}, P)$, etc, now correspond to this choice of the sequences $\{\delta_j\}$, $\{t_j\}$. Then, it is easy to verify that for all $\delta \geq \sigma$

$$\frac{U_n(\delta)}{\delta} \leq 2q\left[\sup_{\delta_j \geq \sigma} \frac{\phi_n(\delta_j)}{\delta_j} + \sup_{\delta_j \geq \sigma} \frac{D(\delta_j)}{\sqrt{\delta_j}}\sqrt{\frac{t\delta_j}{n\sigma\delta_j}} + \frac{t\delta_j}{n\sigma\delta_j}\right]$$

$$\leq 2q\left[\sup_{\delta \geq \sigma} \frac{\phi_n(\delta)}{\delta} + \sup_{\delta \geq \sigma} \frac{D(\delta)}{\sqrt{\delta}}\sqrt{\frac{t}{n\sigma}} + \frac{t}{n\sigma}\right]$$

$$= 2q\left[\phi_n^\flat(\sigma) + \sqrt{(D^2)^\flat(\sigma)}\sqrt{\frac{t}{n\sigma}} + \frac{t}{n\sigma}\right] = V_n^t(\sigma). \tag{4.3}$$

Since $\sigma > \sigma_n^t$ and the function $V_n^t$ is strictly decreasing, we have $V_n^t(\sigma) < 1$ and, for all $\delta > \sigma_n^t$,

$$U_n(\delta) \leq V_n^t(\sigma)\delta < \delta.$$

Therefore, $\sigma_n^t \geq \delta_n(\mathscr{F}; P)$. It follows from Theorem 4.1 that

$$\mathbb{P}\{\mathscr{E}(\hat{f}_n) \geq \sigma\} \leq \sum_{\delta_j \geq \sigma} e^{-tj}.$$

The right hand side can be now bounded as follows:

$$\sum_{\delta_j \geq \sigma} e^{-tj} = \sum_{\delta_j \geq \sigma} \exp\left\{-t\frac{\delta_j}{\sigma}\right\} \leq \sum_{j \geq 0} e^{-tq^j}$$

$$= e^{-t} + \frac{q}{q-1} \sum_{j=1}^{\infty} q^{-j} e^{-tq^j} (q^j - q^{j-1}) \leq e^{-t} + \frac{1}{q-1} \int_1^{\infty} e^{-tx} dx$$

$$= e^{-t} + \frac{1}{q-1}\frac{1}{t}e^{-t} \leq \frac{q}{q-1}e^{-t},\ t \geq 1. \tag{4.4}$$

This implies the first bound for $t \geq 1$ and it is trivial for $t \leq 1$ because of the definition of the constant $C_q$.

To prove the second bound use Theorem 4.2 and note that, by (4.3), $V_n(\sigma) \leq V_n^t(\sigma)$. The result follows from Theorem 4.2 and (4.4).                    □

The result of Lemma 4.1 below is due to Massart [106, 107] (we formulate it in a slightly different form). Suppose that $\mathscr{F}$ is a class of measurable functions from $S$ into $[0, 1]$ and $f_* : S \mapsto [0, 1]$ is a measurable function such that with some numerical constant $D > 0$

$$D(Pf - Pf_*) \geq \rho_P^2(f, f_*) \geq P(f - f_*)^2 - (P(f - f_*))^2, \tag{4.5}$$

where $\rho_P$ is a (pseudo)metric. The assumptions of this type are frequently used in model selection problems (see Sect. 6.3). They describe the link between the excess risk (or the approximation error) $Pf - P_*$ and the variance of the "excess loss" $f - f_*$. This particular form of bound (4.5) is typical in regression problems with $L_2$-loss (see Sect. 5.1): the link function in this case is just the square. In some other problems, such as classification under "low noise" assumption other link functions are also used (see Sect. 5.3).

Assume, for simplicity, that the infimum of $Pf$ over $\mathscr{F}$ is attained at a function $\bar{f} \in \mathscr{F}$ (the result can be easily modified if this is not the case). Let

$$\omega_n(\delta) := \omega_n(\mathscr{F}; \bar{f}; \delta) := \mathbb{E} \sup_{f \in \mathscr{F}, \rho_P^2(f,\bar{f}) \leq \delta} |(P_n - P)(f - \bar{f})|.$$

**Lemma 4.1.** *There exists a constant $K > 0$ such that for all $\varepsilon \in (0, 1]$ and for all $t > 0$*

$$\sigma_n^t(\mathscr{F}; P) \leq \varepsilon(\inf_{\mathscr{F}} Pf - Pf_*) + \frac{1}{D}\omega_n^{\sharp}\left(\frac{\varepsilon}{KD}\right) + \frac{KD}{\varepsilon}\frac{t}{n}.$$

*Proof.* Note that

$$\phi_n(\delta) = \mathbb{E}\|P_n - P\|_{\mathscr{F}'(\delta)} \le 2\mathbb{E} \sup_{f \in \mathscr{F}(\delta)} |(P_n - P)(f - \bar{f})|.$$

For $f \in \mathscr{F}(\delta)$,

$$\rho_P(f, \bar{f}) \le \rho_P(f, f_*) + \rho_P(\bar{f}, f_*) \le \sqrt{D(Pf - Pf_*)} + \sqrt{D(P\bar{f} - Pf_*)} \le$$

$$\le \sqrt{D(Pf - P\bar{f})} + 2\sqrt{D(P\bar{f} - Pf_*)} \le \sqrt{D\delta} + 2\sqrt{D\Delta} \le \sqrt{2D(\delta + 4\Delta)},$$

where

$$\Delta := P\bar{f} - Pf_* = \inf_{\mathscr{F}} Pf - Pf_*.$$

As a result, it follows that

$$D(\delta) \le 2\sqrt{D}(\sqrt{\delta} + 2\sqrt{\Delta}) \le \sqrt{8D(\delta + 4\Delta)}$$

and

$$\phi_n(\delta) \le 2\omega_n\Big(2D(\delta + 4\Delta)\Big).$$

We will now bound the functions $\phi_n^\flat(\sigma)$ and $(D^2)^\flat(\sigma)$ involved in the definition of $V_n^t(\sigma)$ (see the proof of Theorem 4.3). Denote $\tau := \frac{\Delta}{\sigma}$. Then

$$\phi_n^\flat(\sigma) = \sup_{\delta \ge \sigma} \frac{\phi_n(\delta)}{\delta} \le 2\sup_{\delta \ge \sigma} \frac{\omega_n\Big(2D(1 + 4\tau)\delta\Big)}{\delta} = 4D(1 + 4\tau)\omega_n^\flat\Big(2D(1 + 4\tau)\sigma\Big)$$

and also

$$(D^2)^\flat(\sigma) = \sup_{\delta \ge \sigma} \frac{D^2(\delta)}{\delta} \le \sup_{\delta \ge \sigma} \frac{8D(\delta + 4\Delta)}{\delta} \le 8D(1 + 4\tau).$$

Therefore,

$$V_n^t(\sigma) \le 2q\left[4D(1 + 4\tau)\omega_n^\flat\Big(2D(1 + 4\tau)\sigma\Big) + 2\sqrt{2D}\sqrt{1 + 4\tau}\sqrt{\frac{t}{n\sigma}} + \frac{t}{n\sigma}\right].$$

Suppose that, for some $\varepsilon \in (0, 1]$, we have $\sigma \ge \varepsilon\Delta$ implying that $\tau \le \frac{1}{\varepsilon}$. Then we can upper bound $V_n^t(\sigma)$ as follows:

$$V_n^t(\sigma) \le 2q\left[\frac{20D}{\varepsilon}\omega_n^\flat\Big(2D\sigma\Big) + 2\sqrt{10}\sqrt{\frac{tD}{n\varepsilon\sigma}} + \frac{t}{n\sigma}\right].$$

As soon as

$$\sigma \geq \frac{1}{2D}\omega_n^\sharp\left(\frac{\varepsilon}{KD}\right) \vee \frac{KDt}{n\varepsilon}$$

with a sufficiently large $K$, the right hand side of the last bound can be made smaller than 1. Thus, $\sigma_n^t$ is upper bounded either by $\varepsilon\Delta$, or by the expression

$$\frac{1}{2D}\omega_n^\sharp\left(\frac{\varepsilon}{KD}\right) \vee \frac{KDt}{n\varepsilon},$$

which implies the bound of the lemma.                                                                                     □

**Remark.** By increasing the value of the constant $K$ it is easy to upper bound the quantity $\sup\{\sigma : V_n^t(\sigma) \leq 1/2\}$ in exactly the same way.

The next statement follows immediately from Lemma 4.1 and Theorem 4.3.

**Proposition 4.1.** *There exists a large enough constant $K > 0$ such that for all $\varepsilon \in (0, 1]$ and all $t > 0$*

$$\mathbb{P}\left\{P\hat{f} - Pf_* \geq (1+\varepsilon)(\inf_{\mathcal{F}} Pf - Pf_*) + \frac{1}{D}\omega_n^\sharp\left(\frac{\varepsilon}{KD}\right) + \frac{KD}{\varepsilon}\frac{t}{n}\right\} \leq C_q e^{-t}.$$

Let us call $\psi : \mathbb{R}_+ \mapsto \mathbb{R}_+$ a function of concave type if it is nondecreasing and $u \mapsto \frac{\psi(u)}{u}$ is decreasing. If, in addition, for some $\gamma \in (0, 1)$, $u \mapsto \frac{\psi(u)}{u^\gamma}$ is decreasing, $\psi$ will be called a function of strictly concave type (with exponent $\gamma$). In particular, if $\psi(u) := \varphi(u^\gamma)$, or $\psi(u) := \varphi^\gamma(u)$, where $\varphi$ is a nondecreasing strictly concave function with $\varphi(0) = 0$, then $\psi$ is of concave type for $\gamma = 1$ and of strictly concave type for $\gamma < 1$.

**Proposition 4.2.** *Let $\delta_j := q^{-j}$, $j \geq 0$ for some $q > 1$. If $\psi$ is a function of strictly concave type with some exponent $\gamma \in (0, 1)$, then*

$$\sum_{j:\delta_j \geq \delta} \frac{\psi(\delta_j)}{\delta_j} \leq c_{\gamma,q}\frac{\psi(\delta)}{\delta},$$

*where $c_{\gamma,q}$ is a constant depending only on $q, \gamma$.*

*Proof.* Note that

$$\sum_{j:\delta_j \geq \delta} \frac{\psi(\delta_j)}{\delta_j} = \sum_{j:\delta_j \geq \delta} \frac{\psi(\delta_j)}{\delta_j^\gamma \delta_j^{1-\gamma}} \leq \frac{\psi(\delta)}{\delta^\gamma} \sum_{j:\delta_j \geq \delta} \frac{1}{\delta_j^{1-\gamma}} =$$

$$= \frac{\psi(\delta)}{\delta} \sum_{j:\delta_j \geq \delta} \left(\frac{\delta}{\delta_j}\right)^{1-\gamma} \leq \frac{\psi(\delta)}{\delta} \sum_{j \geq 0} q^{-j(1-\gamma)} = c_{\gamma,q}\frac{\psi(\delta)}{\delta},$$

which implies the bound.                                                                                                 □

Assume that $\phi_n(\delta) \leq \check{\phi}_n(\delta)$ and $D(\delta) \leq \check{D}(\delta)$, $\delta > 0$, where $\check{\phi}_n$ is a function of strictly concave type with some exponent $\gamma \in (0, 1)$ and $\check{D}$ is a concave type function. Define

$$\check{U}_n(\delta; t) := \check{U}_{n,t}(\delta) := \check{K}\left(\check{\phi}_n(\delta) + \check{D}(\delta)\sqrt{\frac{t}{n}} + \frac{t}{n}\right)$$

with some numerical constant $\check{K}$. Then $\check{U}_n(\cdot; t)$ is also a function of strictly concave type. In this case, it is natural to define

$$\check{V}_n(\delta; t) := \check{U}_{n,t}^\flat(\delta) = \frac{\check{U}_n(\delta; t)}{\delta} \quad and \quad \check{\delta}_n(t) := \check{U}_{n,t}^\sharp(1).$$

**Theorem 4.4.** *There exists a constant $\check{K}$ in the definition of the function $\check{U}_n(\delta; t)$ such that for all $t > 0$*

$$\mathbb{P}\{\mathscr{E}(\hat{f}_n) \geq \check{\delta}_n(t)\} \leq e^{-t}$$

*and for all $\delta \geq \check{\delta}_n(t)$,*

$$\mathbb{P}\left\{\sup_{f \in \mathscr{F}, \mathscr{E}(f) \geq \delta}\left|\frac{\hat{\mathscr{E}}_n(f)}{\mathscr{E}(f)} - 1\right| \geq \check{V}_n(\delta; t)\right\} \leq e^{-t}.$$

*Proof.* It is similar to the proof of Theorem 4.2, but now our goal is to avoid using the concentration inequality repeatedly for each value of $\delta_j$ since this leads to a logarithmic factor. The trick was previously used in Massart [106] and in the Ph.D. dissertation of Bousquet (see also Bartlett et al. [15]). Define

$$\mathscr{G}_\delta := \bigcup_{\sigma \geq \delta} \frac{\delta}{\sigma}\{f - g : f, g \in \mathscr{F}(\sigma)\}.$$

Then the functions in $\mathscr{G}_\delta$ are bounded by 1 and

$$\sigma_P(\mathscr{G}_\delta) \leq \sup_{\sigma \geq \delta} \frac{\delta}{\sigma} \sup_{f,g \in \mathscr{F}(\sigma)} \sigma_P(f - g) \leq \delta \sup_{\sigma \geq \delta} \frac{\check{D}(\sigma)}{\sigma} \leq \check{D}(\delta),$$

since $\check{D}$ is of concave type. Also, since $\check{\phi}_n$ is of strictly concave type, Proposition 4.2 yields

$$\mathbb{E}\|P_n - P\|_{\mathscr{G}_\delta} = \mathbb{E} \sup_{j:\delta_j \geq \delta} \sup_{\sigma \in (\delta_{j+1}, \delta_j]} \frac{\delta}{\sigma}\|P_n - P\|_{\mathscr{F}'(\sigma)} \leq$$

$$\leq q \sum_{j:\delta_j \geq \delta} \frac{\delta}{\delta_j}\mathbb{E}\|P_n - P\|_{\mathscr{F}'(\delta_j)} \leq q\delta \sum_{j:\delta_j \geq \delta} \frac{\check{\phi}_n(\delta_j)}{\delta_j} \leq qc_{\gamma,q}\check{\phi}_n(\delta).$$

Now, Talagrand's concentration inequality implies that there exists an event $E$ of probability $\mathbb{P}(E) \geq 1 - e^{-t}$ such that on this event $\|P_n - P\|_{\mathscr{G}_\delta} \leq \check{U}_n(\delta; t)$ (the constant $\check{K}$ in the definition of $\check{U}_n(\delta; t)$ should be chosen properly). Then, on the event $E$, for all $\sigma \geq \delta$,

$$\|P_n - P\|_{\mathscr{F}'(\sigma)} \leq \frac{\sigma}{\delta} \check{U}_n(\delta; t) \leq \check{V}_n(\delta; t)\sigma.$$

The rest repeats the proof of Theorems 4.1 and 4.2.                              □

In the next theorem, we consider empirical risk minimization problems over Donsker classes of functions under the assumption that, roughly speaking, the true risk has unique minimum and, as a consequence, the $\delta$-minimal sets $\mathscr{F}(\delta)$ shrink to a set consisting of a single function as $\delta \to 0$. It will be shown that in such cases the excess risk is of the order $o_\mathbb{P}(n^{-1/2})$.

**Theorem 4.5.** *If $\mathscr{F}$ is a $P$-Donsker class and*

$$D_P(\mathscr{F}; \delta) \to 0 \text{ as } n \to \infty,$$

*then*

$$\mathscr{E}_P(\hat{f}_n) = o_\mathbb{P}(n^{-1/2}) \text{ as } n \to \infty.$$

*Proof.* If $\mathscr{F}$ is a $P$-Donsker class, then the sequence of empirical processes

$$Z_n(f) := n^{1/2}(P_n f - Pf), f \in \mathscr{F}$$

is asymptotically equicontinuous, that is, for all $\varepsilon > 0$

$$\lim_{\delta \to 0} \limsup_{n \to \infty} \mathbb{P}\left\{ \sup_{\rho_P(f,g) \leq \delta, f, g \in \mathscr{F}} \left| Z_n(f) - Z_n(g) \right| \geq \varepsilon \right\} = 0.$$

(see, e.g., van der Vaart and Wellner [148], Sect. 2.1.2). This also implies (in the case of uniformly bounded classes, by an application of Talagrand's concentration inequality) that

$$\lim_{\delta \to 0} \limsup_{n \to \infty} \mathbb{E} \sup_{\rho_P(f,g) \leq \delta, f, g \in \mathscr{F}} \left| Z_n(f) - Z_n(g) \right| = 0.$$

Since $D_P(\mathscr{F}; \delta) \to 0$ as $\delta \to 0$, it follows that

$$\lim_{\delta \to 0} \limsup_{n \to \infty} n^{1/2} \phi_n(\mathscr{F}; P; \delta) = \lim_{\delta \to 0} \limsup_{n \to \infty} n^{1/2} \mathbb{E}\|P_n - P\|_{\mathscr{F}'(\delta)}$$

$$\leq \lim_{\delta \to 0} \limsup_{n \to \infty} \mathbb{E} \sup_{\rho_P(f,g) \leq D(\mathscr{F};\delta), f, g \in \mathscr{F}} \left| Z_n(f) - Z_n(g) \right| = 0. \tag{4.6}$$

Without loss of generality, assume that $D(\delta) \geq \delta$ (otherwise, in what follows, replace $D(\delta)$ by $D(\delta) \vee \delta$). Let now $\{\delta_j\}$ be a decreasing sequence such that $\delta_0 = 1$, $\delta_j \to 0$ as $j \to \infty$ and $D(\delta_j) \geq e^{-(j+1)}$. Define

$$t_j := t + 2 \log \log \frac{1}{D(\delta_j)} \leq t + 2 \log(j+1)$$

and

$$U_n^t(\delta) := 2\left[\phi_n(\delta_j) + D(\delta_j)\sqrt{\frac{t_j}{n}} + \frac{t_j}{n}\right], \quad \delta \in (\delta_{j+1}, \delta_j], \quad j \geq 0.$$

Clearly, $U_n^t$ is an upper bound on the function $U_n$ (used in Theorem 4.1) provided that $U_n$ is based on the same sequences $\{\delta_j\}, \{t_j\}$. Denote

$$\delta_n^t := \sup\{\delta \in (0, 1] : \delta \leq U_n^t(\delta)\}.$$

Then $\delta_n^t \geq \delta_n(\mathscr{F}; P)$ and also $\delta_n^t \geq \frac{t}{n}$. It follows from Theorem 4.1 that

$$\mathbb{P}\{\mathscr{E}_P(\hat{f}_n) > \delta_n^t\} \leq \sum_{\delta_j \geq \delta_n^t} e^{-t_j} \leq \sum_{j \geq 0} e^{-t_j} \leq \sum_{j \geq 0} e^{-t - 2\log(j+1)} = \sum_{j \geq 1} j^{-2} e^{-t} \leq 2e^{-t}.$$

The definitions of $\delta_n^t$ and $U_n^t$ easily imply that

$$\delta_n^t \leq U_n^t(\delta_n^t) \leq 2\left[\phi_n(1) + D(1)\sqrt{\frac{t + 2\log\log(n/t)}{n}} + \frac{t + 2\log\log(n/t)}{n}\right],$$

which tends to 0 as $n \to \infty$ since

$$\phi_n(1) \leq 2\mathbb{E}\|P_n - P\|_{\mathscr{F}} = O(n^{-1/2}) \to 0$$

for a Donsker class $\mathscr{F}$ and $D(1) < +\infty$. Denote by $j_n$ the number for which

$$\delta_n^t \in (\delta_{j_n+1}, \delta_{j_n}].$$

Then, clearly, $j_n \to \infty$ and $\delta_{j_n} \to 0$ as $n \to \infty$. Now, we have

$$n^{1/2}\delta_n^t \leq n^{1/2}U_n^t(\delta_n^t) \leq 2\left[n^{1/2}\phi_n(\delta_{j_n})\right.$$

$$\left. + D(\delta_{j_n})\sqrt{t + 2\log\log(1/D(\delta_{j_n}))} + \frac{t + 2\log\log(n/t)}{n^{1/2}}\right],$$

and, in view of (4.6) and the assumption that $D(\delta) \to 0, \delta \to 0$, it is easy to conclude that, for all $t > 0$,

$$n^{1/2}\delta_n^t \to 0 \text{ as } n \to \infty.$$

It remains to show that there is a choice of $t = \tau_n \to \infty$ (slowly enough) such that

$$\delta_n^{\tau_n} = o(n^{-1/2}).$$

The claim of the theorem now follows from the bound

$$\mathbb{P}\{\mathscr{E}_P(\hat{f}_n) > \delta_n^{\tau_n}\} \leq 2e^{-\tau_n} \to 0 \text{ as } n \to \infty. \qquad \square$$

There is another version of the proof that is based on Theorem 4.3.

The condition $D(\mathscr{F}; \delta) \to 0$ as $\delta \to 0$ is quite natural when the true risk minimization problem (1.1) has unique solution. In this case, such quantities as $\delta_n(\mathscr{F}; P)$ often give correct (in a minimax sense) convergence rate for the excess risk in risk minimization problems. However, if the minimum in (1.1) is not unique, the diameter $D(\delta)$ of the $\delta$-minimal set is bounded away from 0. In such cases, $\delta_n(\mathscr{F}; P)$ is bounded from below by $c\sqrt{\frac{1}{n}}$. At the same time, the optimal convergence rate of the excess risk to 0 is often better than this (in fact, it can be close to $n^{-1}$, e.g., in classification problems).

## 4.2  Rademacher Complexities and Data Dependent Bounds on Excess Risk

In a variety of statistical problems, it is crucial to have data dependent upper and lower confidence bounds on the sup-norm of the empirical process $\|P_n - P\|_{\mathscr{F}}$ for a given function class $\mathscr{F}$. This random variable is a natural measure of the accuracy of approximation of an unknown distribution $P$ by its empirical distribution $P_n$. However, $\|P_n - P\|_{\mathscr{F}}$ depends on the unknown distribution $P$ and, hence, it can not be used directly. It happens that it is easy to construct rather simple upper and lower bounds on $\|P_n - P\|_{\mathscr{F}}$ in terms of the sup-norm of Rademacher process $\|R_n\|_{\mathscr{F}}$. The last random variable depends only on the data $X_1, \ldots, X_n$ and on random signs $\varepsilon_1, \ldots, \varepsilon_n$ that are independent of $X_1, \ldots, X_n$ and are easy to simulate. Thus, $\|R_n\|_{\mathscr{F}}$ can be used as a data dependent complexity measure of the class $\mathscr{F}$ that allows one to estimate the accuracy of approximation of $P$ by $P_n$ based on the data. This bootstrap type approach was introduced independently by Koltchinskii [81] and Bartlett et al. [14] and it was used to develop a general method of model selection and complexity regularization in learning theory. It is based on the following simple bounds. Their proof is very elementary and relies only on the symmetrization and bounded difference inequalities.

Assume that the functions in the class $\mathscr{F}$ are uniformly bounded by a constant $U > 0$.

**Theorem 4.6.** *For all $t > 0$,*

$$\mathbb{P}\left\{ \|P_n - P\|_{\mathscr{F}} \geq 2\|R_n\|_{\mathscr{F}} + \frac{3tU}{\sqrt{n}} \right\} \leq \exp\left\{ -\frac{t^2}{2} \right\}$$

*and*

$$\mathbb{P}\left\{ \|P_n - P\|_{\mathscr{F}} \leq \frac{1}{2}\|R_n\|_{\mathscr{F}} - \frac{2tU}{\sqrt{n}} - \frac{U}{2\sqrt{n}} \right\} \leq \exp\left\{ -\frac{t^2}{2} \right\}.$$

*Proof.* Denote

$$Z_n := \|P_n - P\|_{\mathscr{F}} - 2\|R_n\|_{\mathscr{F}}.$$

Then, by symmetrization inequality, $\mathbb{E}Z_n \leq 0$ and applying bounded difference inequality to the random variable $Z_n$ easily yields

$$\mathbb{P}\left\{ Z_n \geq \mathbb{E}Z_n + \frac{3tU}{\sqrt{n}} \right\} \leq \exp\left\{ -\frac{t^2}{2} \right\},$$

which implies the first bound.

The second bound is proved similarly by considering the random variable

$$Z_n := \|P_n - P\|_{\mathscr{F}} - \frac{1}{2}\|R_n\|_{\mathscr{F}} - \frac{U}{2\sqrt{n}}$$

and using symmetrization and bounded difference inequalities.                        $\square$

Note that other versions of bootstrap, most notably, the classical Efron's bootstrap, can be also used in a similar way (see Fromont [61]).

The major drawback of Theorem 4.6 is that the error term does not take into account the size of the variance of functions in the class $\mathscr{F}$. In some sense, this is a data dependent version of uniform Hoeffding inequality and what is often needed is a data dependent version of uniform Bernstein type inequality. We provide such a result below. It can be viewed as a *statistical version* of Talagrand's concentration inequality. Recently, Giné and Nickl [67] used some inequalities of similar nature in adaptive density estimation.

Denote

$$\sigma_P^2(\mathscr{F}) := \sup_{f \in \mathscr{F}} Pf^2 \quad \text{and} \quad \sigma_n^2(\mathscr{F}) := \sup_{f \in \mathscr{F}} P_n f^2.$$

**Theorem 4.7.** *There exists a numerical constant $K > 0$ such that for all $t \geq 1$ with probability at least $1 - e^{-t}$ the following bounds hold:*

$$\left| \|R_n\|_{\mathscr{F}} - \mathbb{E}\|R_n\|_{\mathscr{F}} \right| \leq K\left[ \sqrt{\frac{t}{n}\left( \sigma_n^2(\mathscr{F}) + U\|R_n\|_{\mathscr{F}} \right)} + \frac{tU}{n} \right], \tag{4.7}$$

$$\mathbb{E}\|R_n\|_{\mathscr{F}} \leq K\left[ \|R_n\|_{\mathscr{F}} + \sigma_n(\mathscr{F})\sqrt{\frac{t}{n}} + \frac{tU}{n} \right], \tag{4.8}$$

$$\sigma_P^2(\mathcal{F}) \le K\left(\sigma_n^2(\mathcal{F}) + U\|R_n\|_{\mathcal{F}} + \frac{tU}{n}\right) \tag{4.9}$$

*and*

$$\sigma_n^2(\mathcal{F}) \le K\left(\sigma_P^2(\mathcal{F}) + U\mathbb{E}\|R_n\|_{\mathcal{F}} + \frac{tU}{n}\right). \tag{4.10}$$

*Also, for all $t \ge 1$ with probability at least $1 - e^{-t}$*

$$\mathbb{E}\|P_n - P\|_{\mathcal{F}} \le K\left[\|R_n\|_{\mathcal{F}} + \sigma_n(\mathcal{F})\sqrt{\frac{t}{n}} + \frac{tU}{n}\right] \tag{4.11}$$

*and*

$$\left|\|P_n - P\|_{\mathcal{F}} - \mathbb{E}\|P_n - P\|_{\mathcal{F}}\right| \le K\left[\sqrt{\frac{t}{n}\left(\sigma_n^2(\mathcal{F}) + U\|R_n\|_{\mathcal{F}}\right)} + \frac{tU}{n}\right]. \tag{4.12}$$

*Proof.* It is enough to consider the case when $U = 1/2$. The general case then follows by rescaling. Using Talagrand's concentration inequality (to be specific, Klein–Rio bound, see Sect. 2.3), we claim that on an event $E$ of probability at least $1 - e^{-t}$

$$\mathbb{E}\|R_n\|_{\mathcal{F}} \le \|R_n\|_{\mathcal{F}} + \sqrt{\frac{2t}{n}\left(\sigma_P^2(\mathcal{F}) + 2\mathbb{E}\|R_n\|_{\mathcal{F}}\right)} + \frac{t}{n}, \tag{4.13}$$

which implies that

$$\mathbb{E}\|R_n\|_{\mathcal{F}} \le \|R_n\|_{\mathcal{F}} + \sigma_P(\mathcal{F})\sqrt{\frac{2t}{n}} + \frac{t}{n} + 2\sqrt{\frac{1}{2}\mathbb{E}\|R_n\|_{\mathcal{F}}\frac{2t}{n}} \le$$

$$\le \|R_n\|_{\mathcal{F}} + \sigma_P(\mathcal{F})\sqrt{\frac{2t}{n}} + \frac{t}{n} + \frac{1}{2}\mathbb{E}\|R_n\|_{\mathcal{F}} + \frac{2t}{n},$$

or

$$\mathbb{E}\|R_n\|_{\mathcal{F}} \le 2\|R_n\|_{\mathcal{F}} + 2\sqrt{2}\sigma_P(\mathcal{F})\sqrt{\frac{t}{n}} + \frac{6t}{n}. \tag{4.14}$$

We will now upper bound $\sigma_P^2(\mathcal{F})$ in terms of $\sigma_n^2(\mathcal{F})$. Denote $\mathcal{F}^2 := \{f^2 : f \in \mathcal{F}\}$. Again, we apply Talagrand's concentration inequality (namely, Bousquet's bound, Sect. 2.3) and show that on an event $F$ of probability at least $1 - e^{-t}$

$$\sigma_P^2(\mathcal{F}) = \sup_{f \in \mathcal{F}} Pf^2 \le \sup_{f \in \mathcal{F}} P_n f^2 + \|P_n - P\|_{\mathcal{F}^2} \le$$

$$\le \sigma_n^2(\mathcal{F}) + \mathbb{E}\|P_n - P\|_{\mathcal{F}^2} + \sqrt{\frac{2t}{n}\left(\sigma_P^2(\mathcal{F}) + 2\mathbb{E}\|P_n - P\|_{\mathcal{F}^2}\right)} + \frac{t}{3n},$$

where we also used the fact that

$$\sup_{f \in \mathscr{F}^2} \operatorname{Var}_P(f^2) \le \sup_{f \in \mathscr{F}} Pf^4 < \sup_{f \in \mathscr{F}} Pf^2 = \sigma_P^2(\mathscr{F})$$

since the functions from $\mathscr{F}$ are uniformly bounded by $U = 1/2$. Using symmetrization inequality and then contraction inequality for Rademacher processes, we get

$$\mathbb{E}\|P_n - P\|_{\mathscr{F}^2} \le 2\mathbb{E}\|R_n\|_{\mathscr{F}^2} \le 8\mathbb{E}\|R_n\|_{\mathscr{F}}.$$

Hence,

$$\sigma_P^2(\mathscr{F}) \le \sigma_n^2(\mathscr{F}) + 8\mathbb{E}\|R_n\|_{\mathscr{F}} + \sigma_P(\mathscr{F})\sqrt{\frac{2t}{n}} + 2\sqrt{\frac{8t}{n}\mathbb{E}\|R_n\|_{\mathscr{F}}} + \frac{t}{3n} \le$$
$$\le \sigma_n^2(\mathscr{F}) + 9\mathbb{E}\|R_n\|_{\mathscr{F}} + \sigma_P(\mathscr{F})\sqrt{\frac{2t}{n}} + \frac{9t}{n},$$

where the inequality $2\sqrt{ab} \le a + b$, $a, b \ge 0$ was applied. Next we use bound (4.14) on $\mathbb{E}\|R_n\|_{\mathscr{F}}$ to get

$$\sigma_P^2(\mathscr{F}) \le \sigma_n^2(\mathscr{F}) + 18\|R_n\|_{\mathscr{F}} + 19\sigma_P(\mathscr{F})\sqrt{\frac{2t}{n}} + \frac{100t}{n}.$$

As before, we bound the term $19\sigma_P(\mathscr{F})\sqrt{\frac{2t}{n}} = 2 \times 19\frac{\sigma_P(\mathscr{F})}{\sqrt{2}}\sqrt{\frac{t}{n}}$ using the inequality $2ab \le a^2 + b^2$, which gives

$$\sigma_P^2(\mathscr{F}) \le \frac{1}{2}\sigma_P^2(\mathscr{F}) + \sigma_n^2(\mathscr{F}) + 18\|R_n\|_{\mathscr{F}} + \frac{500t}{n}.$$

As a result, the following bound holds on the event $E \cap F$:

$$\sigma_P^2(\mathscr{F}) \le 2\sigma_n^2(\mathscr{F}) + 36\|R_n\|_{\mathscr{F}} + \frac{1000t}{n}. \tag{4.15}$$

It also implies that

$$\sigma_P(\mathscr{F}) \le \sqrt{2}\sigma_n(\mathscr{F}) + 6\sqrt{\|R_n\|_{\mathscr{F}}} + 32\sqrt{\frac{t}{n}}.$$

We use this bound on $\sigma_P(\mathscr{F})$ in terms of $\sigma_n(\mathscr{F})$ to derive from (4.14) that

$$\mathbb{E}\|R_n\|_{\mathscr{F}} \le 2\|R_n\|_{\mathscr{F}} + 4\sigma_n(\mathscr{F})\sqrt{\frac{t}{n}} +$$
$$+ 12\sqrt{2}\sqrt{\|R_n\|_{\mathscr{F}}}\sqrt{\frac{t}{n}} + \frac{100t}{n} \le 3\|R_n\|_{\mathscr{F}} + 4\sigma_n(\mathscr{F})\sqrt{\frac{t}{n}} + \frac{172t}{n}.$$

The last bound holds on the same event $E \cap F$ of probability at least $1 - 2e^{-t}$. This implies inequalities (4.8) and (4.9) of the theorem. Inequality (4.7) follows from Talagrand's inequality, specifically, from combination of Klein–Rio inequality (4.13), the following application of Bousquet's inequality

$$\|R_n\|_{\mathscr{F}} \leq \mathbb{E}\|R_n\|_{\mathscr{F}} + \sqrt{\frac{2t}{n}\left(\sigma_P^2(\mathscr{F}) + 2\mathbb{E}\|R_n\|_{\mathscr{F}}\right)} + \frac{t}{3n} \qquad (4.16)$$

and bounds (4.8), (4.9) that have been already proved. The proof of the next inequality (4.10) is another application of symmetrization, contraction and Talagrand's concentration and is similar to the proof of (4.9). The last two bounds follow from the inequalities for the Rademacher process and symmetrization inequality.

Under the assumption $t \geq 1$, the exponent in the expression for probability can be written as $e^{-t}$ without a constant in front of it. The constant can be removed by increasing the value of $K$.    □

We will use the above tools to construct data dependent bounds on the excess risk. As in the previous section, we assume that the functions in the class $\mathscr{F}$ are uniformly bounded by 1. First we show that the $\delta$-minimal sets of the risk can be estimated by the $\delta$-minimal sets of the empirical risk provided that $\delta$ is not too small, which is a consequence of Theorem 4.2. Let

$$\hat{\mathscr{F}}_n(\delta) := \mathscr{F}_{P_n}(\delta)$$

be the $\delta$-minimal set of $P_n$.

**Lemma 4.2.** *Let $\delta_n^\diamond$ be a number such that $\delta_n^\diamond \geq U_n^\sharp\left(\frac{1}{2}\right)$. There exists an event of probability at least $1 - \sum_{\delta_j \geq \delta_n^\diamond} e^{-t_j}$ such that on this event, for all $\delta \geq \delta_n^\diamond$,*

$$\mathscr{F}(\delta) \subset \hat{\mathscr{F}}_n(3\delta/2) \text{ and } \hat{\mathscr{F}}_n(\delta) \subset \mathscr{F}(2\delta).$$

*Proof.* It easily follows from the definitions that $\delta_n^\diamond \geq \delta_n(\mathscr{F}; P)$. Denote

$$E_n := \bigcap_{\delta_j \geq \delta_n^\diamond} E_{n,j},$$

where $E_{n,j}$ are the events defined in the proof of Theorem 4.1. Then

$$\mathbb{P}(E_n) \geq 1 - \sum_{\delta_j \geq \delta_n^\diamond} e^{-t_j}.$$

It follows from the proof of Theorem 4.2, that, on the event $E_n$, for all $f \in \mathscr{F}$ with $\mathscr{E}(f) \geq \delta_n^\diamond$,

$$\frac{1}{2} \leq \frac{\hat{\mathscr{E}}_n(f)}{\mathscr{E}(f)} \leq \frac{3}{2}.$$

By the proof of Theorem 4.2, on the same event

$$\|P_n - P\|_{\mathscr{F}'(\delta_n^\diamond)} \leq U_n(\delta_n^\diamond).$$

Therefore, on the event $E_n$,

$$\mathscr{E}(f) \leq 2\hat{\mathscr{E}}_n(f) \vee \delta_n^\diamond, \quad f \in \mathscr{F}, \tag{4.17}$$

which implies that, for all $\delta \geq \delta_n^\diamond$, $\hat{\mathscr{F}}_n(\delta) \subset \mathscr{F}(2\delta)$. On the other hand, on the same event $E_n$, for all $f \in \mathscr{F}$, the assumption $\mathscr{E}(f) \geq \delta_n^\diamond$ implies that $\hat{\mathscr{E}}_n(f) \leq \frac{3}{2}\mathscr{E}(f)$ and the assumption $\mathscr{E}(f) \leq \delta_n^\diamond$ implies that

$$\hat{\mathscr{E}}_n(f) \leq \mathscr{E}(f) + \|P_n - P\|_{\mathscr{F}'(\delta_n^\diamond)} \leq \mathscr{E}(f) + U_n(\delta_n^\diamond) \leq \delta_n^\diamond + V_n(\delta_n^\diamond)\delta_n^\diamond \leq \frac{3}{2}\delta_n^\diamond.$$

Thus, for all $f \in \mathscr{F}$,

$$\hat{\mathscr{E}}_n(f) \leq \frac{3}{2}\left(\mathscr{E}(f) \vee \delta_n^\diamond\right), \tag{4.18}$$

which implies that on the event $E_n$, for all $\delta \geq \delta_n^\diamond$, $\mathscr{F}(\delta) \subset \hat{\mathscr{F}}_n(3\delta/2)$. □

Now we are ready to define an empirical version of excess risk bounds. It will be convenient to use the following definition of $\rho_P$:

$$\rho_P^2(f, g) := P(f - g)^2.$$

Given a decreasing sequence $\{\delta_j\}$ of positive numbers with $\delta_0 = 1$ and a sequence $\{t_j\}$ of real numbers, $t_j \geq 1$, define

$$\bar{U}_n(\delta) := \bar{K}\left(\phi_n(\delta_j) + D(\delta_j)\sqrt{\frac{t_j}{n}} + \frac{t_j}{n}\right), \quad \delta \in (\delta_{j+1}, \delta_j], j \geq 0,$$

where $\bar{K} = 2$. Comparing this with the definition (4.2) of the function $U_n$, it is easy to check that $U_n(\delta) \leq \bar{U}_n(\delta), \delta \in (0, 1]$. As a consequence, if we define $\bar{\delta}_n := \bar{U}_n^\#(1/2)$, then $\delta_n(\mathscr{F}; P) \leq \bar{\delta}_n$.

Empirical versions of the functions $D$ and $\phi_n$ are defined by the following relationships:

$$\hat{D}_n(\delta) := \sup_{f,g \in \hat{\mathscr{F}}_n(\delta)} \rho_{P_n}(f, g) \quad \text{and} \quad \hat{\phi}_n(\delta) := \|R_n\|_{\hat{\mathscr{F}}'_n(\delta)}.$$

Also, let

$$\hat{U}_n(\delta) := \hat{K}\left(\hat{\phi}_n(\hat{c}\delta_j) + \hat{D}_n(\hat{c}\delta_j)\sqrt{\frac{t_j}{n}} + \frac{t_j}{n}\right), \quad \delta \in (\delta_{j+1}, \delta_j], j \geq 0,$$

$$\tilde{U}_n(\delta) := \tilde{K}\left(\phi_n(\tilde{c}\delta_j) + D(\tilde{c}\delta_j)\sqrt{\frac{t_j}{n}} + \frac{t_j}{n}\right), \quad \delta \in (\delta_{j+1}, \delta_j], j \geq 0,$$

where $2 \leq \hat{K} \leq \tilde{K}$, $\hat{c}, \tilde{c} \geq 1$ are numerical constants. Define

$$\bar{V}_n(\delta) := \bar{U}_n^\flat(\delta), \quad \hat{V}_n(\delta) := \hat{U}_n^\flat(\delta), \quad \tilde{V}_n(\delta) := \tilde{U}_n^\flat(\delta)$$

and

$$\hat{\delta}_n := \hat{U}_n^\sharp(1/2), \quad \tilde{\delta}_n := \tilde{U}_n^\sharp(1/2).$$

The constants in the definitions of the functions $\bar{U}_n$ and $\tilde{U}_n$ can be chosen in such a way that for all $\delta$ $U_n(\delta) \leq \bar{U}_n(\delta) \leq \tilde{U}_n(\delta)$, which yields the bound $\delta_n(\mathscr{F}; P) \leq \bar{\delta}_n \leq \tilde{\delta}_n$. Since the definitions of the functions $U_n, \bar{U}_n, \tilde{U}_n$ differ only in the constants, it is plausible that the quantities $\delta_n(\mathscr{F}; P), \bar{\delta}_n, \tilde{\delta}_n$ are of the same order (in fact, it can be checked in numerous examples).

We will prove that with a high probability, for all $\delta$, $\bar{U}_n(\delta) \leq \hat{U}_n(\delta) \leq \tilde{U}_n(\delta)$, so, $\hat{U}_n$ provides a data-dependent upper bound on $\bar{U}_n$ and $\tilde{U}_n$ provides a distribution dependent upper bound on $\hat{U}_n$. This implies that, with a high probability, $\bar{\delta}_n \leq \hat{\delta}_n \leq \tilde{\delta}_n$, which provides a data dependent bound $\hat{\delta}_n$ on the excess risk $\mathscr{E}_P(\hat{f}_n)$ which is of correct size (up to a constant) in many cases.

**Theorem 4.8.** *With the above notations,*

$$\mathbb{P}\left\{\bar{\delta}_n \leq \hat{\delta}_n \leq \tilde{\delta}_n\right\} \geq 1 - 3 \sum_{\delta_j \geq \bar{\delta}_n} \exp\{-t_j\}.$$

*Proof.* The proof follows from the inequalities of Theorem 4.7 and Lemma 4.2 in a rather straightforward way. Note that $\bar{\delta}_n \geq U_n^\sharp(1/2)$, so we can use it as $\delta_n^\diamond$ in Lemma 4.2. Denote $H$ the event introduced in the proof of this lemma (it was called $E_n$ in the proof). Then

$$\mathbb{P}(H) \geq 1 - \sum_{\delta_j \geq \bar{\delta}_n} e^{-t_j}$$

and, on the event $H$,

$$\mathscr{F}(\delta) \subset \hat{\mathscr{F}}_n(3\delta/2) \quad \text{and} \quad \hat{\mathscr{F}}_n(\delta) \subset \mathscr{F}(2\delta)$$

for all $\delta \geq \bar{\delta}_n$.

First, the values of $\delta$ and $t$ will be fixed. At the end, the resulting bounds will be used for $\delta = \delta_j$ and $t = t_j$. We will apply the inequalities of Theorem 4.7 to the function class $\mathscr{F}'(\delta)$. It easily follows from bound (4.11) that there exists an event $F = F(\delta)$ of probability at least $1 - e^{-t}$ such that, on the event $H \cap F$,

$$\mathbb{E}\|P_n - P\|_{\mathscr{F}'(\delta)} \leq K\left[\|R_n\|_{\hat{\mathscr{F}}'_n(3/2\delta)} + \hat{D}_n\left(\frac{3}{2}\delta\right)\sqrt{\frac{t}{n}} + \frac{t}{n}\right]$$

with a properly chosen $K$. Recalling the definition of $\bar{U}_n$ and $\hat{U}_n$, the last bound immediately implies that with a straightforward choice of numerical constants $\hat{K}, \hat{c}$, the inequality $\bar{U}_n(\delta) \leq \hat{U}_n(\delta)$ holds on the event $H \cap F$.

Quite similarly, using the inequalities of Theorem 4.7 (in particular, using bound (4.10) to control the "empirical" diameter $\hat{D}(\delta)$ in terms of the "true" diameter $D(\delta)$) and also the desymmetrization inequality, it is easy to see that there exists an event $G = G(\delta)$ of probability at least $1 - e^{-t}$ such that the inequality $\hat{U}_n(\delta) \leq \tilde{U}_n(\delta)$ holds on $H \cap G$ with properly chosen numerical constants $\tilde{K}, \tilde{c}$ in the definition of $\tilde{U}_n$.

Using the resulting inequalities for $\delta = \delta_j \geq \bar{\delta}_n$ yields

$$\mathbb{P}(E) \geq 1 - 3 \sum_{\delta_j \geq \bar{\delta}_n} \exp\{-t_j\},$$

where

$$E := \left\{\forall \delta_j \geq \bar{\delta}_n : \bar{U}_n(\delta_j) \leq \hat{U}_n(\delta_j) \leq \tilde{U}_n(\delta_j)\right\} \supset \bigcup_{j:\delta_j \geq \bar{\delta}_n} (H \cap F(\delta_j) \cap G(\delta_j)).$$

By the definitions of $\bar{U}_n, \hat{U}_n$ and $\tilde{U}_n$, this also implies that, on the event $E$,

$$\bar{U}_n(\delta) \leq \hat{U}_n(\delta) \leq \tilde{U}_n(\delta)$$

for all $\delta \geq \bar{\delta}_n$. By simple properties of $\sharp$-transform, we conclude that $\bar{\delta}_n \leq \hat{\delta}_n \leq \tilde{\delta}_n$ on the event $E$, which completes the proof. $\qquad\square$

It is easily seen from the proof of Theorem 4.8 and from the definitions and constructions of the events involved in this proof as well as in the proofs of Theorem 4.2 and Lemma 4.2 that on an event $E$ of probability at least $1 - p$, where $p = 3\sum_{\delta_j \geq \bar{\delta}_n} e^{-t_j}$, the following conditions hold:

(i) $\bar{\delta}_n \leq \hat{\delta}_n \leq \tilde{\delta}_n$

(ii) $\mathscr{E}(\hat{f}) \leq \bar{\delta}_n$

(iii) for all $f \in \mathscr{F}$,

$$\mathscr{E}(f) \leq 2\hat{\mathscr{E}}_n(f) \vee \bar{\delta}_n \quad \text{and} \quad \hat{\mathscr{E}}_n(f) \leq \frac{3}{2}\left(\mathscr{E}(f) \vee \bar{\delta}_n\right);$$

(iv)  for all $\delta \geq \bar{\delta}_n$,

$$\|P_n - P\|_{\mathscr{F}'(\delta)} \leq U_n(\delta).$$

Sometimes it is convenient to deal with different triples $(\bar{\delta}_n, \hat{\delta}_n, \tilde{\delta}_n)$ (defined in terms of various complexity measures of the class $\mathscr{F}$) that still satisfy conditions (i)–(iv) with a high probability. In fact, to satisfy conditions (ii)–(iv) it is enough to choose $\bar{\delta}_n$ in such a way that

(v)  $\bar{\delta}_n \geq U_n^\sharp(1/2)$.

This is reflected in the following definition.

**Definition 4.1.** Suppose sequences $\{\delta_j\}$, $\{t_j\}$ and the corresponding function $U_n$ are given. We will call $\bar{\delta}_n$ that depends on $\mathscr{F}$ and $P$ an *admissible distribution dependent bound* on the excess risk iff it satisfies condition (v), and, as a consequence, also conditions (ii)–(iv). If (ii)–(iv) hold on an event $E$ such that $\mathbb{P}(E) \geq 1 - p$, then $\bar{\delta}_n$ will be called an admissible bound of confidence level $1 - p$. A triple $(\bar{\delta}_n, \hat{\delta}_n, \tilde{\delta}_n)$, such that $\bar{\delta}_n$ and $\tilde{\delta}_n$ depend on $\mathscr{F}$ and $P$, $\hat{\delta}_n$ depends on $\mathscr{F}$ and $X_1, \ldots, X_n$, and, for some $p \in (0, 1)$, conditions (i)–(v) hold on an event $E$ with $\mathbb{P}(E) \geq 1 - p$, will be called a *triple bound* on the excess risk of confidence level $1 - p$.

Such triple bounds will be used later in model selection methods based on penalized empirical risk minimization.

## 4.3  Further Comments

Distribution dependent excess risk bounds of Sect. 4.1 are closely related to ratio type empirical processes studied in the 1980s by many authors (notably, by Alexander [5]). This connection was emphasized by Giné and Koltchinskii [66] (see also Giné et al. [65]). It was understood long ago that convergence rates of statistical estimators based on empirical risk minimization could often be found as solutions of certain fixed point equations defined in terms of proper complexities of underlying function classes and that such complexities are related to continuity moduli of empirical processes (see, e.g., van der Vaart and Wellner [148], Sect. 3.2, van de Geer [62], Shen and Wong [132]). Massart [106] and Koltchinskii and Panchenko [92] started defining such fixed point based complexities in terms of continuity moduli of empirical and Rademacher processes in a variety of problems of statistical learning theory. This approach was developed further by Bartlett et al. [15], Bousquet et al. [34] and Koltchinskii [83]. In the last paper, the data dependent Rademacher complexities were defined in terms of $\delta$-minimal sets of the true risk and the $L_2(\Pi)$-diameters of these sets play an important role in the analysis of the problem. We followed this approach here. Bartlett and Mendelson [17] introduced different definitions of localized Rademacher complexities that provided a way to distinguish between bounding excess risk of empirical risk minimizers and estimating the level sets of the true risk. Boucheron and Massart [32] obtained concentration inequalities for the excess empirical risk in terms of fixed point complexities.

The idea to use Rademacher processes ("Rademacher bootstrap") in order to construct data dependent excess risk bounds was introduced by Koltchinskii [81] and Bartlett et al. [14]. Koltchinskii and Panchenko [92] suggested a localized version of such complexities (in the "zero error case"). This idea was developed by Bartlett et al. [15], Bousquet et al. [34] and Koltchinskii [83]. "The statistical version" of Talagrand's concentration inequality (Theorem 4.7) was essentially used (without stating it) in Koltchinskii [83]. We follow the approach of this paper in our construction of data dependent Rademacher complexities. This construction provides reasonable excess risk bounds only when the $L_2(P)$-diameters of the $\delta$-minimal sets are small for small values of $\delta$. This is not the case when the true risk has multiple minima. Koltchinskii [83] gives a simple example showing that, in the multiple minima case, the distribution dependent excess risk bounds developed in the previous section are not always sharp. Moreover, there is a difficulty in estimation of the level sets of the risk (the $\delta$-minimal sets), which is of importance in constructing data dependent excess risk bounds. Some more subtle geometric characteristics of the class $\mathscr{F}$ that can be used in such cases to recover the correct convergence rates were suggested in Koltchinskii [83]. However, the extension of the theory of data dependent excess risk bounds to the multiple minima case remains an open problem.

Rademacher complexities have been also used in other problems of statistical learning theory, in particular, in margin type bounds on generalization error (see [93–95]). Recently, Hanneke [73] and Koltchinskii [87] used localized Rademacher complexities in the development of active learning algorithms.

Boucheron et al. [29] provide an excellent review of excess risk bounds in empirical risk minimization and their role in classification problems. Some further references can be found in this paper and in lecture notes by Massart [107]. Recent results on lower bounds in empirical risk minimization can be found in Mendelson [114].

# Chapter 5
# Examples of Excess Risk Bounds in Prediction Problems

Let $(X, Y)$ be a random couple in $S \times T, T \subset \mathbb{R}$ with distribution $P$. The distribution of $X$ will be denoted by $\Pi$. Assume that the random variable $X$ is "observable" and $Y$ is to be predicted based on an observation of $X$. Let $\ell : T \times \mathbb{R} \mapsto \mathbb{R}$ be a loss function. Given a function $g : S \mapsto \mathbb{R}$, the quantity $(\ell \bullet g)(x, y) := \ell(y, g(x))$ is interpreted as a loss suffered when $g(x)$ is used to predict $y$. The problem of optimal prediction can be viewed as a risk minimization

$$\mathbb{E}\ell(Y, g(X)) = P(\ell \bullet g) \longrightarrow \min, \ g : S \mapsto \mathbb{R}.$$

Since the distribution $P$ and the risk function $g \mapsto P(\ell \bullet g)$ are unknown, the risk minimization problem is usually replaced by the empirical risk minimization

$$P_n(\ell \bullet g) = n^{-1} \sum_{j=1}^{n} \ell(Y_j, g(X_j)) \longrightarrow \min, \ g \in \mathscr{G},$$

where $\mathscr{G}$ is a given class of functions $g : S \mapsto \mathbb{R}$ and $(X_1, Y_1), \ldots, (X_n, Y_n)$ is a sample of i.i.d. copies of $(X, Y)$ ("training data"). Obviously, this can be viewed as a special case of abstract empirical risk minimization problems discussed in Chap. 4. In this case, the class $\mathscr{F}$ is the "loss class" $\mathscr{F} := \{\ell \bullet g : g \in \mathscr{G}\}$ and the goal of this chapter is to derive excess risk bounds for concrete examples of loss functions and function classes frequently used in statistics and learning theory.

Let $\mu_x$ denote a version of conditional distribution of $Y$ given $X = x$. The following representation of the risk holds under very mild regularity assumptions:

$$P(\ell \bullet g) = \int_S \int_T \ell(y; g(x)) \mu_x(dy) \Pi(dx).$$

Given a probability measure $\mu$ on $T$, let

$$u_\mu \in \mathrm{Argmin}_{u \in \bar{\mathbb{R}}} \int_T \ell(y; u) \mu(dy).$$

V. Koltchinskii, *Oracle Inequalities in Empirical Risk Minimization and Sparse Recovery Problems*, Lecture Notes in Mathematics 2033, DOI 10.1007/978-3-642-22147-7_5, © Springer-Verlag Berlin Heidelberg 2011

Denote

$$g_*(x) := u_{\mu_x} = \mathrm{argmin}_{u \in \bar{\mathbb{R}}} \int_T \ell(y; u)\mu_x(dy).$$

Assume that the function $g_*$ is well defined and properly measurable. Then, for all $g$, $P(\ell \bullet g) \geq P(\ell \bullet g_*)$, which implies that $g_*$ is a point of *global* minimum of $P(\ell \bullet g)$.

Let

$$\hat{g}_n := \mathrm{argmin}_{g \in \mathscr{G}} P_n(\ell \bullet g)$$

be a solution of the corresponding empirical risk minimization problem (for simplicity, assume its existence).

The following assumption on the loss function $\ell$ is often used in the analysis of the problem: there exists a function $D(u, \mu) \geq 0$ such that for all measures $\mu = \mu_x$, $x \in S$

$$\int_T (\ell(y, u) - \ell(y, u_\mu))^2 \mu(dy) \leq D(u, \mu) \int_T (\ell(y, u) - \ell(y, u_\mu))\mu(dy). \quad (5.1)$$

In the case when the functions in the class $\mathscr{G}$ take their values in $[-M/2, M/2]$ and

$$D(u, \mu_x), \ |u| \leq M/2, x \in S$$

is uniformly bounded by a constant $D > 0$, it immediately follows from (5.1) (just by plugging in $u = g(x)$, $\mu = \mu_x$ and integrating with respect to $\Pi$) that, for all $g \in \mathscr{G}$,

$$P(\ell \bullet g - \ell \bullet g_*)^2 \leq DP(\ell \bullet g - \ell \bullet g_*). \quad (5.2)$$

As a consequence, if $g_* \in \mathscr{G}$, then the $L_2(P)$-diameter of the $\delta$-minimal set of $\mathscr{F}$ is bounded as follows:

$$D(\mathscr{F}; \delta) \leq 2(D\delta)^{1/2}.$$

Moreover, even if $g_* \notin \mathscr{G}$, condition (4.5) might still hold for the loss class $\mathscr{F}$ with $f_* = \ell \bullet g_*$, providing a link between the excess risk (approximation error) and the variance of the "excess loss" and opening a way for Massart's type penalization methods (see Sects. 4.1, 6.3).

## 5.1   Regression with Quadratic Loss

We start with regression problems with bounded response and with quadratic loss. To be specific, assume that $Y$ takes values in $T = [0, 1]$ and $\ell(y, u) := (y - u)^2$, $y \in T, u \in \mathbb{R}$. The minimum of the risk

$$P(\ell \bullet g) = \mathbb{E}(Y - g(X))^2$$

over the set of all measurable functions $g : S \mapsto \mathbb{R}$ is attained at the regression function

$$g_*(x) := \eta(x) := \mathbb{E}(Y \mid X = x).$$

If $\mathscr{G}$ is a class of measurable functions from $S$ into $[0, 1]$ such that $g_* \in \mathscr{G}$, then it is easy to check that for all $g \in \mathscr{G}$

$$\mathscr{E}_P(\ell \bullet g) = \|g - g_*\|^2_{L_2(\Pi)}.$$

In general, the excess risk is given by

$$\mathscr{E}_P(\ell \bullet g) = \|g - g_*\|^2_{L_2(\Pi)} - \inf_{h \in \mathscr{G}} \|h - g_*\|^2_{L_2(\Pi)}.$$

The following lemma provides an easy way to bound the excess risk from below in the case of a *convex class* $\mathscr{G}$ and $\bar{g} := \operatorname{argmin}_{g \in \mathscr{G}} \|g - g_*\|^2_{L_2(\Pi)}$.

**Lemma 5.1.** *If $\mathscr{G}$ is a convex class of functions, then*

$$2\mathscr{E}_P(\ell \bullet g) \geq \|g - \bar{g}\|^2_{L_2(\Pi)}.$$

*Proof.* The identity

$$\frac{u^2 + v^2}{2} - \left(\frac{u + v}{2}\right)^2 = \frac{(u - v)^2}{4}$$

implies that

$$\frac{(g - g_*)^2 + (\bar{g} - g_*)^2}{2} = \left(\frac{g + \bar{g}}{2} - g_*\right)^2 + \frac{(g - \bar{g})^2}{4}.$$

Integrating the last identity with respect to $\Pi$ yields

$$\frac{\|g - g_*\|^2_{L_2(\Pi)} + \|\bar{g} - g_*\|^2_{L_2(\Pi)}}{2} = \left\|\frac{g + \bar{g}}{2} - g_*\right\|^2_{L_2(\Pi)} + \frac{\|g - \bar{g}\|^2_{L_2(\Pi)}}{4}.$$

Since $\mathscr{G}$ is convex and $g, \bar{g} \in \mathscr{G}$, we have $\frac{g + \bar{g}}{2} \in \mathscr{G}$ and

$$\left\|\frac{g + \bar{g}}{2} - g_*\right\|^2_{L_2(\Pi)} \geq \|\bar{g} - g_*\|^2_{L_2(\Pi)}.$$

Therefore,

$$\frac{\|g - g_*\|^2_{L_2(\Pi)} + \|\bar{g} - g_*\|^2_{L_2(\Pi)}}{2} \geq \|\bar{g} - g_*\|^2_{L_2(\Pi)} + \frac{\|g - \bar{g}\|^2_{L_2(\Pi)}}{4},$$

implying the claim.                                                               $\square$

As before, we denote $\mathscr{F} := \{\ell \bullet g : g \in \mathscr{G}\}$. It follows from Lemma 5.1 that

$$\mathscr{F}(\delta) \subset \{\ell \bullet g : \|g - \bar{g}\|^2_{L_2(\Pi)} \leq 2\delta\}.$$

Also, for all functions $g_1, g_2 \in \mathscr{G}$ and all $x \in S, y \in T$,

$$\left|(\ell \bullet g_1)(x, y) - (\ell \bullet g_2)(x, y)\right| = \left|(y - g_1(x))^2 - (y - g_2(x))^2\right|$$

$$= |g_1(x) - g_2(x)||2y - g_1(x) - g_2(x)| \leq 2|g_1(x) - g_2(x)|,$$

which implies

$$P\left(\ell \bullet g_1 - \ell \bullet g_2\right)^2 \leq 4\|g_1 - g_2\|^2_{L_2(\Pi)}.$$

Hence

$$D(\delta) \leq 2 \sup\left\{\|g_1 - g_2\|_{L_2(\Pi)} : \|g_1 - \bar{g}\|^2_{L_2(\Pi)} \leq 2\delta, \|g_2 - \bar{g}\|^2_{L_2(\Pi)} \leq 2\delta\right\} \leq 4\sqrt{2}\sqrt{\delta}.$$

In addition, by symmetrization inequality,

$$\phi_n(\delta) = \mathbb{E}\|P_n - P\|_{\mathscr{F}'(\delta)} \leq 2\mathbb{E}\|R_n\|_{\mathscr{F}'(\delta)}$$

$$\leq 2\mathbb{E} \sup\left\{\left|R_n(\ell \bullet g_1 - \ell \bullet g_2)\right| : g_1, g_2 \in \mathscr{G}, \|g_1 - \bar{g}\|^2_{L_2(\Pi)}\right.$$

$$\left. \vee \|g_2 - \bar{g}\|^2_{L_2(\Pi)} \leq 2\delta\right\}$$

$$\leq 4\mathbb{E} \sup\left\{\left|R_n(\ell \bullet g - \ell \bullet \bar{g})\right| : g \in \mathscr{G}, \|g - \bar{g}\|^2_{L_2(\Pi)} \leq 2\delta\right\},$$

and since $\ell(y, \cdot)$ is Lipschitz with constant 2 on the interval $[0, 1]$ one can use the contraction inequality to get

$$\phi_n(\delta) \leq 16\mathbb{E} \sup\{|R_n(g - \bar{g})| : g \in \mathscr{G}, \|g - \bar{g}\|^2_{L_2(\Pi)} \leq 2\delta\} =: \psi_n(\delta).$$

As a result, we get

$$\phi_n^\flat(\sigma) \leq \psi_n^\flat(\sigma) \quad \text{and} \quad \sqrt{(D^2)^\flat(\sigma)} \leq 4\sqrt{2}.$$

This yields an upper bound on the quantity $\sigma_n^t$ involved in Theorem 4.3:

$$\sigma_n^t \leq K\left(\psi_n^\sharp\left(\frac{1}{2q}\right) + \frac{t}{n}\right),$$

and the following statement is a corollary of this theorem.

**Theorem 5.1.** *Let $\mathscr{G}$ be a convex class of functions from $S$ into $[0,1]$ and let $\hat{g}$ denotes the least squares estimator of the regression function*

$$\hat{g} := \mathrm{argmin}_{g \in \mathscr{G}} n^{-1} \sum_{j=1}^{n} (Y_j - g(X_j))^2.$$

*Then, there exist constants $K > 0, C > 0$ such that for all $t > 0$,*

$$\mathbb{P}\left\{ \|\hat{g} - g_*\|^2_{L_2(\Pi)} \geq \inf_{g \in \mathscr{G}} \|g - g_*\|^2_{L_2(\Pi)} + K\left(\psi_n^{\#}\left(\frac{1}{2q}\right) + \frac{t}{n}\right) \right\} \leq Ce^{-t}.$$

A slightly weaker result holds in the case when the class $\mathscr{G}$ is not necessarily convex. It follows from Lemma 4.1. Note that the condition

$$4(P(\ell \bullet g) - P(\ell \bullet g_*)) = 4\|g - g_*\|^2_{L_2(\Pi)} =: \rho_P^2(\ell \bullet g, \ell \bullet g_*) \geq P(\ell \bullet g - \ell \bullet g_*)^2$$

is satisfied for all functions $g : S \mapsto [0,1]$. Also,

$$\omega_n(\delta) = \mathbb{E} \sup_{4\|g - \bar{g}\|^2_{L_2(\Pi)} \leq \delta} \left| (P_n - P)(\ell \bullet g - \ell \bullet \bar{g}) \right| \leq \frac{1}{2}\psi_n(\delta/8)$$

(by symmetrization and contraction inequalities). Therefore, the following result holds.

**Theorem 5.2.** *Let $\mathscr{G}$ be a class of functions from $S$ into $[0,1]$ and let $\hat{g}$ denote the least squares estimator of the regression function. Then, there exist constants $K > 0, C > 0$ such that for all $t > 0$,*

$$\mathbb{P}\left\{ \|\hat{g} - g_*\|^2_{L_2(\Pi)} \geq (1 + \varepsilon) \inf_{g \in \mathscr{G}} \|g - g_*\|^2_{L_2(\Pi)} + \frac{1}{4}\psi_n^{\#}\left(\frac{\varepsilon}{K}\right) + \frac{Kt}{n\varepsilon} \right\} \leq Ce^{-t}.$$

Clearly, similar results hold (with different constants) if the functions in $\mathscr{G}$ take their values in an arbitrary bounded interval.

Several more specific examples are discussed below.

- **Example 1. Finite dimensional classes.** Suppose that $L \subset L_2(\Pi)$ is a finite dimensional linear space with $\dim(L) = d < \infty$ and let $\mathscr{G} \subset L$ be a convex class of functions taking values in a bounded interval (for simplicity, $[0,1]$). It follows from Proposition 3.2 that

$$\psi_n(\delta) \leq C\sqrt{\frac{d\delta}{n}}$$

with some constant $C > 0$. Hence,

$$\psi_n^{\#}\left(\frac{1}{2q}\right) \leq K\frac{d}{n},$$

and Theorem 5.1 implies that

$$\mathbb{P}\left\{\|\hat{g} - g_*\|^2_{L_2(\Pi)} \geq \inf_{g \in \mathcal{G}} \|g - g_*\|^2_{L_2(\Pi)} + K\left(\frac{d}{n} + \frac{t}{n}\right)\right\} \leq Ce^{-t}$$

with some constant $K > 0$.

- **Example 2. Reproducing kernel Hilbert spaces (RKHS).** Suppose $\mathcal{G}$ is the unit ball in RKHS $\mathcal{H}_K$:

$$\mathcal{G} := \{h : \|h\|_{\mathcal{H}_K} \leq 1\}.$$

Denote $\{\lambda_k\}$ the eigenvalues of the integral operator from $L_2(\Pi)$ into $L_2(\Pi)$ with kernel $K$. Then Proposition 3.3 implies that

$$\psi_n(\delta) \leq C\left(n^{-1} \sum_{j=1}^{\infty} (\lambda_j \wedge \delta)\right)^{1/2}.$$

The function

$$\delta \mapsto \left(n^{-1} \sum_{j=1}^{\infty} (\lambda_j \wedge \delta)\right)^{1/2} =: \gamma_n(\delta)$$

is strictly concave with $\gamma_n(0) = 0$, and, as a result,

$$\gamma_n^{\flat}(\delta) = \frac{\gamma_n(\delta)}{\delta}$$

is strictly decreasing. By a simple computation, Theorem 5.1 yields

$$\mathbb{P}\left\{\|\hat{g} - g_*\|^2_{L_2(\Pi)} \geq \inf_{g \in \mathcal{G}} \|g - g_*\|^2_{L_2(\Pi)} + K\left(\gamma_n^{\sharp}(1) + \frac{t}{n}\right)\right\} \leq Ce^{-t}$$

with some constant $K > 0$.

- **Example 3. VC-subgraph classes.** Suppose that $\mathcal{G}$ is a VC-subgraph class of functions $g : S \mapsto [0, 1]$ of VC-dimension $V$. Then the function $\psi_n(\delta)$ can be upper bounded using (3.17):

$$\psi_n(\delta) \leq C\left[\sqrt{\frac{V\delta}{n} \log \frac{1}{\delta}} \bigvee \frac{V}{n} \log \frac{1}{\delta}\right].$$

Therefore

$$\psi_n^{\sharp}(\varepsilon) \leq \frac{CV}{n\varepsilon^2} \log \frac{n\varepsilon^2}{V}.$$

Theorem 5.2 implies

$$\mathbb{P}\left\{\|\hat{g} - g_*\|^2_{L_2(\Pi)} \geq (1+\varepsilon) \inf_{g \in \mathcal{G}} \|g - g_*\|^2_{L_2(\Pi)} + K\left(\frac{V}{n\varepsilon^2} \log \frac{n\varepsilon^2}{V} + \frac{t}{n\varepsilon}\right)\right\} \leq Ce^{-t}.$$

- **Example 4. Entropy conditions**. In the case when the entropy of the class $\mathscr{G}$ (random, uniform, bracketing, etc.) is bounded by $O(\varepsilon^{-2\rho})$ for some $\rho \in (0, 1)$, we typically have

$$\psi_n^{\#}(\varepsilon) = O\left(n^{-1/(1+\rho)}\right).$$

For instance, if (3.18) holds, then it follows from (3.19) (with $F \equiv U = 1$ for simplicity) that

$$\psi_n(\delta) \leq K\left(\frac{A^\rho}{\sqrt{n}}\delta^{(1-\rho)/2} \bigvee \frac{A^{2\rho/(\rho+1)}}{n^{1/(1+\rho)}}\right).$$

Therefore,

$$\psi_n^{\#}(\varepsilon) \leq \frac{CA^{2\rho/(1+\rho)}}{(n\varepsilon^2)^{1/(1+\rho)}}.$$

In this case Theorem 5.2 gives the bound

$$\mathbb{P}\left\{\|\hat{g} - g_*\|_{L_2(\Pi)}^2 \geq (1+\varepsilon) \inf_{g \in \mathscr{G}} \|g - g_*\|_{L_2(\Pi)}^2 + K\left(\frac{A^{2\rho/(1+\rho)}}{(n\varepsilon^2)^{1/(1+\rho)}} + \frac{t}{n\varepsilon}\right)\right\}$$
$$\leq Ce^{-t}.$$

- **Example 5. Convex hulls**. If

$$\mathscr{G} := \operatorname{conv}(\mathscr{H}) := \left\{\sum_j \lambda_j h_j : \sum_j |\lambda_j| \leq 1, h_j \in \mathscr{H}\right\}$$

is the symmetric convex hull of a given VC-type class $\mathscr{H}$ of measurable functions from $S$ into $[0, 1]$, then the condition of the previous example is satisfied with $\rho := \frac{V}{V+2}$. This yields

$$\psi_n^{\#}(\varepsilon) \leq \left(\frac{K(V)}{n\varepsilon^2}\right)^{\frac{1}{2}\frac{2+V}{1+V}}$$

and Theorem 5.1 yields

$$\mathbb{P}\left\{\|\hat{g} - g_*\|_{L_2(\Pi)}^2 \geq \inf_{g \in \mathscr{G}} \|g - g_*\|_{L_2(\Pi)}^2 + K\left(\left(\frac{1}{n}\right)^{\frac{1}{2}\frac{2+V}{1+V}} + \frac{t}{n}\right)\right\} \leq Ce^{-t}$$

with some constant $K > 0$ depending on $V$.

## 5.2   Empirical Risk Minimization with Convex Loss

A standard assumption on the loss function $\ell$ that makes the empirical risk minimization problem computationally tractable is that $\ell(y, \cdot)$ is a convex function for all $y \in T$. Assuming, in addition, that $\mathcal{G}$ is a convex class of functions, the convexity of the loss implies that the empirical risk $\mathcal{G} \ni g \mapsto P_n(\ell \bullet g)$ is a convex functional and the empirical risk minimization is a convex minimization problem. We will call the problems of this type *convex risk minimization*. The least squares and the $L_1$-regression as well as some of the methods of large margin classification (such as boosting) are examples of convex risk minimization.

The convexity assumption also simplifies the analysis of empirical risk minimization problems. In particular, it makes easier proving the existence of the minimal point $g_*$, checking condition (5.1), etc. In this section, we extend the results for $L_2$-regression to this more general framework.

Assume the functions in $\mathcal{G}$ take their values in $[-M/2, M/2]$. We will need the following assumptions on the loss function $\ell$ :

- $\ell$ satisfies the Lipschitz condition with some $L > 0$

$$\forall y \in T \; \forall u, v \in [-M/2, M/2] \;\; |\ell(y, u) - \ell(y, v)| \le L|u - v|; \qquad (5.3)$$

- The following assumption on convexity modulus of $\ell$ holds for some $\Lambda > 0$ :

$$\forall y \in T \; \forall u, v \in [-M/2, M/2] \;\; \frac{\ell(y, u) + \ell(y, v)}{2} - \ell\left(y; \frac{u+v}{2}\right) \ge \Lambda|u - v|^2. \qquad (5.4)$$

Note that, if $g_*$ is bounded by $M/2$, conditions (5.3) and (5.4) imply (5.1) with $D(u, \mu) \le \frac{L^2}{2\Lambda}$. To see this, it is enough to use (5.4) with $v = u_\mu$, $\mu = \mu_x$ and to integrate it with respect to $\mu$. As a result, for the function

$$L(u) := \int_T \ell(y, u)\mu(dy),$$

whose minimum is attained at $u_\mu$, the following bound holds:

$$\frac{L(u) - L(u_\mu)}{2} = \frac{L(u) + L(u_\mu)}{2} - L(u_\mu)$$

$$\ge \frac{L(u) + L(u_\mu)}{2} - L\left(\frac{u + u_\mu}{2}\right) \ge \Lambda|u - u_\mu|^2. \qquad (5.5)$$

On the other hand, the Lipschitz condition implies that

$$\int_T |\ell(y, u) - \ell(y, u_\mu)|^2 \mu(dy) \le L^2|u - u_\mu|^2, \qquad (5.6)$$

and (5.1) follows from (5.5) and (5.6). This nice and simple convexity argument has been used repeatedly in the theory of excess risk bounds (see, for instance, Bartlett et al. [16]). We will also use it in the proof of Theorem 5.3.

**Theorem 5.3.** *Suppose that $\mathscr{G}$ is a convex class of functions taking values in $[-M/2, M/2]$. Assume that the minimum of $P(\ell \bullet g)$ over $\mathscr{G}$ is attained at $\bar{g} \in \mathscr{G}$ and*

$$\omega_n(\delta) := \mathbb{E} \sup_{g \in \mathscr{G}, \|g - \bar{g}\|^2_{L_2(\Pi)} \leq \delta} |R_n(g - \bar{g})|.$$

*Denote*

$$\hat{g} := \operatorname{argmin}_{g \in \mathscr{G}} P_n(\ell \bullet g).$$

*Then there exist constants $K > 0, C > 0, c > 0$ such that*

$$\mathbb{P}\left\{ P(\ell \bullet \hat{g}) \geq \inf_{g \in \mathscr{G}} P(\ell \bullet g) + K\left( \Lambda \omega_n^{\sharp}\left(\frac{c\Lambda}{L}\right) + \frac{L^2}{\Lambda} \frac{t}{n} \right) \right\} \leq Ce^{-t}, \ t > 0.$$

*Proof.* Note that by Lipschitz condition (5.3), for all $g_1, g_2 \in \mathscr{G}$,

$$P|\ell \bullet g_1 - \ell \bullet g_2|^2 \leq L^2 \|g_1 - g_2\|^2_{L_2(\Pi)}.$$

On the other hand, by (5.4), for all $g \in \mathscr{G}, x \in S, y \in T$,

$$\frac{\ell(y, g(x)) + \ell(y, \bar{g}(x))}{2} \geq \ell\left(y; \frac{g(x) + \bar{g}(x)}{2}\right) + \Lambda|g(x) - \bar{g}(x)|^2.$$

Integrating this inequality and observing that $\frac{g + \bar{g}}{2} \in \mathscr{G}$ and hence

$$P\left( \ell \bullet \left( \frac{g + \bar{g}}{2} \right) \right) \geq P(\ell \bullet \bar{g}),$$

yields

$$\frac{P(\ell \bullet g) + P(\ell \bullet \bar{g})}{2} \geq P(\ell \bullet \bar{g}) + \Lambda\Pi|g - \bar{g}|^2,$$

or

$$P(\ell \bullet g) - P(\ell \bullet \bar{g}) \geq 2\Lambda\Pi|g - \bar{g}|^2.$$

For the loss class $\mathscr{F} = \{\ell \bullet g : g \in \mathscr{G}\}$, this gives the following upper bound on the $L_2(P)$-diameter of the $\delta$-minimal set $\mathscr{F}(\delta) : D^2(\delta) \leq \frac{2\delta}{\Lambda}$. By symmetrization and contraction inequalities, it is easy to bound

$$\phi_n(\delta) = \mathbb{E}\|P_n - P\|_{\mathscr{F}'(\delta)}$$

in terms of $\omega_n(\delta)$ :

$$\phi_n(\delta) \leq CL\omega_n\left(\frac{\delta}{2\Lambda}\right).$$

By a simple computation, the quantity $\sigma_n^t$ used in Theorem 4.3 is bounded as follows:

$$\sigma_n^t \leq K\left(\Lambda\omega_n^{\sharp}\left(\frac{c\Lambda}{L}\right) + \frac{L^2}{\Lambda}\frac{t}{n}\right).$$

Under the additional assumption that $\ell$ is uniformly bounded by 1 in $T \times [-M/2, M/2]$, Theorem 4.3 implies the result. To get rid of the extra assumption, suppose that $\ell$ is uniformly bounded by $D$ on $T \times [-M/2, M/2]$. Then the result holds for the loss function $\ell/D$. For this loss function, $L$ and $\Lambda$ are replaced by $L/D$ and $\Lambda/D$, and the expression

$$\Lambda\omega_n^{\sharp}\left(\frac{c\Lambda}{L}\right) + \frac{L^2}{\Lambda}\frac{t}{n}$$

becomes

$$\Lambda/D\omega_n^{\sharp}\left(\frac{c\Lambda/D}{L/D}\right) + \frac{L^2/D^2}{\Lambda/D}\frac{t}{n} = \frac{1}{D}\left(\Lambda\omega_n^{\sharp}\left(\frac{c\Lambda}{L}\right) + \frac{L^2}{\Lambda}\frac{t}{n}\right),$$

so the result follows by rescaling.    $\square$

As an example, consider the case when $\mathscr{G} := M\mathrm{conv}(\mathscr{H})$ for a base class $\mathscr{H}$ of functions from $S$ into $[-1/2, 1/2]$. There are many powerful functional gradient descent type algorithms (such as boosting) that provide an implementation of convex empirical risk minimization over a convex hull or a linear span of a given base class. Assume that condition (3.16) holds for the class $\mathscr{H}$ with some $V > 0$, i.e., $\mathscr{H}$ is a VC-type class. Define

$$\pi_n(M, L, \Lambda; t) := K\left[\Lambda M^{V/(V+1)}\left(\frac{L}{\Lambda}\bigvee 1\right)^{(V+2)/(V+1)} n^{-\frac{1}{2}\frac{V+2}{V+1}} + \frac{L^2}{\Lambda}\frac{t}{n}\right]$$

with a numerical constant $K$. The next result is a slightly generalized version of a theorem due to Bartlett et al. [16].

**Theorem 5.4.** *Under the conditions (5.3) and (5.4),*

$$\mathbb{P}\left\{P(\ell \bullet \hat{g}_n) \geq \min_{g \in \mathscr{G}} P(\ell \bullet g) + \pi_n(M, L, \Lambda; t)\right\} \leq Ce^{-t}.$$

*Proof.* To apply Theorem 5.3, it is enough to bound the function $\omega_n$. Since $\mathscr{G} := M\,\mathrm{conv}(\mathscr{H})$, where $\mathscr{H}$ is a VC-type class of functions from $S$ into $[-1/2, 1/2]$, condition (3.16) holds for $\mathscr{H}$ with envelope $F \equiv 1$ (see Theorem 3.13). Together with (3.19), this gives

$$\omega_n(\delta) \leq C\left[\frac{M^\rho}{\sqrt{n}}\delta^{(1-\rho)/2} \bigvee \frac{M^{2\rho/(\rho+1)}}{n^{1/(1+\rho)}}\right]$$

with $\rho := \frac{V}{V+2}$. Hence,

$$\omega_n^\sharp(\varepsilon) \leq C\,\frac{M^{2\rho/(1+\rho)}}{n^{1/(1+\rho)}}\varepsilon^{-2/(1+\rho)}$$

for $\varepsilon \leq 1$. If $\ell(y, \cdot)$ is bounded by 1 in $T \times [-M/2, M/2]$, then Theorem 5.3 yields

$$\mathbb{P}\left\{P(\ell \bullet \hat{g}) \geq \min_{g \in \mathscr{G}} P(\ell \bullet g) + \pi_n(M, L, \Lambda; t)\right\} \leq Ce^{-t}.$$

To remove the assumption that $\ell$ is bounded by 1, one should use the same rescaling argument as in the proof of Theorem 5.3.                                    □


## 5.3  Binary Classification Problems

Binary classification is a prediction problem with $T = \{-1, 1\}$ and $\ell(y, u) := I(y \neq u)$, $y, u \in \{-1, 1\}$ (binary loss). It is a simple example of risk minimization with a nonconvex loss function.

Measurable functions $g : S \mapsto \{-1, 1\}$ are called *classifiers*. The risk of a classifier $g$ with respect to the binary loss

$$L(g) := P(\ell \bullet g) = \mathbb{E}I(Y \neq g(X)) = \mathbb{P}\{Y \neq g(X)\}$$

is called *the generalization error*. It is well known that the minimum of the generalization error over the set of all classifiers is attained at the classifier

$$g_*(x) = \mathrm{sign}(\eta(x)),$$

where $\eta(x) = \mathbb{E}(Y|X = x)$ is the regression function. The function $g_*$ is called *the Bayes classifier*. It is also well known that for all classifiers $g$

$$L(g) - L(g_*) = \int_{\{x:g(x)\neq g_*(x)\}} |\eta(x)|\Pi(dx) \tag{5.7}$$

(see, e.g., [51]).

Suppose there exists $h \in (0, 1]$ such that for all $x \in S$

$$|\eta(x)| \geq h. \tag{5.8}$$

The parameter $h$ characterizes the level of noise in classification problems: for small values of $h$, $\eta(x)$ can get close to 0 and, in such cases, correct classification is harder to achieve. The following condition provides a more flexible way to describe the level of the noise:

$$\Pi\{x : |\eta(x)| \leq t\} \leq Ct^\alpha \tag{5.9}$$

for some $\alpha > 0$. It is often referred to as "Tsybakov's low noise assumption" or "Tsybakov's margin assumption" (sometimes, condition (5.8) is called "Massart's low noise assumption").

**Lemma 5.2.** *Under condition (5.8),*

$$L(g) - L(g_*) \geq h\Pi(\{x : g(x) \neq g_*(x)\}).$$

*Under condition (5.9),*

$$L(g) - L(g_*) \geq c\Pi^\kappa(\{x : g(x) \neq g_*(x)\}),$$

*where $\kappa = \frac{1+\alpha}{\alpha}$ and $c > 0$ is a constant.*

*Proof.* The first bound follows immediately from formula (5.7). To prove the second bound, use the same formula to get

$$L(g) - L(g_*) \geq t\Pi \{x : g(x) \neq g_*(x), |\eta(x)| > t\}$$
$$\geq t\Pi \{x : g(x) \neq g_*(x)\} - t\Pi\{x : |\eta(x)| \leq t\}$$
$$\geq t\Pi \{x : g(x) \neq g_*(x)\} - Ct^{1+\alpha}.$$

It remains to substitute in the last bound the value of $t$ that solves the equation

$$\Pi \{x : g(x) \neq g_*(x)\} = 2Ct^\alpha$$

to get the result.                                                         □

Let $\mathcal{G}$ be a class of binary classifiers. Denote

$$\hat{g} := \mathrm{argmin}_{g \in \mathcal{G}} L_n(g),$$

where

$$L_n(g) := n^{-1} \sum_{j=1}^{n} I(Y_j \neq g(X_j))$$

is the empirical risk with respect to the binary loss (*the training error*).

First we obtain upper bounds on the excess risk $L(\hat{g}) - L(g_*)$ of $\hat{g}$ in terms of random shattering numbers

$$\Delta^{\mathcal{G}}(X_1, \ldots, X_n) := \mathrm{card}\left\{ (g(X_1), \ldots, g(X_n)) : g \in \mathcal{G} \right\}$$

and parameter $h$ involved in condition (5.8).

**Theorem 5.5.** *Suppose condition (5.8) holds with some $h \in (0, 1]$. If $g_* \in \mathcal{G}$, then*

$$\mathbb{P}\left\{ L(\hat{g}) - L(g_*) \geq K\left( \frac{\mathbb{E} \log \Delta^{\mathcal{G}}(X_1, \ldots, X_n)}{nh} + \frac{t}{nh} \right) \right\} \leq Ce^{-t}$$

*with some constants $K, C > 0$. In the general case, when $g_*$ does not necessarily belong to $\mathcal{G}$, the following bound holds for all $\varepsilon \in (0, 1)$ :*

$$\mathbb{P}\left\{ L(\hat{g}) - L(g_*) \geq (1 + \varepsilon)\left( \inf_{g \in \mathcal{G}} L(g) - L(g_*) \right) \right.$$

$$\left. + K\left( \frac{\mathbb{E} \log \Delta^{\mathcal{G}}(X_1, \ldots, X_n)}{nh\varepsilon^2} + \frac{t}{nh\varepsilon} \right) \right\} \leq Ce^{-t}$$

*Proof.* Note that

$$|(\ell \bullet g)(x, y) - (\ell \bullet g_*)(x, y)| = I(g(x) \neq g_*(x)),$$

which implies

$$\left\| \ell \bullet g - \ell \bullet g_* \right\|_{L_2(P)}^2 = P|(\ell \bullet g) - (\ell \bullet g_*)|^2 = \Pi\{x : g(x) \neq g_*(x)\}.$$

As always, denote $\mathscr{F} := \{\ell \bullet g : g \in \mathcal{G}\}$. Under the assumption $g_* \in \mathcal{G}$, the first inequality of Lemma 5.2 implies that

$$\mathscr{F}(\delta) = \left\{ \ell \bullet g : \mathscr{E}(\ell \bullet g) = L(g) - L(g_*) \leq \delta \right\} \subset \left\{ \ell \bullet g : \left\| \ell \bullet g - \ell \bullet g_* \right\|_{L_2(P)} \leq \sqrt{\frac{\delta}{h}} \right\},$$

so the $L_2(P)$-diameter $D(\delta)$ of the class $\mathscr{F}(\delta)$ satisfies $D(\delta) \leq 2\sqrt{\frac{\delta}{h}}$. Next we have

$$\phi_n(\delta) = \mathbb{E}\| P_n - P \|_{\mathscr{F}'(\delta)} \leq 2\mathbb{E} \sup_{g \in \mathcal{G}, \Pi(\{g \neq g_*\}) \leq \delta/h} |(P_n - P)(\ell \bullet g - \ell \bullet g_*)|. \quad (5.10)$$

Denote

$$\mathscr{D} := \left\{ \{(x, y) : y \neq g(x)\} : g \in \mathcal{G} \right\} \text{ and } D_* := \{(x, y) : y \neq g_*(x)\}.$$

It is easy to check that for

$$D_1 := \{(x, y) : y \neq g_1(x)\}, \quad D_2 := \{(x, y) : y \neq g_2(x)\},$$

we have

$$\Pi(\{g_1 \neq g_2\}) = P(D_1 \triangle D_2).$$

It follows from (5.10) that

$$\phi_n(\delta) \leq 2\mathbb{E} \sup_{D \in \mathcal{D}, P(D \triangle D_*) \leq \delta/h} |(P_n - P)(D \setminus D_*)|$$

$$+ 2\mathbb{E} \sup_{D \in \mathcal{D}, P(D \triangle D_*) \leq \delta/h} |(P_n - P)(D_* \setminus D)|.$$

Theorem 3.9 yields

$$\phi_n(\delta) \leq K \left[ \sqrt{\frac{\delta}{h}} \sqrt{\frac{\mathbb{E} \log \Delta^{\mathcal{D}}((X_1, Y_1), \ldots, (X_n, Y_n))}{n}} \right.$$

$$\left. \bigvee \frac{\mathbb{E} \log \Delta^{\mathcal{D}}((X_1, Y_1), \ldots, (X_n, Y_n))}{n} \right]$$

with some constant $K > 0$. Also, it is easy to observe that

$$\Delta^{\mathcal{D}}((X_1, Y_1), \ldots, (X_n, Y_n)) = \Delta^{\mathcal{G}}(X_1, \ldots, X_n)$$

which gives the bound

$$\phi_n(\delta) \leq K \left[ \sqrt{\frac{\delta}{h}} \sqrt{\frac{\mathbb{E} \log \Delta^{\mathcal{G}}(X_1, \ldots, X_n)}{n}} \bigvee \frac{\mathbb{E} \log \Delta^{\mathcal{G}}(X_1, \ldots, X_n)}{n} \right].$$

The bounds on $\phi_n(\delta)$ and $D(\delta)$ provide a way to control the quantity $\sigma_n^t$ involved in Theorem 4.3:

$$\sigma_n^t \leq K \left[ \frac{\mathbb{E} \log \Delta^{\mathcal{G}}(X_1, \ldots, X_n)}{nh} + \frac{t}{nh} \right]$$

with some constant $K > 0$, which implies the first bound of the theorem.

The proof of the second bound follows the same lines and it is based on Lemma 4.1. $\qquad \square$

The next theorem provides bounds on excess risk in terms of shattering numbers under Tsybakov's condition (5.9). We skip the proof which is similar to that of Theorem 5.5.

**Theorem 5.6.** *Suppose condition (5.9) holds with some $\alpha > 0$. Let $\kappa := \frac{1+\alpha}{\alpha}$. If $g_* \in \mathscr{G}$, then*

$$\mathbb{P}\left\{L(\hat{g}) - L(g_*) \geq K\left(\left(\frac{\mathbb{E}\log\Delta^{\mathscr{G}}(X_1,\ldots,X_n)}{n}\right)^{\kappa/(2\kappa-1)} + \left(\frac{t}{n}\right)^{\kappa/(2\kappa-1)}\right)\right\} \leq Ce^{-t}$$

*with some constants $K, C > 0$.*

We will also mention the following result in spirit of Tsybakov [144].

**Theorem 5.7.** *Suppose, for some $A > 0, \rho \in (0,1)$*

$$\log N(\mathscr{G}; L_2(P_n); \varepsilon) \leq \left(\frac{A}{\varepsilon}\right)^{2\rho}, \quad \varepsilon > 0, \tag{5.11}$$

*and condition (5.9) holds with some $\alpha > 0$. Let $\kappa := \frac{1+\alpha}{\alpha}$. If $g_* \in \mathscr{G}$, then*

$$\mathbb{P}\left\{L(\hat{g}) - L(g_*) \geq K\left(\left(\frac{1}{n}\right)^{\kappa/(2\kappa+\rho-1)} + \left(\frac{t}{n}\right)^{\kappa/(2\kappa-1)}\right)\right\} \leq Ce^{-t}$$

*with some constant $K, C > 0$ depending on $A$.*

The proof is very similar to the proofs of the previous results except that now (3.19) is used to bound the empirical process. One can also use other notions of entropy such as entropy with bracketing and obtain very similar results.

We conclude this section with a theorem by Giné and Koltchinskii [66] that refines an earlier result by Massart and Nedelec [108]. To formulate it, let

$$\mathscr{C} := \{\{g = 1\} : g \in \mathscr{G}\}, \quad C_* := \{g_* = 1\},$$

and define the following local version of Alexander's capacity function of the class $\mathscr{C}$ (see [5]):

$$\tau(\delta) := \frac{\Pi\left(\bigcup_{C \in \mathscr{C}, \Pi(C \triangle C_*) \leq \delta}(C \triangle C_*)\right)}{\delta}.$$

**Theorem 5.8.** *Suppose condition (5.8) holds with some $h \in (0,1]$. Suppose also that $\mathscr{C}$ is a VC-class of VC-dimension $V$. If $g_* \in \mathscr{G}$, then*

$$\mathbb{P}\left\{L(\hat{g}) - L(g_*) \geq K\left(\frac{V}{nh}\log\tau\left(\frac{V}{nh^2}\right) + \frac{t}{nh}\right)\right\} \leq Ce^{-t}$$

*with some constants $K, C > 0$. In the general case, when $g_*$ does not necessarily belong to $\mathscr{G}$, the following bound holds for all $\varepsilon \in (0, 1)$ :*

$$\mathbb{P}\left\{ L(\hat{g}) - L(g_*) \geq (1 + \varepsilon)\left( \inf_{g \in \mathscr{G}} L(g) - L(g_*) \right) \right.$$
$$\left. + K\left( \frac{V}{nh\varepsilon^2} \log \tau\left( \frac{V}{nh^2\varepsilon^2} \right) + \frac{t}{nh\varepsilon} \right) \right\} \leq Ce^{-t}.$$

*Proof.* We give only a sketch of the proof that relies on bound (3.17). For instance, to prove the second inequality this bound is used to control

$$\omega_n(\delta) = \mathbb{E} \sup_{g \in \mathscr{G}, \|\ell \bullet g - \ell \bullet \bar{g}\|_{L_2(P)}^2 \leq \delta} |(P_n - P)(\ell \bullet g - \ell \bullet \bar{g})|,$$

where $\bar{g}$ is a minimal point of $P(\ell \bullet g)$ on $\mathscr{G}$. To use (3.17) one has to find the envelope

$$F_\delta(x, y) := \sup_{g \in \mathscr{G}, \|\ell \bullet g - \ell \bullet \bar{g}\|_{L_2(P)}^2 \leq \delta} |\ell \bullet g(x, y) - \ell \bullet \bar{g}(x, y)|.$$

Easy computations show that

$$\|F_\delta\|_{L_2(\Pi)} = 2\sqrt{\delta \tau(\delta)}$$

and an application of (3.17) yields

$$\omega_n(\delta) \leq K\left[ \sqrt{\frac{V\delta}{n} \log \tau(\delta)} \bigvee \frac{V}{n} \log \tau(\delta) \right]$$

with some constant $K$. This implies that, for all $\varepsilon \in (0, 1)$,

$$\omega_n^\sharp(\varepsilon) \leq K\frac{V}{n\varepsilon^2} \log \tau\left( \frac{V}{n\varepsilon^2} \right)$$

with some constant $K > 0$. Now we can use Lemma 4.1 to complete the proof of the second bound of the theorem (condition (4.5) of this lemma holds with $D = \frac{1}{h}$).
□

A straightforward upper bound on the capacity function $\tau(\delta) \leq \frac{1}{\delta}$ leads to the result of Massart and Nedelec [108] in which the main part of the error term is $\frac{V}{nh} \log\left( \frac{nh^2}{V} \right)$. However, it is easy to find examples in which the capacity $\tau(\delta)$ is uniformly bounded. For instance, suppose that $S = [0, 1]^d$, $\Pi$ is the Lebesgue measure on $S$, $\mathscr{C}$ is a VC-class of convex sets, $C_* \in \mathscr{C}$ and $\Pi(C_*) > 0$. Suppose also that with some constant $L > 0$

$$L^{-1}h(C, C_*) \leq \Pi(C \triangle C_*) \leq Lh(C, C_*), C \in \mathscr{C},$$

where $h$ is the Hausdorff distance. Then the boundedness of $\tau$ easily follows. In such cases, the main part of the error is of the order $\frac{V}{nh}$ (without a logarithmic factor).

## 5.4 Further Comments

The idea to control the variance of a loss in terms of its expectation has been extensively used by Massart [106] (and even in a much earlier work of Birgé and Massart) as well as in the learning theory literature Mendelson [112], Bartlett et al. [16], Blanchard et al. [26], Bartlett et al. [15] (and even earlier, see [6, 19]).

$L_2(\Pi)$-error bounds in regression problems with quadratic loss, given in Examples 1–5 of Sect. 5.1, are well known. In particular, the bound of Example 2 (regression in RKHS) goes back to Mendelson [113] and the bound of Example 5 (regression in convex hulls) to Blanchard et al. [26]. Other important references on $L_2$-error bounds in regression problems include [11, 62, 72, 79].

Empirical risk minimization with convex loss was studied in detail by Blanchard et al. [26] and Bartlett et al. [16]. In the last paper, rather subtle bounds relating excess risks with respect to a "surrogate" convex loss and with respect to the binary classification loss were also studied. Earlier, Zhang [155] provided initial versions of such bounds in connection with his study of consistency of classification algorithms (see also [103, 134] for other important results on this problem).

Classification problems under condition (5.9) ("Tsybakov's low noise assumption") have been intensively studied by Mammen and Tsybakov [105] and, especially, by Tsybakov [144] (see also [145]). Condition (5.8) was later suggested by Massart and used in a number of papers (see, e.g., [108]). Koltchinskii [83] provided an interpretation of assumptions of this type as special cases of conditions on the $L_2(\Pi)$-diameters of $\delta$-minimal sets of the true risk (see Chap. 4).

In the recent years, the capacity function $\tau$ used in Theorem 5.8 (see also Giné and Koltchinskii [66]) started playing an important role in the analysis of active learning algorithms (see Hanneke [73] and Koltchinskii [87]).

# Chapter 6
# Penalized Empirical Risk Minimization and Model Selection Problems

Let $\mathscr{F}$ be a class of measurable functions on $(S, \mathscr{A})$ and let $\{\mathscr{F}_k : k \geq 1\}$ be a family of its subclasses $\mathscr{F}_k \subset \mathscr{F}, k \geq 1$. The subclasses $\mathscr{F}_k$ will be used to approximate a solution of the problem of risk minimization (1.1) over a large class $\mathscr{F}$ by a family of solutions of "smaller" empirical risk minimization problems

$$\hat{f}_k := \hat{f}_{n,k} := \operatorname{argmin}_{f \in \mathscr{F}_k} P_n f.$$

For simplicity, we assume that the solutions $\{\hat{f}_{n,k}\}$ exist.

In what follows, we call

$$\mathscr{E}_P(\mathscr{F}; f) = Pf - \inf_{f \in \mathscr{F}} Pf$$

*the global excess risk* of $f \in \mathscr{F}$. Given $k \geq 1$, we call $\mathscr{E}_P(\mathscr{F}_k; f) = Pf - \inf_{f \in \mathscr{F}_k} Pf$ *the local excess risk of* $f \in \mathscr{F}_k$.

Usually, the classes $\mathscr{F}_k, k \geq 1$ represent losses associated with certain statistical models and the problem is to use the estimators $\{\hat{f}_{n,k}\}$ to construct a function $\hat{f} \in \mathscr{F}$ (for instance, to choose one of the estimators $\hat{f}_{n,k}$) with a small value of the global excess risk $\mathscr{E}_P(\mathscr{F}; \hat{f})$. To be more precise, suppose that there exists an index $k(P)$ such that

$$\inf_{\mathscr{F}_{k(P)}} Pf = \inf_{\mathscr{F}} Pf.$$

In other words, the risk minimizer over the whole class $\mathscr{F}$ belongs to a subclass $\mathscr{F}_{k(P)}$. A statistician does not know the distribution $P$ and, hence, the index $k(P)$ of the correct model. Let $\tilde{\delta}_n(k)$ be an upper bound on the local excess risk $\mathscr{E}_P(\mathscr{F}_k; \hat{f}_{n,k})$ of $\hat{f}_{n,k}$ that provides an "optimal", or just a "desirable" accuracy of solution of empirical risk minimization problem on the class $\mathscr{F}_k$. If there were an oracle who could tell the statistician that, say, $k(P) = 5$ is the correct index of the model, then the risk minimization problem could be solved with an accuracy at least $\tilde{\delta}_n(5)$. The *model selection problem* deals with constructing a data dependent index

V. Koltchinskii, *Oracle Inequalities in Empirical Risk Minimization and Sparse Recovery Problems*, Lecture Notes in Mathematics 2033, DOI 10.1007/978-3-642-22147-7_6, © Springer-Verlag Berlin Heidelberg 2011

$\hat{k} = \hat{k}(X_1, \ldots, X_n)$ of the model such that the excess risk of $\hat{f} := \hat{f}_{n,\hat{k}}$ is within a constant from $\tilde{\delta}_n(k(P))$ with a high probability. More generally, in the case when the global minimum of the risk $Pf$, $f \in \mathcal{F}$ is not attained in any of the classes $\mathcal{F}_k$, one can still try to show that with a high probability

$$\mathcal{E}_P(\mathcal{F}; \hat{f}) \leq C \inf_k \left[ \inf_{\mathcal{F}_k} Pf - Pf_* + \tilde{\pi}_n(k) \right],$$

where

$$f_* := \operatorname{argmin}_{f \in \mathcal{F}} Pf.$$

For simplicity, assume the existence of a function $f_* \in \mathcal{F}$ at which the global minimum of the risk $Pf$, $f \in \mathcal{F}$ is attained. The quantities $\tilde{\pi}_n(k)$ involved in the above bound are "ideal" distribution dependent complexity penalties associated with risk minimization over $\mathcal{F}_k$ and $C$ is a constant (preferably, $C = 1$ or at least close to 1). The inequalities that express such a property are often called *oracle inequalities*.

Among the most popular approaches to model selection are *penalization methods*, in which $\hat{k}$ is defined as a solution of the following minimization problem

$$\hat{k} := \operatorname{argmin}_{k \geq 1} \left\{ P_n \hat{f}_k + \hat{\pi}_n(k) \right\} \tag{6.1}$$

where $\hat{\pi}_n(k)$ is a *complexity penalty* (generally, data dependent) associated with the class (the model) $\mathcal{F}_k$. In other words, instead of minimizing the empirical risk on the whole class $\mathcal{F}$ we now minimize a penalized empirical risk.

We discuss below penalization strategies with the penalties based on data dependent bounds on excess risk developed in the previous sections. Penalization methods have been widely used in a variety of statistical problems, in particular, in nonparametric regression. At the same time, there are difficulties in extending penalization method of model selection to some other problems, such as nonparametric classification.

To provide some motivation for the approach discussed below, note that ideally one would want to find $\hat{k}$ by minimizing the global excess risk $\mathcal{E}_P(\mathcal{F}; \hat{f}_{n,k})$ of the solutions of ERM problems with respect to $k$. This is impossible without the help of the oracle. Instead, data dependent upper confidence bounds on the excess risk have to be developed. The following trivial representation (that plays the role of "bias-variance decomposition")

$$\mathcal{E}_P(\mathcal{F}; \hat{f}_{n,k}) = \inf_{\mathcal{F}_k} Pf - Pf_* + \mathcal{E}_P(\mathcal{F}_k; \hat{f}_{n,k})$$

shows that a part of the problem is to come up with data dependent upper bounds on *the local excess risk* $\mathcal{E}_P(\mathcal{F}_k; \hat{f}_{n,k})$. This was precisely the question studied in the previous sections. Another part of the problem is to bound $\inf_{\mathscr{F}_k} Pf - Pf_*$ in terms of $\inf_{\mathscr{F}_k} P_n f - P_n f_*$, which is what will be done in Lemma 6.3 below. Combining these two bounds provides an upper bound on *the global excess risk* that can be now minimized with respect to $k$ (the term $P_n f_*$ can be dropped since it does not depend on $k$).

Suppose that for each class $\mathscr{F}_k$, the function $U_n(\cdot) = U_{n,k}(\cdot)$ is given (it was defined in Sect. 4.1 in terms of sequences $\{\delta_j\}$ $\{t_j\}$ that, in this case, might also depend on $k$). In what follows, we will assume that, for each $k \geq 1$, $(\bar{\delta}_n(k), \hat{\delta}_n(k), \tilde{\delta}_n(k))$ is a triple bound on the excess risk for the class $\mathscr{F}_k$ of confidence level $1 - p_k$ (see Definition 4.1). Suppose $p := \sum_{k=1}^{\infty} p_k < 1$. Then, there exists an event $E$ of probability at least $1 - p$ such that on this event the following properties hold for all $k \geq 1$:

(i) $U_{n,k}^{\sharp}\left(\frac{1}{2}\right) \leq \bar{\delta}_n(k) \leq \hat{\delta}_n(k) \leq \tilde{\delta}_n(k)$

(ii) $\mathscr{E}(\mathscr{F}_k, \hat{f}_{n,k}) \leq \bar{\delta}_n(k)$

(iii) for all $f \in \mathscr{F}_k$,

$$\mathscr{E}_P(\mathscr{F}_k, f) \leq 2\mathscr{E}_{P_n}(\mathscr{F}_k; f) \vee \bar{\delta}_n(k)$$

and

$$\mathscr{E}_{P_n}(\mathscr{F}_k; f) \leq \frac{3}{2}\left(\mathscr{E}_P(\mathscr{F}_k; f) \vee \bar{\delta}_n(k)\right);$$

(iv) for all $\delta \geq \bar{\delta}_n(k)$, $\|P_n - P\|_{\mathscr{F}_k'(\delta)} \leq U_{n,k}(\delta)$.

In the next sections, we study several special cases of general penalized empirical risk minimization problem in which it will be possible to prove oracle inequalities.

## 6.1 Penalization in Monotone Families $\mathscr{F}_k$

In this section, we make a simplifying assumption that $\{\mathscr{F}_k\}$ is a monotone family, that is, $\mathscr{F}_k \subset \mathscr{F}_{k+1}$, $k \geq 1$. Let $\mathscr{F} := \bigcup_{j \geq 1} \mathscr{F}_j$. Define

$$\hat{k} := \operatorname{argmin}_{k \geq 1}\left[\inf_{f \in \mathscr{F}_k} P_n f + 4\hat{\delta}_n(k)\right]$$

and $\hat{f} := \hat{f}_{\hat{k}}$. The next statement is akin to the result of Bartlett [13].

**Theorem 6.1.** *The following oracle inequality holds with probability at least* $1 - p$ :

$$\mathscr{E}_P(\mathscr{F}; \hat{f}) \leq \inf_{j \geq 1}\left[\inf_{\mathscr{F}_j} Pf - \inf_{\mathscr{F}} Pf + 9\tilde{\delta}_n(j)\right].$$

*Proof.* We will consider the event $E$ of probability at least $1-p$ on which properties (i)–(iv) hold. Then, for all $j \geq \hat{k}$,

$$\mathcal{E}_P(\mathscr{F}_j; \hat{f}) \leq 2\mathcal{E}_{P_n}(\mathscr{F}_j; \hat{f}) \vee \bar{\delta}_n(j) \leq 2\left[ \inf_{f \in \mathscr{F}_{\hat{k}}} P_n f - \inf_{f \in \mathscr{F}_j} P_n f \right] + \bar{\delta}_n(j)$$

$$\leq 2\left[ \inf_{f \in \mathscr{F}_{\hat{k}}} P_n f + 4\hat{\delta}_n(\hat{k}) - \inf_{f \in \mathscr{F}_j} P_n f - 4\hat{\delta}_n(j) \right] + 9\hat{\delta}_n(j),$$

which is bounded by $9\tilde{\delta}_n(j)$ since, by the definition of $\hat{k}$, the term in the bracket is nonpositive and $\hat{\delta}_n(j) \leq \bar{\delta}_n(j)$. This implies

$$P\hat{f} \leq \inf_{f \in \mathscr{F}_j} Pf + 9\tilde{\delta}_n(j).$$

The next case is when $j < \hat{k}$ and $\hat{\delta}_n(j) \geq \hat{\delta}_n(\hat{k})/9$. Then $\mathcal{E}_P(\mathscr{F}_{\hat{k}}; \hat{f}_{\hat{k}}) \leq \bar{\delta}_n(\hat{k})$, and, as a consequence,

$$P\hat{f} \leq \inf_{f \in \mathscr{F}_{\hat{k}}} Pf + \hat{\delta}_n(\hat{k}) \leq \inf_{f \in \mathscr{F}_j} Pf + 9\tilde{\delta}_n(j).$$

The last case to consider is when $j < \hat{k}$ and $\hat{\delta}_n(j) < \hat{\delta}_n(\hat{k})/9$. In this case, the definition of $\hat{k}$ implies that

$$\inf_{f \in \mathscr{F}_j} \mathcal{E}_{P_n}(\mathscr{F}_{\hat{k}}; f) = \inf_{f \in \mathscr{F}_j} P_n f - \inf_{f \in \mathscr{F}_{\hat{k}}} P_n f \geq 4(\hat{\delta}_n(\hat{k}) - \hat{\delta}_n(j)) \geq 3\hat{\delta}_n(\hat{k}).$$

Hence,

$$\frac{3}{2}\left( \inf_{f \in \mathscr{F}_j} \mathcal{E}_P(\mathscr{F}_{\hat{k}}; f) \vee \bar{\delta}_n(\hat{k}) \right) \geq \inf_{f \in \mathscr{F}_j} \mathcal{E}_{P_n}(\mathscr{F}_{\hat{k}}; f) \geq 3\hat{\delta}_n(\hat{k}),$$

which yields

$$3 \inf_{f \in \mathscr{F}_j} \mathcal{E}_P(\mathscr{F}_{\hat{k}}; f) + 3\bar{\delta}_n(\hat{k}) \geq 6\hat{\delta}_n(\hat{k}).$$

Therefore

$$\inf_{f \in \mathscr{F}_j} \mathcal{E}_P(\mathscr{F}_{\hat{k}}; f) \geq \hat{\delta}_n(\hat{k}) \geq \mathcal{E}_P(\mathscr{F}_{\hat{k}}; \hat{f}).$$

As a consequence,

$$P\hat{f} \leq \inf_{f \in \mathscr{F}_j} Pf \leq \inf_{f \in \mathscr{F}_j} Pf + 9\tilde{\delta}_n(j).$$

This completes the proof.                                                         □

*Example 6.1.* Consider a regression problem with quadratic loss and with a bounded response variable $Y \in [0, 1]$ (see Sect. 5.1). Let $\mathscr{G}_k$, $k \geq 1$ be convex classes of functions $g$ taking values in $[0, 1]$ such that $\mathscr{G}_k \subset \mathscr{G}_{k+1}$, $k \geq 1$. Moreover, suppose that for all $k \geq 1$ $\mathscr{G}_k \subset L_k$, where $L_k$ is a finite dimensional space of dimension $d_k$. Let

$$\hat{g}_{n,k} := \mathrm{argmin}_{g \in \mathscr{G}_k} n^{-1} \sum_{j=1}^{n} (Y_j - g(X_j))^2.$$

Take a nondecreasing sequence $\{t_k\}$ of positive numbers such that

$$\sum_{k \geq 1} e^{-t_k} = p \in (0, 1).$$

Define

$$\bar{\delta}_n(k) = \hat{\delta}_n(k) = \tilde{\delta}_n(k) = K \frac{d_k + t_k}{n}, \quad k \geq 1.$$

It is straightforward to see that, for a large enough constant $K$, $(\bar{\delta}_n(k), \hat{\delta}_n(k), \tilde{\delta}_n(k))$ is a triple bound of level $1 - e^{-t_k}$ (see Example 1, Sect. 5.1). Hence, if we define

$$\hat{k} := \mathrm{argmin}_{k \geq 1} \left[ \inf_{g \in \mathscr{G}_k} n^{-1} \sum_{j=1}^{n} (Y_j - g(X_j))^2 + 4K \frac{d_k + t_k}{n} \right]$$

with a sufficiently large constant $K$ and set $\hat{g} := \hat{g}_{n,\hat{k}}$, then it follows from Theorem 6.1 that with probability at least $1 - p$

$$\|\hat{g} - g_*\|_{L_2(\Pi)}^2 \leq \inf_{k \geq 1} \left[ \inf_{g \in \mathscr{G}_k} \|g - g_*\|_{L_2(\Pi)}^2 + 9K \frac{d_k + t_k}{n} \right].$$

Clearly, one can also construct triple bounds and implement this penalization method in more complicated situations (see Examples 2–5 in Sect. 5.1) and for other loss functions (for instance, for convex losses discussed in Sect. 5.2). Moreover, one can use a general construction of triple bounds in Theorem 4.8 that provides a universal approach to complexity penalization (which, however, is more of theoretical interest).

Despite the fact that it is possible to prove nice and simple oracle inequalities under the monotonicity assumption, this assumption might be restrictive and, in what follows, we explore what can be done without it.

## 6.2   Penalization by Empirical Risk Minima

In this section, we study a simple penalization technique in spirit of the work of Lugosi and Wegkamp [104] in which the infimum of empirical risk $\inf_{\mathscr{F}_k} P_n f$ is explicitly involved in the penalty. It will be possible to prove rather natural oracle inequalities for this penalization method. However, the drawback of this approach is that, in most of the cases, it yields only suboptimal convergence rates.

Given triple bounds $(\bar{\delta}_n(k), \hat{\delta}_n(k), \tilde{\delta}_n(k))$ of level $1 - p_k$ for classes $\mathscr{F}_k$, define the following penalties:

$$\hat{\pi}(k) := \hat{\pi}_n(k) := \hat{K}\left[\hat{\delta}_n(k) + \sqrt{\frac{t_k}{n} \inf_{\mathscr{F}_k} P_n f} + \frac{t_k}{n}\right]$$

and

$$\tilde{\pi}(k) := \tilde{\pi}_n(k) := \tilde{K}\left[\tilde{\delta}_n(k) + \sqrt{\frac{t_k}{n} \inf_{\mathscr{F}_k} P f} + \frac{t_k}{n}\right],$$

where $\hat{K}, \tilde{K}$ are sufficiently large numerical constants. Here $\tilde{\pi}(k)$ represents a "desirable accuracy" of risk minimization on the class $\mathscr{F}_k$.

The index estimate $\hat{k}$ is defined by minimizing the penalized empirical risk

$$\hat{k} := \mathrm{argmin}_{k \geq 1}\left\{P_n \hat{f}_k + \hat{\pi}(k)\right\}$$

and, as always, $\hat{f} := \hat{f}_{\hat{k}}$.

The next theorem provides an upper confidence bound on the risk of $\hat{f}$ and an oracle inequality for the global excess risk $\mathscr{E}_P(\mathscr{F}; \hat{f})$.

**Theorem 6.2.** *There exists a choice of $\hat{K}, \tilde{K}$ such that for any sequence $\{t_k\}$ of positive numbers, the following bounds hold:*

$$\mathbb{P}\left\{P\hat{f} \geq \inf_{k \geq 1}\left\{P_n \hat{f}_{n,k} + \hat{\pi}(k)\right\}\right\} \leq \sum_{k=1}^{\infty}\left(p_k + e^{-t_k}\right)$$

*and*

$$\mathbb{P}\left\{\mathscr{E}_P(\mathscr{F}; \hat{f}) \geq \inf_{k \geq 1}\left\{\inf_{f \in \mathscr{F}_k} P f - \inf_{f \in \mathscr{F}} P f + \tilde{\pi}(k)\right\}\right\} \leq \sum_{k=1}^{\infty}\left(p_k + e^{-t_k}\right).$$

**Remark.** Note that, unless $\inf_{\mathscr{F}_k} P f = 0$, $\tilde{\pi}(k) = \tilde{\pi}_n(k)$ can not be smaller than const $n^{-1/2}$. In many cases (see Chap. 5), the excess risk bound $\tilde{\delta}_n(k)$ is smaller than this, and the penalization method of this section is suboptimal.

*Proof.* The following lemma is the main tool used in the proof.

**Lemma 6.1.** *Let $\mathscr{F}$ be a class measurable functions from $S$ into $[0, 1]$. If $\bar{\delta}_n$ is an admissible distribution dependent bound of confidence level $1 - p$, $p \in (0, 1)$ (see Definition 4.1), then the following inequality holds for all $t > 0$:*

$$\mathbb{P}\left\{\left|\inf_{\mathscr{F}} P_n f - \inf_{\mathscr{F}} Pf\right| \geq 2\bar{\delta}_n + \sqrt{\frac{2t}{n}\inf_{\mathscr{F}} Pf} + \frac{t}{n}\right\} \leq p + e^{-t}.$$

*If $(\bar{\delta}_n, \hat{\delta}_n, \tilde{\delta}_n)$ is a triple bound of confidence level $1 - p$, then*

$$\mathbb{P}\left\{\left|\inf_{\mathscr{F}} P_n f - \inf_{\mathscr{F}} Pf\right| \geq 4\hat{\delta}_n + 2\sqrt{\frac{2t}{n}\inf_{\mathscr{F}} P_n f} + \frac{8t}{n}\right\} \leq p + e^{-t}.$$

*Proof.* Let $E$ be the event where conditions (i)–(iv) of Definition 4.1 hold. Then $\mathbb{P}(E) \geq 1 - p$. On the event $E$, $\mathscr{E}(\hat{f}_n) \leq \bar{\delta}_n$ and, for all $\varepsilon < \bar{\delta}_n$ and $g \in \mathscr{F}(\varepsilon)$

$$\left|\inf_{\mathscr{F}} P_n f - \inf_{\mathscr{F}} Pf\right| = \left|P_n \hat{f}_n - \inf_{\mathscr{F}} Pf\right|$$

$$\leq P\hat{f}_n - \inf_{\mathscr{F}} Pf + |(P_n - P)(\hat{f}_n - g)| + |(P_n - P)(g)| \leq$$

$$\leq \bar{\delta}_n + \|P_n - P\|_{\mathscr{F}'(\bar{\delta}_n)} + |(P_n - P)(g)|. \tag{6.2}$$

Also, on the same event $E$,

$$\|P_n - P\|_{\mathscr{F}'(\bar{\delta}_n)} \leq U_n(\bar{\delta}_n(t)) \leq \bar{V}_n(\bar{\delta}_n)\bar{\delta}_n \leq \bar{\delta}_n. \tag{6.3}$$

By Bernstein's inequality, with probability at least $1 - e^{-t}$

$$|(P_n - P)(g)| \leq \sqrt{2\frac{t}{n}\mathrm{Var}_P g} + \frac{2t}{3n} \leq \sqrt{2\frac{t}{n}\left(\inf_{\mathscr{F}} Pf + \varepsilon\right)} + \frac{2t}{3n}, \tag{6.4}$$

since $g$ takes values in $[0, 1]$, $g \in \mathscr{F}(\varepsilon)$, and $\mathrm{Var}_P g \leq Pg^2 \leq Pg \leq \inf_{\mathscr{F}} Pf + \varepsilon$. It follows from (6.2), (6.3) and (6.4) that, on the event

$$E(\varepsilon) := E \bigcap \left\{|(P_n - P)(g)| \leq \sqrt{2\frac{t}{n}\left(\inf_{\mathscr{F}} Pf + \varepsilon\right)} + \frac{2t}{3n}\right\}, \tag{6.5}$$

the following inequality holds:

$$\left|\inf_{\mathscr{F}} P_n f - \inf_{\mathscr{F}} Pf\right| \leq 2\bar{\delta}_n + \sqrt{2\frac{t}{n}\left(\inf_{\mathscr{F}} Pf + \varepsilon\right)} + \frac{t}{n}. \tag{6.6}$$

Since the events $E(\varepsilon)$ are monotone in $\varepsilon$, let $\varepsilon \to 0$ to get

$$\mathbb{P}(E(0)) \geq 1 - p - e^{-t}.$$

This yields the first bound of the lemma.

For the proof of the second bound, note that on the event $E(0)$,

$$\left|\inf_{\mathscr{F}} P_n f - \inf_{\mathscr{F}} Pf\right| \leq \sqrt{2\frac{t}{n}|\inf_{\mathscr{F}} P_n f - \inf_{\mathscr{F}} Pf|} + 2\bar{\delta}_n + \sqrt{2\frac{t}{n}\inf_{\mathscr{F}} P_n f} + \frac{t}{n}. \quad (6.7)$$

Thus, either

$$|\inf_{\mathscr{F}} P_n f - \inf_{\mathscr{F}} Pf| \leq \frac{8t}{n}, \quad \text{or} \quad \frac{2t}{n} \leq \frac{|\inf_{\mathscr{F}} P_n f - \inf_{\mathscr{F}} Pf|}{4}.$$

In the last case (6.7) implies that

$$\left|\inf_{\mathscr{F}} P_n f - \inf_{\mathscr{F}} Pf\right| \leq 4\bar{\delta}_n + 2\sqrt{2\frac{t}{n}\inf_{\mathscr{F}} P_n f} + \frac{2t}{n}.$$

The condition of the lemma allows us to replace (on the event $E$) $\bar{\delta}_n$ by $\hat{\delta}_n$ and to get the following bound that holds with probability at least $1 - p - e^{-t}$ :

$$\left|\inf_{\mathscr{F}} P_n f - \inf_{\mathscr{F}} Pf\right| \leq 4\hat{\delta}_n + 2\sqrt{2\frac{t}{n}\inf_{\mathscr{F}} P_n f} + \frac{8t}{n}. \qquad \square$$

Now, we return to the proof of the theorem. For each class $\mathscr{F}_k$ and $t = t_k$, define the event $E_k(0)$ as in (6.5) with $\varepsilon = 0$. Clearly,

$$\mathbb{P}(E_k(0)) \geq 1 - p_k - e^{-t_k}.$$

Let

$$F := \bigcap_{k \geq 1} E_k(0).$$

Then

$$\mathbb{P}(F^c) \leq \sum_{k=1}^{\infty}\left(p_k + e^{-t_k}\right).$$

We use the following consequence of Lemma 6.1 and the definition of the triple bounds: on the event $F$ for all $k \geq 1$,

$$P\hat{f}_k - \inf_{f \in \mathscr{F}_k} Pf \leq \bar{\delta}_n(k) \leq \hat{\delta}_n(k) \leq \tilde{\delta}_n(k)$$

and

$$\left| \inf_{\mathscr{F}_k} P_n f - \inf_{\mathscr{F}_k} P f \right| \le 2 \bar{\delta}_n(k) + \sqrt{\frac{2t_k}{n} \inf_{\mathscr{F}_k} P f} + \frac{t_k}{n},$$

$$\left| \inf_{\mathscr{F}_k} P_n f - \inf_{\mathscr{F}_k} P f \right| \le 4 \hat{\delta}_n(k) + 2 \sqrt{\frac{2t_k}{n} \inf_{\mathscr{F}_k} P_n f} + \frac{8t_k}{n}.$$

Therefore,

$$P \hat{f} = P \hat{f}_{\hat{k}} \le \inf_{\mathscr{F}_{\hat{k}}} P f + \bar{\delta}_n(\hat{k}) \le \inf_{\mathscr{F}_{\hat{k}}} P_n f + 5 \hat{\delta}_n(\hat{k}) + 2 \sqrt{\frac{2t_{\hat{k}}}{n} \inf_{\mathscr{F}_{\hat{k}}} P_n f} + \frac{8t_{\hat{k}}}{n} \le$$

$$\le \inf_{\mathscr{F}_{\hat{k}}} P_n f + \hat{\pi}(\hat{k}) = \inf_k \left[ \inf_{\mathscr{F}_k} P_n f + \hat{\pi}(k) \right],$$

provided that the constant $\hat{K}$ in the definition of $\hat{\pi}$ was chosen properly. The first bound of the theorem has been proved.

To prove the second bound, note that

$$\sqrt{\frac{t_k}{n} \inf_{\mathscr{F}_k} P_n f} \le \sqrt{\frac{t_k}{n} \inf_{\mathscr{F}_k} P f} + \sqrt{\frac{t_k}{n} | \inf_{\mathscr{F}_k} P_n f - \inf_{\mathscr{F}_k} P f |}$$

$$\le \sqrt{\frac{t_k}{n} \inf_{\mathscr{F}_k} P f} + \frac{t_k}{2n} + \frac{1}{2} | \inf_{\mathscr{F}_k} P_n f - \inf_{\mathscr{F}_k} P f |.$$

Therefore, on the event $F$, for all $k$,

$$\hat{\pi}(k) = \hat{K} \left[ \hat{\delta}_n(k) + \sqrt{\frac{t_k}{n} \inf_{\mathscr{F}_k} P_n f} + \frac{t_k}{n} \right]$$

$$\le \frac{\tilde{K}}{2} \left[ \bar{\delta}_n(k) + \sqrt{\frac{t_k}{n} \inf_{\mathscr{F}_k} P f} + \frac{t_k}{n} \right] = \tilde{\pi}(k)/2$$

and

$$\left| \inf_{\mathscr{F}_k} P_n f - \inf_{\mathscr{F}_k} P f \right| \le 2 \bar{\delta}_n(k) + \sqrt{\frac{2t_k}{n} \inf_{\mathscr{F}_k} P f} + \frac{t_k}{n}$$

$$\le \frac{\tilde{K}}{2} \left[ \bar{\delta}_n(k) + \sqrt{\frac{t_k}{n} \inf_{\mathscr{F}_k} P f} + \frac{t_k}{n} \right] = \tilde{\pi}(k)/2,$$

provided that the constant $\tilde{K}$ in the definition of $\tilde{\pi}(k)$ is large enough. As a result, on the event $F$,

$$P\hat{f} \le \inf_k\left[\inf_{\mathscr{F}_k} P_n f + \hat{\pi}(k)\right] \le \inf_k\left[\inf_{\mathscr{F}_k} Pf + \tilde{\pi}(k)\right],$$

proving the second bound.                                                             □

*Example 6.2.*  As an example, we derive some of the results of Lugosi and Wegkamp [104] (in a slightly modified form). Suppose that $\mathscr{F}$ is a class of measurable functions on $S$ taking values in $\{0, 1\}$ (binary functions). As before, let $\Delta^{\mathscr{F}}(X_1, \ldots, X_n)$ be the shattering number of the class $\mathscr{F}$ on the sample $(X_1, \ldots, X_n)$ :

$$\Delta^{\mathscr{F}}(X_1, \ldots, X_n) := \operatorname{card}\left(\left\{(f(X_1), \ldots, f(X_n)) : f \in \mathscr{F}\right\}\right).$$

Given a sequence $\{\mathscr{F}_k\}$, $\mathscr{F}_k \subset \mathscr{F}$ of classes of binary functions, define the following complexity penalties

$$\hat{\pi}(k) := \hat{K}\left[\sqrt{\inf_{f \in \mathscr{F}_k} P_n f \frac{\log \Delta^{\mathscr{F}_k}(X_1, \ldots, X_n) + t_k}{n}} + \frac{\log \Delta^{\mathscr{F}_k}(X_1, \ldots, X_n) + t_k}{n}\right]$$

and

$$\tilde{\pi}(k) := \tilde{K}\left[\sqrt{\inf_{f \in \mathscr{F}_k} Pf \frac{\mathbb{E}\log \Delta^{\mathscr{F}_k}(X_1, \ldots, X_n) + t_k}{n}} + \frac{\mathbb{E}\log \Delta^{\mathscr{F}_k}(X_1, \ldots, X_n) + t_k}{n}\right].$$

Let $\hat{k}$ be a solution of the penalized empirical risk minimization problem

$$\hat{k} := \operatorname{argmin}_{k \ge 1}\left[\min_{\mathscr{F}_k} P_n f + \hat{\pi}(k)\right].$$

Denote $\hat{f} := \hat{f}_{n,\hat{k}}$.

**Theorem 6.3.** *There exists a choice of $\hat{K}, \tilde{K}$ such that for all $t_k > 0$,*

$$\mathbb{P}\left\{\mathscr{E}_P(\mathscr{F}; \hat{f}) \ge \inf_{k \ge 1}\left\{\inf_{f \in \mathscr{F}_k} Pf - \inf_{f \in \mathscr{F}} Pf + \tilde{\pi}(k)\right\}\right\} \le \sum_{k=1}^{\infty} e^{-t_k}.$$

Note that penalization based on random shattering numbers is natural in classification problems and the result of Theorem 6.3 can be easily stated in classification setting. The result follows from Theorem 6.2 and the next lemma that provides a version of triple bound on excess risk for classes of binary functions.

**Lemma 6.2.** *Given a class of binary functions $\mathscr{F}$ and $t > 0$, define*

$$\bar{\delta}_n := \bar{K}\left[\sqrt{\inf_{f \in \mathscr{F}} Pf \frac{\mathbb{E} \log \Delta^{\mathscr{F}}(X_1, \ldots, X_n) + t}{n}} + \frac{\mathbb{E} \log \Delta^{\mathscr{F}}(X_1, \ldots, X_n) + t}{n}\right],$$

$$\hat{\delta}_n := \hat{K}\left[\sqrt{\inf_{f \in \mathscr{F}} P_n f \frac{\log \Delta^{\mathscr{F}}(X_1, \ldots, X_n) + t}{n}} + \frac{\log \Delta^{\mathscr{F}}(X_1, \ldots, X_n) + t}{n}\right]$$

*and*

$$\tilde{\delta}_n := \tilde{K}\left[\sqrt{\inf_{f \in \mathscr{F}} Pf \frac{\mathbb{E} \log \Delta^{\mathscr{F}}(X_1, \ldots, X_n) + t}{n}} + \frac{\mathbb{E} \log \Delta^{\mathscr{F}}(X_1, \ldots, X_n) + t}{n}\right].$$

*There exists a choice of constants $\bar{K}, \hat{K}, \tilde{K}$ such that $(\bar{\delta}_n, \hat{\delta}_n, \tilde{\delta}_n)$ is a triple bound of level $1 - e^{-t}$ for the class $\mathscr{F}$.*

*Proof.* The following upper bounds on the $L_2(P)$-diameter of the $\delta$-minimal set $\mathscr{F}(\delta)$ and on the function $\phi_n(\delta)$ hold:

$$D^2(\mathscr{F}; \delta) = \sup_{f,g \in \mathscr{F}(\delta)} P(f - g)^2 \leq \sup_{f,g \in \mathscr{F}(\delta)} (Pf + Pg) \leq 2(\inf_{f \in \mathscr{F}} Pf + \delta).$$

By Theorem 3.9,

$$\phi_n(\delta) \leq K\left[\sqrt{2\left(\inf_{f \in \mathscr{F}} Pf + \delta\right) \frac{\mathbb{E} \log \Delta^{\mathscr{F}}(X_1, \ldots, X_n)}{n}} + \frac{\mathbb{E} \log \Delta^{\mathscr{F}}(X_1, \ldots, X_n)}{n}\right].$$

A straightforward computation implies the next bound on the quantity $\sigma_n^t$ from Theorem 4.3:

$$\sigma_n^t \leq \bar{\delta}_n = \bar{K}\left[\sqrt{\inf_{f \in \mathscr{F}} Pf \frac{\mathbb{E} \log \Delta^{\mathscr{F}}(X_1, \ldots, X_n) + t}{n}} + \frac{\mathbb{E} \log \Delta^{\mathscr{F}}(X_1, \ldots, X_n) + t}{n}\right],$$

provided that the constant $\bar{K}$ is large enough. Moreover, with a proper choice of this constant, $\bar{\delta}_n$ is an admissible bound of level $1 - e^{-t}$.

The following deviation inequality for shattering numbers is due to Boucheron et al. [31]: with probability at least $1 - e^{-t}$

$$\log \Delta^{\mathscr{F}}(X_1, \ldots, X_n) \leq 2\mathbb{E} \log \Delta^{\mathscr{F}}(X_1, \ldots, X_n) + 2t$$

and

$$\mathbb{E} \log \Delta^{\mathscr{F}}(X_1, \ldots, X_n) \leq 2 \log \Delta^{\mathscr{F}}(X_1, \ldots, X_n) + 2t.$$

Together with the first bound of Lemma 6.1, this easily implies that with probability at least $1 - 8e^{-t}$, $\bar{\delta}_n \leq \hat{\delta}_n \leq \tilde{\delta}_n$. First we prove that $\bar{\delta}_n \leq \hat{\delta}_n$. To this end, we use the first bound of Lemma 6.1 and the inequality $2ab \leq a^2 + b^2$ to show that with probability at least $1 - 2e^{-t}$

$$\inf_{\mathscr{F}} Pf \leq \inf_{\mathscr{F}} P_n f + 2\bar{\delta}_n + 2\sqrt{\frac{t}{2n} \inf_{\mathscr{F}} Pf} + \frac{t}{3n} \leq \inf_{\mathscr{F}} P_n f + 2\bar{\delta}_n + \frac{\inf_{\mathscr{F}} Pf}{2} + \frac{2t}{n}.$$

Therefore,

$$\inf_{\mathscr{F}} Pf \leq 2 \inf_{\mathscr{F}} P_n f + 4\bar{\delta}_n + \frac{4t}{n}.$$

We substitute this inequality into the definition of $\bar{\delta}_n$ and replace $\mathbb{E} \log \Delta^{\mathscr{F}}$ $(X_1, \ldots, X_n)$ by the upper bound $2 \log \Delta^{\mathscr{F}}(X_1, \ldots, X_n) + 2t$ that holds with probability at least $1 - e^{-t}$. It follows that, with some constant $K$,

$$\bar{\delta}_n \leq K \left[ \sqrt{\inf_{f \in \mathscr{F}} P_n f \frac{\log \Delta^{\mathscr{F}}(X_1, \ldots, X_n) + t}{n}} + \frac{\log \Delta^{\mathscr{F}}(X_1, \ldots, X_n) + t}{n} \right] +$$

$$+ 2\sqrt{\frac{\bar{\delta}_n}{2} \frac{K^2 \log \Delta^{\mathscr{F}}(X_1, \ldots, X_n) + t}{2n}},$$

Again, using the inequality $2ab \leq a^2 + b^2$, we get the following bound that holds with some constant $\hat{K}$ and with probability at least $1 - 4e^{-t}$ :

$$\bar{\delta}_n \leq \hat{K} \left[ \sqrt{\inf_{f \in \mathscr{F}} P_n f \frac{\log \Delta^{\mathscr{F}}(X_1, \ldots, X_n) + t}{n}} + \frac{\log \Delta^{\mathscr{F}}(X_1, \ldots, X_n) + t}{n} \right] =: \hat{\delta}_n.$$

The proof of the second inequality $\hat{\delta}_n \leq \tilde{\delta}_n$ is similar. By increasing the values of the constants $\bar{K}, \hat{K}, \tilde{K}$, it is easy to eliminate the numerical factor in front of $e^{-t}$ and to obtain a triple bound of level $1 - e^{-t}$, as it was claimed.    □

## 6.3  Linking Excess Risk and Variance in Penalization

In a variety of regression and classification problems, the following assumption plays the crucial role: for all $f \in \mathscr{F}$,

$$Pf - Pf_* \geq \varphi\left(\sqrt{\mathrm{Var}_P(f - f_*)}\right), \tag{6.8}$$

where $\varphi$ is a convex nondecreasing function on $[0, +\infty)$ with $\varphi(0) = 0$. In Chap. 5, we have already dealt with several examples of this condition. For instance, in the case of regression with quadratic loss $\ell(y, u) = (y-u)^2$ and with bounded response $Y \in [0, 1]$, condition (6.8) is satisfied for the loss class $\mathscr{F} = \{\ell \bullet g : g \in \mathscr{G}\}$, where $\mathscr{G}$ is a class of functions from $S$ into $[0, 1]$. In this case, one can take $\varphi(u) = u^2/2$, so the function $\varphi$ does not depend on the unknown distribution $P$ (except that the assumption $Y \in [0, 1]$ is already a restriction on the class of distributions $P$). On the other hand, in classification problems, $\varphi$ is related to the parameters of the noise such as parameter $\alpha$ in Tsybakov's low noise assumption (5.9) or parameter $h$ in Massart's low noise assumption (5.8). So, in this case, $\varphi$ does depend on $P$. The function $\varphi$ describes the relationship between the excess risk $Pf - P_*$ and the variance $\mathrm{Var}_P(f - f_*)$ of the "excess loss" $f - f_*$. In what follows, we will call $\varphi$ the *link function*. It happens that the link function is involved in a rather natural way in the construction of complexity penalties that provide optimal convergence rates in many problems. Since the link function is generally distribution dependent, the development of adaptive penalization methods of model selection is a challenge, for instance, in classification setting.

We will assume that with some $\gamma > 0$

$$\varphi(uv) \leq \gamma\varphi(u)\varphi(v), \quad u, v \geq 0. \tag{6.9}$$

Denote by

$$\varphi^*(v) := \sup_{u \geq 0}[uv - \varphi(u)]$$

the conjugate of $\varphi$. Then

$$uv \leq \varphi(u) + \varphi^*(v), \quad u, v \geq 0.$$

Let $(\bar{\delta}_n(k), \hat{\delta}_n(k), \tilde{\delta}_n(k))$ be a triple bound of level $1 - p_k$ for the class $\mathscr{F}_k, k \geq 1$. For a fixed $\varepsilon > 0$, define the penalties as follows:

$$\hat{\pi}(k) := A(\varepsilon)\hat{\delta}_n(k) + \varphi^*\left(\sqrt{\frac{2t_k}{\varepsilon n}}\right) + \frac{t_k}{n}$$

and

$$\tilde{\pi}(k) := \frac{A(\varepsilon)}{1 + \gamma\varphi(\sqrt{\varepsilon})}\tilde{\delta}_n(k) + \frac{2}{1 + \gamma\varphi(\sqrt{\varepsilon})}\varphi^*\left(\sqrt{\frac{2t_k}{\varepsilon n}}\right) + \frac{2}{1 + \gamma\varphi(\sqrt{\varepsilon})}\frac{t_k}{n},$$

where

$$A(\varepsilon) := \frac{5}{2} - \gamma\varphi(\sqrt{\varepsilon}).$$

As before, $\hat{k}$ is defined by

$$\hat{k} := \operatorname{argmin}_{k \geq 1} \left\{ P_n \hat{f}_k + \hat{\pi}(k) \right\}$$

and $\hat{f} := \hat{f}_{n,\hat{k}}$.

**Theorem 6.4.** *For any sequence $\{t_k\}$ of positive numbers,*

$$\mathbb{P}\left\{ \mathscr{E}_P(\mathscr{F}; \hat{f}) \geq C(\varepsilon) \inf_{k \geq 1} \left\{ \inf_{f \in \mathscr{F}_k} Pf - \inf_{f \in \mathscr{F}} Pf + \tilde{\pi}(k) \right\} \right\} \leq \sum_{k=1}^{\infty} \left( p_k + e^{-t_k} \right),$$

*where*

$$C(\varepsilon) := \frac{1 + \gamma\varphi(\sqrt{\varepsilon})}{1 - \gamma\varphi(\sqrt{\varepsilon})}.$$

*Proof.* The following lemma is needed in the proof.

**Lemma 6.3.** *Let $\mathscr{G} \subset \mathscr{F}$ and let $(\bar{\delta}_n, \hat{\delta}_n, \tilde{\delta}_n)$ be a triple bound of level $1 - p$ for the class $\mathscr{G}$. For all $t > 0$, there exists an event $E$ with probability at least $1 - p - e^{-t}$ such that on this event*

$$\inf_{\mathscr{G}} P_n f - P_n f_* \leq (1 + \gamma\varphi(\sqrt{\varepsilon}))(\inf_{\mathscr{G}} Pf - Pf_*) + \varphi^*\left(\sqrt{\frac{2t}{\varepsilon n}}\right) + \frac{t}{n} \quad (6.10)$$

*and*

$$\inf_{\mathscr{G}} Pf - Pf_* \leq (1 - \gamma\varphi(\sqrt{\varepsilon}))^{-1}\left[ \inf_{\mathscr{G}} P_n f - P_n f_* + \frac{3}{2}\bar{\delta}_n + \varphi^*\left(\sqrt{\frac{2t}{\varepsilon n}}\right) + \frac{t}{n} \right]. \quad (6.11)$$

*In addition, if there exists $\bar{\delta}_n^\varepsilon$ such that*

$$\bar{\delta}_n \leq \varepsilon(\inf_{\mathscr{G}} Pf - Pf_*) + \bar{\delta}_n^\varepsilon,$$

*then*

$$\inf_{\mathscr{G}} Pf - Pf_* \leq \left( 1 - \varphi(\sqrt{\varepsilon}) - \frac{3}{2}\varepsilon \right)^{-1}\left[ \inf_{\mathscr{G}} P_n f - P_n f_* \right.$$

$$\left. + \frac{3}{2}\bar{\delta}_n^\varepsilon + \varphi^*\left(\sqrt{\frac{2t}{\varepsilon n}}\right) + \frac{t}{n} \right]. \quad (6.12)$$

*Proof.* We assume, for simplicity, that $Pf$ attains its minimum over $\mathscr{G}$ at some $\bar{f} \in \mathscr{G}$ (the proof can be easily modified if the minimum is not attained). Let $E'$ be the event from the Definition 4.1 of the triple bound and let

$$E := \left\{ |(P_n - P)(\bar{f} - f_*)| \leq \sqrt{\frac{2t}{n} \mathrm{Var}_P(\bar{f} - f_*)} + \frac{t}{n} \right\} \bigcap E'.$$

It follows from Bernstein inequality and the definition of the triple bound that

$$\mathbb{P}(E) \geq 1 - p - e^{-t}.$$

On the event $E$,

$$|(P_n - P)(\bar{f} - f_*)| \leq \sqrt{\frac{2t}{n} \mathrm{Var}_P(\bar{f} - f_*)} + \frac{t}{n}$$

and

$$\forall f \in \mathcal{G} \quad \hat{\mathcal{E}}_n(\mathcal{G}; f) = \mathcal{E}_{P_n}(\mathcal{G}; f) \leq \frac{3}{2}\left( \mathcal{E}_P(\mathcal{G}; f) \vee \bar{\delta}_n \right).$$

Also,

$$\mathrm{Var}_P^{1/2}(\bar{f} - f_*) \leq \varphi^{-1}(P\bar{f} - Pf_*).$$

Hence, on the event $E$,

$$|(P - P_n)(\bar{f} - f_*)| \leq \varphi(\sqrt{\varepsilon}\varphi^{-1}(P\bar{f} - Pf_*)) + \varphi^*\left(\sqrt{\frac{2t}{\varepsilon n}}\right) + \frac{t}{n} \leq$$

$$\leq \gamma\varphi(\sqrt{\varepsilon})(P\bar{f} - Pf_*) + \varphi^*\left(\sqrt{\frac{2t}{\varepsilon n}}\right) + \frac{t}{n},$$

implying

$$P_n(\bar{f} - f_*) \leq (1 + \gamma\varphi(\sqrt{\varepsilon}))P(\bar{f} - f_*) + \varphi^*\left(\sqrt{\frac{2t}{\varepsilon n}}\right) + \frac{t}{n} \qquad (6.13)$$

and

$$P(\bar{f} - f_*) \leq (1 - \gamma\varphi(\sqrt{\varepsilon}))^{-1}\left[ P_n(\bar{f} - f_*) + \varphi^*\left(\sqrt{\frac{2t}{\varepsilon n}}\right) + \frac{t}{n} \right]. \qquad (6.14)$$

(6.13) immediately yields the first bound of the lemma.

Since, in addition, on the event $E$

$$P_n(\bar{f} - f_*) = P_n\bar{f} - \inf_{\mathcal{G}} P_n f + \inf_{\mathcal{G}} P_n f - P_n f_* = \hat{\mathcal{E}}_n(\mathcal{G}; \bar{f}) + \inf_{\mathcal{G}} P_n f - P_n f_*$$

$$\leq \inf_{\mathcal{G}} P_n f - P_n f_* + \frac{3}{2}\left(\mathcal{E}_P(\mathcal{G}; \bar{f}) \vee \bar{\delta}_n\right),$$

and since $\mathcal{E}_P(\mathcal{G}; \bar{f}) = 0$, we get

$$P_n(\bar{f} - f_*) \leq \inf_{\mathcal{G}} P_n f - P_n f_* + \frac{3}{2}\bar{\delta}_n.$$

Along with (6.14), this implies

$$\inf_{\mathcal{G}} Pf - Pf_* = P(\bar{f} - f_*)$$

$$\leq (1 - \gamma\varphi(\sqrt{\varepsilon}))^{-1}\left[\inf_{\mathcal{G}} P_n f - P_n f_* + \frac{3}{2}\bar{\delta}_n + \varphi^*\left(\sqrt{\frac{2t}{\varepsilon n}}\right) + \frac{t}{n}\right],$$

which is the second bound of the lemma.

Finally, to prove the third bound it is enough to substitute the bound on $\bar{\delta}_n$ into (6.11) and to solve the resulting inequality with respect to $\inf_{\mathcal{G}} Pf - Pf_*$.    □

Let $E_k$ be the event defined in Lemma 6.3 for $\mathcal{G} = \mathcal{F}_k$ and $t = t_k$, so that

$$\mathbb{P}(E_k) \geq 1 - p_k - e^{-t_k}.$$

Let $E := \bigcap_{k \geq 1} E_k$. Then

$$\mathbb{P}(E) \geq 1 - \sum_{k \geq 1}\left(p_k + e^{-t_k}\right).$$

On the event $E$, for all $k \geq 1$

$$\mathcal{E}_P(\mathcal{F}_k; \hat{f}_k) = P\hat{f}_k - \inf_{\mathcal{F}_k} Pf \leq \bar{\delta}_n(k)$$

and

$$\bar{\delta}_n(k) \leq \hat{\delta}_n(k) \leq \tilde{\delta}_n(k).$$

On the same event, first using bound (6.11) and then bound (6.10) of Lemma 6.3, we get

$$\mathcal{E}_P(\mathcal{F}; \hat{f}) = P\hat{f} - \inf_{\mathcal{F}} Pf = P\hat{f}_{\hat{k}} - Pf_*$$

$$= P\hat{f}_{\hat{k}} - \inf_{\mathcal{F}_{\hat{k}}} Pf + \inf_{\mathcal{F}_{\hat{k}}} Pf - Pf_* \leq \bar{\delta}_n(\hat{k}) + \inf_{\mathcal{F}_{\hat{k}}} Pf - Pf_*$$

$$\leq (1 - \gamma\varphi(\sqrt{\varepsilon}))^{-1}\Big[(1 - \gamma\varphi(\sqrt{\varepsilon}))\bar{\delta}_n(\hat{k}) + \inf_{\mathscr{F}_{\hat{k}}} P_n f - P_n f_*$$

$$+ \frac{3}{2}\bar{\delta}_n(\hat{k}) + \varphi^*\Big(\sqrt{\frac{2t_{\hat{k}}}{\varepsilon n}}\Big) + \frac{t_{\hat{k}}}{n}\Big]$$

$$\leq (1 - \gamma\varphi(\sqrt{\varepsilon}))^{-1}\Big\{\inf_k\Big[\inf_{\mathscr{F}_k} P_n f + (5/2 - \gamma\varphi(\sqrt{\varepsilon}))\bar{\delta}_n(k)$$

$$+ \varphi^*\Big(\sqrt{\frac{2t_k}{\varepsilon n}}\Big) + \frac{t_k}{n}\Big] - P_n f_*\Big\}$$

$$= (1 - \gamma\varphi(\sqrt{\varepsilon}))^{-1}\Big\{\inf_k\Big[\inf_{\mathscr{F}_k} P_n f + \hat{\pi}(k)\Big] - P_n f_*\Big\}$$

$$\leq \frac{1 + \gamma\varphi(\sqrt{\varepsilon})}{1 - \gamma\varphi(\sqrt{\varepsilon})} \inf_k\Big[\inf_{\mathscr{F}_k} P f - \inf_{\mathscr{F}} P f + \frac{5/2 - \gamma\varphi(\sqrt{\varepsilon})}{1 + \gamma\varphi(\sqrt{\varepsilon})}\bar{\delta}_n(k)$$

$$+ \frac{2}{1 + \gamma\varphi(\sqrt{\varepsilon})}\varphi^*\Big(\sqrt{\frac{2t_k}{\varepsilon n}}\Big) + \frac{2}{(1 + \gamma\varphi(\sqrt{\varepsilon}))}\frac{t_k}{n}\Big]$$

$$= \frac{1 + \gamma\varphi(\sqrt{\varepsilon})}{1 - \gamma\varphi(\sqrt{\varepsilon})} \inf_k\Big[\inf_{\mathscr{F}_k} P f - \inf_{\mathscr{F}} P f + \tilde{\pi}(k)\Big],$$

and the result follows. □

*Remark 6.1.* Suppose that, for each $k$, $\bar{\delta}_n(k)$ is an admissible excess risk bound for the class $\mathscr{F}_k$ on an event $E_k$ with $\mathbb{P}(E_k) \geq 1 - p_k$ (see Definition 4.1). It is easily seen from the proof of Theorem 6.4 that the same oracle inequality holds for arbitrary penalties $\hat{\pi}(k)$ and $\tilde{\pi}(k)$ such that on the event $E_k$

$$\hat{\pi}(k) \geq A(\varepsilon)\bar{\delta}_n(k) + \varphi^*\Big(\sqrt{\frac{2t_k}{\varepsilon n}}\Big) + \frac{t_k}{n}$$

and

$$\tilde{\pi}(k) \geq \frac{\hat{\pi}(k)}{1 + \gamma\varphi(\sqrt{\varepsilon})} + \frac{\varphi^*\Big(\sqrt{\frac{2t_k}{\varepsilon n}}\Big)}{1 + \gamma\varphi(\sqrt{\varepsilon})} + \frac{t_k}{(1 + \gamma\varphi(\sqrt{\varepsilon}))n}.$$

As it has been already mentioned, the dependence of the penalty on the link function $\varphi$ is the most troubling aspect of this approach since in such problems as classification this function depends on the unknown parameters of the distribution $P$ (such as "low noise" constants $\alpha$ in (5.9) and $h$ in (5.8), see Sect. 5.3). Because of this, it is of importance to know that, using Remark 6.1, it is easy to construct a version of the penalties that *do not depend on* $\varphi$ directly. Suppose that the number of classes $\mathscr{F}_k$ is finite, say, $N$. Take

$$t_k := t + \log N, \ k = 1, \ldots, N.$$

Define

$$\hat{k} := \text{argmin}_{1 \le k \le N} \left[ \min_{f \in \mathscr{F}_k} P_n f + \frac{5}{2} \hat{\delta}_n(k) \right]$$

and $\hat{f} := \hat{f}_{\hat{k}}$. Note that we also have

$$\hat{k} := \text{argmin}_{1 \le k \le N} \left[ \min_{f \in \mathscr{F}_k} P_n f + \hat{\pi}(k) \right],$$

where

$$\hat{\pi}(k) := \frac{5}{2} \hat{\delta}_n(k) + \varphi^* \left( \sqrt{\frac{2t_k}{\varepsilon n}} \right) + \frac{t_k}{n}$$

$$= \frac{5}{2} \hat{\delta}_n(k) + \varphi^* \left( \sqrt{\frac{2(t + \log N)}{\varepsilon n}} \right) + \frac{t + \log N}{n},$$

since $t_k$ in the additional two terms of the definition of $\hat{\pi}(k)$ does not depend on $k$. Denote

$$\tilde{\pi}(k) := \frac{5}{2} \hat{\delta}_n(k) + 2\varphi^* \left( \sqrt{\frac{2(t + \log N)}{\varepsilon n}} \right) + 2\frac{t + \log N}{n}.$$

Then it follows from Theorem 6.4 and from Remark 6.1 that

$$\mathbb{P} \left\{ \mathscr{E}_P(\mathscr{F}; \hat{f}) \ge C(\varepsilon) \inf_{1 \le k \le N} \left\{ \inf_{f \in \mathscr{F}_k} Pf - \inf_{f \in \mathscr{F}} Pf + \tilde{\pi}(k) \right\} \right\} \le e^{-t} + \sum_{k=1}^{N} p_k.$$
$$(6.15)$$

*Example 6.3.* Consider, for instance, model selection in binary classification problems (see Sect. 5.3). Suppose that condition (5.8) holds with some $h > 0$ and, as a result, condition (6.8) holds with $\varphi(u) = hu^2, u \ge 0$, for $f = \ell \bullet g$ and $f_* = \ell \bullet g_*$, where $g$ is a binary classifier, $g_*$ is the Bayes classifier and $\ell(y, u) = I(y \ne u)$ is the binary loss. In this case, $\varphi^*(v) = v^2/(4h), v \ge 0$.

Let $\{\mathscr{G}_k\}$ be a family of classes of functions from $S$ into $\{-1, 1\}$ (binary classifiers). For any $k$, define

$$\hat{g}_{n,k} := \text{argmin}_{g \in \mathscr{G}_k} L_n(g) = \text{argmin}_{g \in \mathscr{G}_k} n^{-1} \sum_{j=1}^{n} I(Y_j \ne g(X_j)).$$

Let $\mathscr{F}_k := \{\ell \bullet g : g \in \mathscr{G}_k\}$. Denote $(\bar{\delta}_n(k), \hat{\delta}_n(k), \tilde{\delta}_n(k))$ the standard triple bound of Theorem 4.8 for the class $\mathscr{F}_k$ of level $1 - p_k$. Suppose that $\sum_{k=1}^{N} p_k = p \in (0, 1)$. Define

$$\hat{k} := \operatorname{argmin}_{1 \leq k \leq N} \left[ \inf_{g \in \mathcal{G}_k} L_n(g) + \frac{5}{2} \hat{\delta}_n(k) \right]$$

and $\hat{g} := \hat{g}_{n,\hat{k}}$. Then it easily follows from bound (6.15) that with probability at least $1 - p - e^{-t}$

$$L(\hat{g}) - L(g_*) \leq C \inf_{1 \leq k \leq N} \left[ \inf_{g \in \mathcal{G}_k} L(g) - L(g_*) + \tilde{\delta}_n(k) + \frac{t + \log N}{nh} \right]$$

(we have fixed $\varepsilon > 0$ and the constant $C$ depends on $\varepsilon$). It is also easy to deduce from the proof of Theorem 5.5 that, for the standard choice of $\tilde{\delta}_n(k)$,

$$\tilde{\delta}_n(k) \leq C \left[ \inf_{g \in \mathcal{G}_k} L(g) - L(g_*) + \frac{\mathbb{E} \log \Delta^{\mathcal{G}_k}(X_1, \ldots, X_n)}{nh} + \frac{t_k}{nh} \right].$$

This leads to the following oracle inequality that holds with probability at least $1 - p - e^{-t}$ and with some constant $C > 0$ :

$$L(\hat{g}) - L(g_*) \leq C \inf_{1 \leq k \leq N} \left[ \inf_{g \in \mathcal{G}_k} L(g) - L(g_*) + \frac{\mathbb{E} \log \Delta^{\mathcal{G}_k}(X_1, \ldots, X_n)}{nh} \right]$$
$$+ C \frac{t + \log N}{nh}.$$

Thus, this penalization method is adaptive to the unknown noise parameter $h$.

We conclude this section with stating a result of Massart [106, 107] that can be derived using the approach of Theorem 6.4. Suppose that $\{\mathcal{F}_k\}$ is a sequence of function classes such that condition (4.5) holds for each class $\mathcal{F}_k$ with some constant $D_k \geq 1$, that is,

$$D_k(Pf - Pf_*) \geq \rho_P^2(f, f_*) \geq \operatorname{Var}_P(f - f_*).$$

We will assume that the sequence $\{D_k\}$ is nondecreasing. Denote

$$\bar{\delta}_n^\varepsilon(k) := D_k^{-1} \omega_n^\sharp \left( \frac{\varepsilon}{KD_k} \right) + \frac{KD_k t_k}{n\varepsilon}.$$

If $K$ is large enough, then Lemma 4.1 implies the following bound:

$$\tilde{\delta}_n(k) := \sigma_n^{t_k}(\mathcal{F}_k; P) \leq \varepsilon(\inf_{\mathcal{F}_k} Pf - Pf_*) + \bar{\delta}_n^\varepsilon(k).$$

Also, it follows from the proof of Theorem 4.3 that $\tilde{\delta}_n(k)$ is an admissible excess risk bound of level $1 - C_q e^{-t_k}$.

Suppose that for each $k$ there exist a data dependent bound $\hat{\delta}_n^\varepsilon(k)$ and a distribution dependent bound $\tilde{\delta}_n^\varepsilon(k)$ such that

$$\mathbb{P}\left\{\bar{\delta}_n^\varepsilon(k) \leq \hat{\delta}_n^\varepsilon(k) \leq \tilde{\delta}_n^\varepsilon(k)\right\} \geq 1 - p_k, \ k \geq 1.$$

Define the following penalties:

$$\hat{\pi}_n^\varepsilon(k) := 3\hat{\delta}_n^\varepsilon(k) + \frac{\hat{K}D_k t_k}{\varepsilon n} \text{ and } \tilde{\pi}_n^\varepsilon(k) := 3\tilde{\delta}_n^\varepsilon(k) + \frac{\tilde{K}D_k t_k}{\varepsilon n}$$

with some numerical constants $\hat{K}, \tilde{K}$. Let

$$\hat{k} := \operatorname{argmin}_{k \geq 1}\left[\min_{f \in \mathscr{F}_k} P_n f + \hat{\pi}_n^\varepsilon(k)\right]$$

and $\hat{f} := f_{\hat{k}}$.

**Theorem 6.5.** *There exist numerical constants $\hat{K}, \tilde{K}, C$ such that for any sequence $\{t_k\}$ of positive numbers,*

$$\mathbb{P}\left\{P\hat{f} - Pf_* \geq \frac{1+\varepsilon}{1-\varepsilon}\inf_{k \geq 1}\left\{\inf_{f \in \mathscr{F}_k} Pf - Pf_* + \tilde{\pi}_n^\varepsilon(k)\right\}\right\} \leq \sum_{k=1}^{\infty}\left(p_k + (C+1)e^{-t_k}\right).$$

To prove this result one has to extend Theorem 6.4 to the case when condition (6.8) holds for each function class $\mathscr{F}_k$ with a different link function $\varphi_k$ and to use this extension for $\varphi_k(u) = u^2/D_k$ and $\varphi_k^*(v) = D_k v^2/4$.

## 6.4  Further Comments

Bartlett [13] suggested a simple and elegant derivation of oracle inequalities in the case of monotone families. Theorem 6.1 is based on this approach. Penalization with empirical risk minima was used by Lugosi and Wegkamp [104]. Section 6.3 is based on the results of Koltchinskii [83]; Theorem 6.5 in this section is essentially due to Massart [106]. Other useful references on oracle inequalities in penalized empirical risk minimization are [7, 25, 26, 29, 107].

Birgé and Massart [24] introduced a concept of minimal penalties and advocated an approach to the problem of calibration of data-dependent penalties based on so called "slope heuristics". So far, this approach has been mathematically justified for several special models by Arlot and Massart [8] with a significant further progress made in the dissertation by Saumard [131]. Concentration inequalities for empirical excess risk obtained by Boucheron and Massart [32] are of importance in this line of work.

Oracle inequalities in penalized empirical risk minimization for kernel machines have been studied by Blanchard et al. [25], Steinwart and Scovel [136], Steinwart and Christmann [135]. Recently, Mendelson and Neeman [116] obtained very subtle oracle inequalities in such problems based on a variety of methods (including, generic chaining bounds).

The methods developed in this and in the previous chapters are also of importance in the study of statistical aggregation procedures, see [35, 119, 143, 152, 153].

# Chapter 7
# Linear Programming in Sparse Recovery

As it was pointed out in the Introduction, many important sparse recovery methods are based on empirical risk minimization with convex loss and convex complexity penalty. Some interesting algorithms, for instance, the Dantzig selector by Candes and Tao [44] can be formulated as linear programs. In this chapter, we develop error bounds for such algorithms that require certain geometric assumptions on the dictionary. They are expressed in terms of restricted isometry constants and other related characteristics that depend both on the dictionary and on the design distribution. Based on these geometric characteristics, we describe the conditions of exact sparse recovery in the noiseless case as well as sparsity oracle inequalities for the Dantzig selector in regression problems with random noise. These results rely on comparison inequalities and exponential bounds for empirical and Rademacher processes.

## 7.1 Sparse Recovery and Neighborliness of Convex Polytopes

Let $\mathcal{H} := \{h_1, \ldots, h_N\}$ be a given finite set of measurable functions from $S$ into $\mathbb{R}$. In what follows, it will be called *a dictionary*. Given $J \subset \{1, \ldots, N\}$, we will write $d(J) := \operatorname{card}(J)$. For $\lambda = (\lambda_1, \ldots, \lambda_N) \in \mathbb{R}^N$, denote

$$f_\lambda = \sum_{j=1}^N \lambda_j h_j, \quad J_\lambda = \operatorname{supp}(\lambda) := \left\{ j : \lambda_j \neq 0 \right\} \text{ and } d(\lambda) := d(J_\lambda).$$

Suppose that a function $f_* \in \text{l.s.}(\mathcal{H}) = \{f_\lambda : \lambda \in \mathbb{R}^N\}$ from the linear span of the dictionary is observed (measured) at points $X_1, \ldots, X_n \in S$. For simplicity, we first assume that there is no noise in the observations:

$$Y_j = f_*(X_j), \quad j = 1, \ldots, n.$$

The goal is to recover a representation of $f_*$ in the dictionary. We are mostly interested in the case when $N > n$ (in fact, $N$ can be much larger than $n$). Define

$$L := \left\{ \lambda \in \mathbb{R}^N : f_\lambda(X_j) = Y_j, \; j = 1, \ldots, n \right\}.$$

Then, $L$ is an affine subspace of dimension at least $N - n$, so, the representation of $f_*$ in the dictionary is not unique. In such cases, it is of interest to find *the sparsest representation*, which means solving the problem

$$\|\lambda\|_{\ell_0} = \sum_{j=1}^{N} I(\lambda_j \neq 0) \longrightarrow \min, \; \lambda \in L. \tag{7.1}$$

If we introduce the following $n \times N$ matrix $A := \left( h_j(X_i) : 1 \leq i \leq n, 1 \leq j \leq N \right)$ and denote $\mathbf{Y}$ the vector with components $Y_1, \ldots, Y_n$, then problem (7.1) can be also rewritten as

$$\|\lambda\|_{\ell_0} = \sum_{j=1}^{N} I(\lambda_j \neq 0) \longrightarrow \min, \; A\lambda = \mathbf{Y}. \tag{7.2}$$

When $N$ is large, such problems are computationally intractable since the function to be minimized is non-smooth and non-convex. Essentially, solving (7.2) would require searching through all $2^N$ coordinate subspaces of $\mathbb{R}^N$. Because of this, the following convex relaxation of the problem is frequently used:

$$\|\lambda\|_{\ell_1} = \sum_{j=1}^{N} |\lambda_j| \longrightarrow \min, \; \lambda \in L, \tag{7.3}$$

or, equivalently,

$$\|\lambda\|_{\ell_1} = \sum_{j=1}^{N} |\lambda_j| \longrightarrow \min, \; A\lambda = \mathbf{Y}. \tag{7.4}$$

The last minimization problem is convex and, moreover, it is a linear programming problem. However, the question is whether solving (7.3) has anything to do with solving (7.1). Next result (due to Donoho [52]) gives an answer to this question by reducing it to some interesting problems in the geometry of convex polytopes. To formulate the result, define

$$P := AU_{\ell_1} = \mathrm{conv}\left( \left\{ a_1, -a_1, \ldots, a_N, -a_N \right\} \right),$$

where $U_{\ell_1} := \{ \lambda \in \mathbb{R}^N : \|\lambda\|_{\ell_1} \leq 1 \}$ is the unit ball in $\ell_1$ and $a_1, \ldots, a_N \in \mathbb{R}^n$ are columns of matrix $A$. Here and in what follows, $U_B$ denotes the closed unit ball of a Banach space $B$ centered at 0.

Clearly, $P$ is a centrally symmetric convex polytope in $\mathbb{R}^n$ with at most $2N$ vertices. Such a centrally symmetric polytope is called $d$-*neighborly* if any set of $d + 1$ vertices that does not contain antipodal vertices (such as $a_k$ and $-a_k$) spans a face of $P$.

**Theorem 7.1.** *Suppose that* $N > n$. *The following two statements are equivalent:*

(i) *The polytope $P$ has $2N$-vertices and is $d$-neighborly*
(ii) *Any solution $\lambda$ of the system of linear equations $A\lambda = \mathbf{Y}$ such that $d(\lambda) \leq d$ is the* **unique** *solution of problem (7.4)*

The unit ball $U_{\ell_1}$ of $\ell_1$ is a trivial example of an $(N - 1)$-neighborly centrally symmetric polytope. However, it is hard to find nontrivial constructive examples of such polytopes with a "high neighborliness". Their existence is usually proved by a probabilistic method, for instance, by choosing the design matrix $A$ at random and showing that the resulting random polytope $P$ is $d$-neighborly for sufficiently large $d$ with a high probability. The problem has been studied for several classes of random matrices (projections on an $n$-dimensional subspace picked at random from the Grassmannian of all $n$-dimensional subspaces; random matrices with i.i.d. Gaussian or Bernoulli entries, etc) both in the case of centrally symmetric polytopes and without the restriction of central symmetry, see Vershik and Sporyshev [150], Affentranger and Schneider [3] and, in connection with sparse recovery, Donoho [52], Donoho and Tanner [57]. The approach taken in these papers is based on rather subtle geometric analysis of the properties of high-dimensional convex polytopes, in particular, on computation of their internal and external angles. This leads to rather sharp estimates of the largest $d$ for which the neighborliness still holds (in other words, for which the phase transition occurs and the polytope starts losing faces). Here we follow another approach that is close to Rudelson and Vershynin [129] and Mendelson et al. [117]. This approach is more probabilistic, it is much simpler and it addresses the sparse recovery problem more directly. On the other hand, it does not give precise bounds on the maximal $d$ for which sparse recovery is possible (although it still provides correct answers up to constants).

## 7.2 Geometric Properties of the Dictionary

In what follows, we introduce several geometric characteristics of the dictionary $\mathcal{H}$ that will be involved in error bounds for sparse recovery methods.

### 7.2.1 Cones of Dominant Coordinates

For $J \subset \{1, \ldots, N\}$ and $b \in [0, +\infty]$, define the following cone consisting of vectors whose "dominant coordinates" are in $J$:

$$C_{b,J} := \left\{ u \in \mathbb{R}^N : \sum_{j \notin J} |u_j| \le b \sum_{j \in J} |u_j| \right\}.$$

Clearly, for $b = +\infty$, $C_{b,J} = \mathbb{R}^N$. For $b = 0$, $C_{b,J}$ is the linear subspace $\mathbb{R}^J$ of vectors $u \in \mathbb{R}^N$ with $\mathrm{supp}(u) \subset J$. For $b = 1$, we will write $C_J := C_{1,J}$. Such cones will be called *cones of dominant coordinates* and some norms in $\mathbb{R}^N$ will be compared on these cones.

Some useful geometric properties of the cones of dominant coordinates will be summarized in the following lemma. It includes several well known facts (see Candes and Tao [44], proof of Theorem 1; Ledoux and Talagrand [101], p. 421; Mendelson et al. [117], Lemma 3.3).

With a minor abuse of notations, we identify in what follows vectors $u \in \mathbb{R}^N$ with $\mathrm{supp}(u) \subset J$, where $J \subset \{1, \ldots, N\}$, with vectors $u = (u_j : j \in J) \in \mathbb{R}^J$.

**Lemma 7.1.** *Let $J \subset \{1, \ldots, N\}$ and let $d := \mathrm{card}(J)$.*

(i) *Take $u \in C_{b,J}$ and denote $J_0 := J$. For $s \ge 1$, $J_1$ will denote the set of $s$ coordinates in $\{1, \ldots, N\} \setminus J_0$ for which $|u_j|$'s are the largest, $J_2$ will be the set of $s$ coordinates in $\{1, \ldots, N\} \setminus (J_0 \cup J_1)$ for which $|u_j|$'s are the largest, etc. (at the end, there might be fewer than $s$ coordinates left). Denote $u^{(k)} := (u_j : j \in J_k)$. Then $u = \sum_{k \ge 0} u^{(k)}$ and*

$$\sum_{k \ge 2} \|u^{(k)}\|_{\ell_2} \le \frac{b}{\sqrt{s}} \sum_{j \in J} |u_j| \le b\sqrt{\frac{d}{s}} \left( \sum_{j \in J} |u_j|^2 \right)^{1/2}.$$

*In addition,*

$$\|u\|_{\ell_2} \le \left( b\sqrt{\frac{d}{s}} + 1 \right) \left( \sum_{j \in J_0 \cup J_1} |u_j|^2 \right)^{1/2}.$$

(ii) *Denote $K_J := C_{b,J} \cap U_{\ell_2}$. There exists a set $\mathcal{M}_d \subset U_{\ell_2}$ such that $d(u) \le d$ for $u \in \mathcal{M}_d$,*

$$\mathrm{card}(\mathcal{M}_d) \le 5^d \binom{N}{\le d}$$

*and*

$$K_J \subset 2(2 + b)\mathrm{conv}(\mathcal{M}_d).$$

*Proof.* To prove (i), note that, for all $j \in J_{k+1}$,

$$|u_j| \le \frac{1}{s} \sum_{i \in J_k} |u_i|,$$

implying that

$$\left(\sum_{j \in J_{k+1}} |u_j|^2\right)^{1/2} \le \frac{1}{\sqrt{s}} \sum_{j \in J_k} |u_j|.$$

Add these inequalities for $k = 1, 2, \ldots$ to get

$$\sum_{k \ge 2} \|u^{(k)}\|_{\ell_2} \le \frac{1}{\sqrt{s}} \sum_{j \notin J} |u_j| \le \frac{b}{\sqrt{s}} \sum_{j \in J} |u_j|$$

$$\le b\sqrt{\frac{d}{s}} \left(\sum_{j \in J} |u_j|^2\right)^{1/2} \le b\sqrt{\frac{d}{s}} \left(\sum_{j \in J \cup J_1} |u_j|^2\right)^{1/2}.$$

Therefore, for $u \in C_J$,

$$\|u\|_{\ell_2} \le \left(b\sqrt{\frac{d}{s}} + 1\right)\left(\sum_{j \in J_0 \cup J_1} |u_j|^2\right)^{1/2}.$$

To prove (ii) denote

$$\mathscr{J}_d := \left\{I \subset \{1, \ldots, N\} : d(I) \le d\right\}$$

and observe that

$$K_J \subset (2 + b) \operatorname{conv}\left(\bigcup_{I \in \mathscr{J}_d} B_I\right),$$

where

$$B_I := \left\{(u_i : i \in I) : \sum_{i \in I} |u_i|^2 \le 1\right\}.$$

Indeed, it is enough to consider $u \in K_J$ and to use statement (i) with $s = d$. Then, we have $u^{(0)} \in B_{J_0}$, $u^{(1)} \in B_{J_1}$ and

$$\sum_{k \ge 2} u^{(k)} \in b \operatorname{conv}\left(\bigcup_{I \in \mathscr{J}_d} B_I\right).$$

It is easy to see that if $B$ is the unit Euclidean ball in $\mathbb{R}^d$ and $M$ is a $1/2$-net of this ball, then

$$B \subset 2 \operatorname{conv}(M).$$

Here is a sketch of the proof of the last claim. For convex sets $C_1, C_2 \subset \mathbb{R}^N$, denote by $C_1 + C_2$ their Minkowski sum

$$C_1 + C_2 = \{x_1 + x_2 : x_1 \in C_1, x_2 \in C_2\}.$$

It follows that

$$B \subset M + \frac{1}{2}B \subset \operatorname{conv}(M) + \frac{1}{2}B \subset \operatorname{conv}(M) + \frac{1}{2}\operatorname{conv}(M) + \frac{1}{4}B \subset \cdots$$

$$\operatorname{conv}(M) + \frac{1}{2}\operatorname{conv}(M) + \frac{1}{4}\operatorname{conv}(M) + \cdots \subset 2\operatorname{conv}(M).$$

For each $I$ with $d(I) = d$, denote $M_I$ a minimal $1/2$-net of $B_I$. Then,

$$K_J \subset 2(2+b) \operatorname{conv}\left(\bigcup_{I \in \mathcal{J}_d} M_I\right) =: 2(2+b) \operatorname{conv}(\mathcal{M}_d).$$

By an easy combinatorial argument,

$$\operatorname{card}(\mathcal{M}_d) \le 5^d \binom{N}{\le d},$$

so, the proof is complete.                                                                      $\square$

### 7.2.2  Restricted Isometry Constants and Related Characteristics

Given a probability measure $\Pi$ on $S$, denote

$$\beta^{(b)}(J;\Pi) := \inf\left\{\beta > 0 : \sum_{j \in J}|\lambda_j| \le \beta\left\|\sum_{j=1}^{N}\lambda_j h_j\right\|_{L_1(\Pi)}, \ \lambda \in C_{b,J}\right\}$$

and

$$\beta_2^{(b)}(J;\Pi) := \inf\left\{\beta > 0 : \sum_{j \in J}|\lambda_j|^2 \le \beta^2\left\|\sum_{j=1}^{N}\lambda_j h_j\right\|_{L_2(\Pi)}^2, \ \lambda \in C_{b,J}\right\}.$$

Let

$$\beta(J,\Pi) := \beta^{(1)}(J,\Pi), \ \beta_2(J,\Pi) := \beta_2^{(1)}(J,\Pi).$$

As soon as the distribution $\Pi$ is fixed, we will often suppress $\Pi$ in our notations and write $\beta(J), \beta_2(J)$, etc. In the case when $J = \emptyset$, we set $\beta^{(b)}(J) = \beta_2^{(b)}(J) = 0$. Note that if $J \ne \emptyset$ and $h_1, \ldots, h_N$ are linearly independent in $L_1(\Pi)$ or in $L_2(\Pi)$, then, for all $b \in (0, +\infty)$, $\beta^{(b)}(J) < +\infty$ or, respectively, $\beta_2^{(b)}(J) < +\infty$. In the case of orthonormal dictionary, $\beta_2^{(b)}(J) = 1$.

We will use several properties of $\beta^{(b)}(J)$ and $\beta_2^{(b)}(J)$ and their relationships with other common characteristics of the dictionary.

Let $\kappa(J)$ denote the minimal eigenvalue of the Gram matrix $\left(\langle h_i, h_j \rangle_{L_2(\Pi)}\right)_{i,j \in J}$. Also denote $L_J$ the linear span of $\{h_j : j \in J\}$ and let

$$\rho(J) := \sup_{f \in L_J, g \in L_{J^c}, f,g \neq 0} \left| \frac{\langle f, g \rangle_{L_2(\Pi)}}{\|f\|_{L_2(\Pi)} \|g\|_{L_2(\Pi)}} \right|.$$

Thus, $\rho(J)$ is the largest "correlation coefficient" (or the largest cosine of the angle) between functions in the linear span of a subset $\{h_j : j \in J\}$ of the dictionary and the linear span of its complement (compare $\rho(J)$ with the notion of canonical correlation in multivariate statistical analysis). In fact, we will rather need a somewhat different quantity defined in terms of the cone $C_{b,J}$:

$$\rho^{(b)}(J) := \sup_{\lambda \in C_{b,J}} \frac{\left| \left\langle \sum_{j \in J} \lambda_j h_j, \sum_{j \notin J} \lambda_j h_j \right\rangle_{L_2(\Pi)} \right|}{\left\| \sum_{j \in J} \lambda_j h_j \right\|_{L_2(\Pi)} \left\| \sum_{j \notin J} \lambda_j h_j \right\|_{L_2(\Pi)}}.$$

Clearly, $\rho^{(b)}(J) \leq \rho(J)$.

**Proposition 7.1.** *The following bound holds:*

$$\beta_2^{(b)}(J) \leq \frac{1}{\sqrt{\kappa(J)(1 - (\rho^{(b)}(J))^2)}}. \tag{7.5}$$

*Proof.* Indeed, let $\lambda \in C_{b,J}$. The next inequality is obvious

$$\left\| \sum_{j \in J} \lambda_j h_j \right\|_{L_2(\Pi)} \leq (1 - (\rho^{(b)}(J))^2)^{-1/2} \left\| \sum_{j=1}^{N} \lambda_j h_j \right\|_{L_2(\Pi)},$$

since for $f = \sum_{j \in J} \lambda_j h_j$ and $g = \sum_{j \notin J} \lambda_j h_j$, we have

$$\|f + g\|_{L_2(\Pi)}^2 = (1 - \cos^2(\alpha))\|f\|_{L_2(\Pi)}^2 + \left( \|f\|_{L_2(\Pi)} \cos(\alpha) + \|g\|_{L_2(\Pi)} \right)^2$$
$$\geq (1 - (\rho^{(b)}(J))^2)\|f\|_{L_2(\Pi)}^2,$$

where $\alpha$ is the angle between $f$ and $g$. This yields

$$\left( \sum_{j \in J} |\lambda_j|^2 \right)^{1/2} \leq \frac{1}{\sqrt{\kappa(J)}} \left\| \sum_{j \in J} \lambda_j h_j \right\|_{L_2(\Pi)}$$

$$\leq \frac{1}{\sqrt{\kappa(J)(1 - (\rho^{(b)}(J))^2)}} \left\| \sum_{j=1}^{N} \lambda_j h_j \right\|_{L_2(\Pi)},$$

which implies (7.5). $\qquad\square$

Lemma 7.1 can be used to provide upper bounds on $\beta_2^{(b)}(J)$. To formulate such bounds, we first introduce so called *restricted isometry constants*.

For $d = 1, \dots, N$, let $\delta_d(\Pi)$ be the smallest $\delta > 0$ such that, for all $\lambda \in \mathbb{R}^N$ with $d(\lambda) \leq d$,

$$(1 - \delta)\|\lambda\|_{\ell_2} \leq \left\| \sum_{j=1}^{N} \lambda_j h_j \right\|_{L_2(\Pi)} \leq (1 + \delta)\|\lambda\|_{\ell_2}.$$

If $\delta_d(\Pi) < 1$, then $d$-dimensional subspaces spanned on subsets of the dictionary and equipped with (a) the $L_2(\Pi)$-norm and (b) the $\ell_2$-norm on vectors of coefficients are "almost" isometric. For a given dictionary $\{h_1, \dots, h_N\}$, the quantity $\delta_d(\Pi)$ will be called *the restricted isometry constant* of dimension $d$ with respect to the measure $\Pi$. The dictionary satisfies *a restricted isometry condition* in $L_2(\Pi)$ if $\delta_d(\Pi)$ is sufficiently small for a sufficiently large value of $d$ (in sparse recovery, this value is usually related to the underlying "sparsity" of the problem).

For $I, J \subset \{1, \dots, N\}$, $I \cap J = \emptyset$, denote

$$r(I; J) := \sup_{f \in L_I, g \in L_J, f, g \neq 0} \left| \frac{\langle f, g \rangle_{L_2(\Pi)}}{\|f\|_{L_2(\Pi)} \|g\|_{L_2(\Pi)}} \right|.$$

Note that $\rho(J) = r(J, J^c)$. Let

$$\rho_d := \max\left\{ r(I, J) : I, J \subset \{1, \dots, N\}, \; I \cap J = \emptyset, \; \mathrm{card}(I) = 2d, \; \mathrm{card}(J) = d \right\}.$$

This quantity measures the correlation between linear spans of disjoint parts of the dictionary of fixed "small cardinalities", in this case, $d$ and $2d$.

Define

$$m_d := \inf\{\|f_u\|_{L_2(\Pi)} : u \in \mathbb{R}^N, \|u\|_{\ell_2} = 1, d(u) \leq d\}$$

and

$$M_d := \sup\{\|f_u\|_{L_2(\Pi)} : u \in \mathbb{R}^N, \|u\|_{\ell_2} = 1, d(u) \leq d\}.$$

If $m_d \leq 1 \leq M_d \leq 2$, the restricted isometry constant can be written as

$$\delta_d = (M_d - 1) \vee (1 - m_d).$$

**Lemma 7.2.** *Suppose* $J \subset \{1, \dots, N\}$, $d(J) = d$ *and* $\rho_d < \frac{m_{2d}}{bM_d}$. *Then*

$$\beta_2^{(b)}(J) \leq \frac{1}{m_{2d} - b\rho_d M_d}.$$

*Proof.* Denote $P_I$ the orthogonal projection on $L_I \subset L_2(\Pi)$. Under the notations of Lemma 7.1 with $s = d$, for all $u \in C_J$,

$$\left\| \sum_{j=1}^{N} u_j h_j \right\|_{L_2(\Pi)} \geq \left\| P_{J_0 \cup J_1} \sum_{j=1}^{N} u_j h_j \right\|_{L_2(\Pi)}$$

$$\geq \left\| \sum_{j \in J_0 \cup J_1} u_j h_j \right\|_{L_2(\Pi)} - \left\| P_{J_0 \cup J_1} \sum_{j \notin J_0 \cup J_1} u_j h_j \right\|_{L_2(\Pi)}$$

$$\geq \left\| \sum_{j \in J_0 \cup J_1} u_j h_j \right\|_{L_2(\Pi)} - \sum_{k \geq 2} \left\| P_{J_0 \cup J_1} \sum_{j \in J_k} u_j h_j \right\|_{L_2(\Pi)}$$

$$\geq \left\| \sum_{j \in J_0 \cup J_1} u_j h_j \right\|_{L_2(\Pi)} - \rho_d \sum_{k \geq 2} \left\| \sum_{j \in J_k} u_j h_j \right\|_{L_2(\Pi)}$$

$$\geq \left\| \sum_{j \in J_0 \cup J_1} u_j h_j \right\|_{L_2(\Pi)} - \rho_d M_d \sum_{k \geq 2} \| u^{(k)} \|_{\ell_2}$$

$$\geq \left\| \sum_{j \in J_0 \cup J_1} u_j h_j \right\|_{L_2(\Pi)} - b \rho_d M_d \left( \sum_{j \in J \cup J_1} |u_j|^2 \right)$$

$$\geq \left\| \sum_{j \in J_0 \cup J_1} u_j h_j \right\|_{L_2(\Pi)} - b \rho_d \frac{M_d}{m_{2d}} \left\| \sum_{j \in J_0 \cup J_1} u_j h_j \right\|_{L_2(\Pi)}$$

$$= \left( 1 - b \rho_d \frac{M_d}{m_{2d}} \right) \left\| \sum_{j \in J_0 \cup J_1} u_j h_j \right\|_{L_2(\Pi)}.$$

On the other hand,

$$\left( \sum_{j \in J} |u_j|^2 \right)^{1/2} \leq \left( \sum_{j \in J_0 \cup J_1} |u_j|^2 \right)^{1/2} \leq m_{2d}^{-1} \left\| \sum_{j \in J_0 \cup J_1} u_j h_j \right\|_{L_2(\Pi)},$$

implying that

$$\left( \sum_{j \in J} |u_j|^2 \right)^{1/2} \leq m_{2d}^{-1} \left( 1 - b \rho_d \frac{M_d}{m_{2d}} \right)^{-1} \left\| \sum_{j=1}^{N} u_j h_j \right\|_{L_2(\Pi)}.$$

Therefore,

$$\beta_2(J) \leq \frac{1}{m_{2d} - b \rho_d M_d}. \qquad \square$$

It is easy to check that

$$\rho_d \leq \frac{1}{2} \left[ \left( \frac{1 + \delta_{3d}}{1 - \delta_{2d}} \right)^2 + \left( \frac{1 + \delta_{3d}}{1 - \delta_d} \right)^2 - 2 \right] \vee \frac{1}{2} \left[ 2 - \left( \frac{1 - \delta_{3d}}{1 + \delta_{2d}} \right)^2 - \left( \frac{1 - \delta_{3d}}{1 + \delta_d} \right)^2 \right].$$

Together with Lemma 7.2 this implies that $\beta_2(J) < +\infty$ for any set $J$ such that $\text{card}(J) \le d$, provided that $\delta_{3d} \le \frac{1}{8}$ (a sharper condition is also possible).

We will give a simple modification of Lemma 7.2 in spirit of [22].

**Lemma 7.3.** *Recall the notations of Lemma 7.1. Suppose $J \subset \{1, \ldots, N\}$, $d(J) = d$ and, for some $s \ge 1$,*

$$\frac{M_s}{m_{d+s}} < \frac{1}{b}\sqrt{\frac{s}{d}}.$$

*Then*

$$\beta_2^{(b)}(J) \le \frac{\sqrt{s}}{\sqrt{s}m_{d+s} - b\sqrt{d}M_s}.$$

*Proof.* For all $u \in C_{b,J}$,

$$\left(\sum_{j \in J}|u_j|^2\right)^{1/2} \le \frac{1}{m_{d+s}}\left\|\sum_{j \in J \cup J_1} u_j h_j\right\|_{L_2(\Pi)}$$

$$\le \frac{1}{m_{d+s}}\left\|\sum_{j=1}^{N} u_j h_j\right\|_{L_2(\Pi)} + \frac{1}{m_{d+s}}\left\|\sum_{j \notin J \cup J_1} u_j h_j\right\|_{L_2(\Pi)}.$$

To bound the last norm in the right hand side, note that

$$\left\|\sum_{j \notin J \cup J_1} u_j h_j\right\|_{L_2(\Pi)} \le \sum_{k \ge 2}\left\|\sum_{j \in J_k} u_j h_j\right\|_{L_2(\Pi)}$$

$$\le M_s \sum_{k \ge 2} \|u^{(k)}\|_{\ell_2} \le M_s \sqrt{\frac{d}{s}}\left(\sum_{j \in J}|u_j|^2\right)^{1/2}.$$

This yields the bound

$$\left(\sum_{j \in J}|u_j|^2\right)^{1/2} \le \frac{1}{m_{d+s}}\left\|\sum_{j=1}^{N} u_j h_j\right\|_{L_2(\Pi)} + \frac{M_s}{m_{d+s}}\sqrt{\frac{d}{s}}\left(\sum_{j \in J}|u_j|^2\right)^{1/2},$$

which implies the result.                                                   $\square$

### 7.2.3   Alignment Coefficients

In what follows, we will use several quantities that describe a way in which vectors in $\mathbb{R}^N$, especially, sparse vectors, are "aligned" with the dictionary. We will use the following definitions. Let $D \subset \mathbb{R}^N$ be a convex set. For $\lambda \in D$, denote by $T_D(\lambda)$

the closure of the set

$$\{v \in \mathbb{R}^N : \exists t > 0 \; \lambda + vt \in D\}.$$

The set $T_D(\lambda)$ will be called *the tangent cone* of convex set $D$ at point $\lambda$. Let

$$H := \left( \langle h_i, h_j \rangle_{L_2(\Pi)} \right)_{i,j=1,\ldots,N}$$

be the Gram matrix of the dictionary in the space $L_2(\Pi)$. Whenever it is convenient, $H$ will be viewed as a linear transformation of $\mathbb{R}^N$.

For a vector $w \in \mathbb{R}^N$ and $b > 0$, we will denote $C_{b,w} := C_{b,\mathrm{supp}(w)}$, which is a cone of vectors whose "dominant" coordinates are in $\mathrm{supp}(w)$. Now define

$$a_H^{(b)}(D, \lambda, w) := \sup\left\{ \langle w, u \rangle_{\ell_2} : u \in -T_D(\lambda) \cap C_{b,w}, \| f_u \|_{L_2(\Pi)} = 1 \right\}, \quad b \in [0, +\infty].$$

The quantities $a_H^{(b)}(D, \lambda, w)$ for $b \in [0, \infty]$ will be called the *alignment coefficients* of vector $w$, matrix $H$ and convex set $D$ at point $\lambda \in D$. In applications that follow, we want the alignment coefficient to be either negative, or, if positive, then small enough.

The geometry of the set $D$ could have an impact on the alignment coefficients for some vectors $w$ that are of interest in sparse recovery problems. For instance, if $L$ is a convex function on $D$ and $\lambda \in D$ is its minimal point, then there exists a subgradient $w \in \partial L(\lambda)$ of $L$ at point $\lambda$ such that, for all $u \in T_D(\lambda)$, $\langle w, u \rangle_{\ell_2} \geq 0$ (that is, the vector $-w$ belongs to *the normal cone* of $D$ at point $\lambda$; see Aubin and Ekeland [9], Chap. 4, Sect. 2, Corollary 6). This implies that $a_H^{(b)}(D, \lambda, w) \leq 0$. If $D = \mathbb{R}^N$, then $T_D(\lambda) = \mathbb{R}^N$, $\lambda \in \mathbb{R}^N$. In this case, we will write

$$a_H^{(b)}(w) := a_H^{(b)}(\mathbb{R}^N, \lambda, w) = \sup\left\{ \langle w, u \rangle_{\ell_2} : u \in C_{b,w}, \| f_u \|_{L_2(\Pi)} = 1 \right\}.$$

Despite the fact that the geometry of set $D$ might be important, in many cases, we are not taking it into account and replace $a_H^{(b)}(D, \lambda, w)$ by its upper bound $a_H^{(b)}(w)$.

Note that

$$\| f_u \|_{L_2(\Pi)}^2 = \langle Hu, u \rangle_{\ell_2} = \langle H^{1/2}u, H^{1/2}u \rangle_{\ell_2}.$$

We will frequently use the following form of alignment coefficient

$$a_H^{(\infty)}(D, \lambda, w) := \sup\left\{ \langle w, u \rangle_{\ell_2} : u \in -T_D(\lambda), \| f_u \|_{L_2(\Pi)} = 1 \right\},$$

or rather a simpler upper bound

$$a_H^{(\infty)}(w) = a_H^{(\infty)}(\mathbb{R}^N, \lambda, w) = \sup\left\{ \langle w, u \rangle_{\ell_2} : \| f_u \|_{L_2(\Pi)} = 1 \right\}.$$

The last quantity is a seminorm in $\mathbb{R}^N$ and, for all $b$, we have

$$a_H^{(b)}(w) \leq a_H^{(\infty)}(w) = \sup_{\|H^{1/2}u\|_{\ell_2}=1} \langle w, u \rangle_{\ell_2} =: \|w\|_H.$$

If $H$ is nonsingular, we can further write

$$\|w\|_H = \sup_{\|H^{1/2}u\|_{\ell_2}=1} \langle H^{-1/2}w, H^{1/2}u \rangle_{\ell_2} = \|H^{-1/2}w\|_{\ell_2}.$$

Even when $H$ is singular, we still have $\|w\|_H \leq \|H^{-1/2}w\|_{\ell_2}$, where, for $w \in \text{Im}(H^{1/2}) = H^{1/2}\mathbb{R}^N$, one defines

$$\|H^{-1/2}w\|_{\ell_2} := \inf\{\|v\|_{\ell_2} : H^{1/2}v = w\}$$

(which means factorization of the space with respect to $\text{Ker}(H^{1/2})$) and, for $w \notin \text{Im}(H^{1/2})$, the norm $\|H^{-1/2}w\|_{\ell_2}$ becomes infinite.

Note also that, for $b = 0$,

$$a_H^{(0)}(w) = a_H^{(0)}(\mathbb{R}^N, \lambda, w) = \sup\left\{\langle w, u \rangle_{\ell_2} : \|f_u\|_{L_2(\Pi)} = 1, \text{supp}(u) = \text{supp}(w)\right\}.$$

This also defines seminorms on subspaces of vectors $w$ with a fixed support, say, $\text{supp}(w) = J$. If $H_J := \left(\langle h_i, h_j \rangle_{L_2(\Pi)}\right)_{i,j \in J}$ is the corresponding submatrix of the Gram matrix $H$ and $H_J$ is nonsingular, then

$$a_H^{(0)}(w) = \|H_J^{-1/2}w\|_{\ell_2},$$

so, in this case, the alignment coefficient depends only on "small" submatrices of the Gram matrix corresponding to the support of $w$ (which is, usually, sparse).

When $0 < b < +\infty$, the definition of alignment coefficients involves cones of dominant coordinates and their values are between the values in the two extreme cases of $b = 0$ and $b = \infty$.

It is easy to bound the alignment coefficient in terms of geometric characteristics of the dictionary introduced earlier in this section. For instance, if $J = \text{supp}(w)$, then

$$\|w\|_H \leq \frac{\|w\|_{\ell_2}}{\sqrt{\kappa(J)(1 - \rho^2(J))}} \leq \frac{\|w\|_{\ell_\infty}\sqrt{d(J)}}{\sqrt{\kappa(J)(1 - \rho^2(J))}},$$

where $\kappa(J)$ is the minimal eigenvalue of the matrix $H_J = \left(\langle h_i, h_j \rangle_{L_2(\Pi)}\right)_{i,j \in J}$ and $\rho(J)$ is the "canonical correlation" defined above.

One can also upper bound the alignment coefficient in terms of the quantity

$$\beta_{2,b}(w; \Pi) := \beta_2^{(b)}(\text{supp}(w); \Pi).$$

Namely, the following bound is straightforward:

$$a_H^{(b)}(w) \leq \|w\|_{\ell_2} \beta_{2,b}(w; \Pi).$$

These upper bounds show that the size of the alignment coefficient is controlled by the "sparsity" of the vector $w$ as well as by some characteristics of the dictionary (or its Gram matrix $H$). For orthonormal dictionaries and for dictionaries that are close enough to being orthonormal (so that, for instance, $\kappa(J)$ is bounded away from 0 and $\rho^2(J)$ is bounded away from 1), the alignment coefficient is bounded from above by a quantity of the order $\|w\|_{\ell_\infty} \sqrt{d(J)}$. However, this is only an upper bound and the alignment coefficient itself is a more flexible characteristic of rather complicated geometric relationships between the vector $w$ and the dictionary. Even the quantity $\|H^{-1/2}w\|_{\ell_2}$ (a rough upper bound on the alignment coefficient not taking into account the geometry of the cone of dominant coordinates), depends not only on the sparsity of $w$, but also on the way in which this vector is aligned with the eigenspaces of $H$. If $w$ belongs to the linear span of the eigenspaces that correspond to large eigenvalues of $H$, then $\|H^{-1/2}w\|_{\ell_2}$ can be of the order $\|w\|_{\ell_2}$.

Note that the geometry of the problem is the geometry of the Hilbert space $L_2(\Pi)$, so it strongly depends on the unknown distribution $\Pi$ of the design variable.

## 7.3  Sparse Recovery in Noiseless Problems

Let $\Pi_n$ denote the empirical measure based on the points $X_1, \ldots, X_n$ (at the moment, not necessarily random).

**Proposition 7.2.** *Let $\hat{\lambda}$ be a solution of (7.3). If $\lambda^* \in L$ and $\beta_2(J_{\lambda^*}; \Pi_n) < +\infty$, then $\hat{\lambda} = \lambda^*$.*

*Proof.* Since $\hat{\lambda} \in L$ and $\lambda^* \in L$, we have

$$f_{\hat{\lambda}}(X_j) = f_{\lambda^*}(X_j), \quad j = 1, \ldots, n$$

implying that $\|f_{\hat{\lambda}} - f_{\lambda^*}\|_{L_2(\Pi_n)} = 0$. On the other hand, since $\hat{\lambda}$ is a solution of (7.3), we have $\|\hat{\lambda}\|_{\ell_1} \leq \|\lambda^*\|_{\ell_1}$. This yields

$$\sum_{j \notin J_{\lambda^*}} |\hat{\lambda}_j| \leq \sum_{j \in J_{\lambda^*}} (|\hat{\lambda}_j| - |\lambda_j^*|) \leq \sum_{j \in J_{\lambda^*}} |\hat{\lambda}_j - \lambda_j^*|.$$

Therefore, $\hat{\lambda} - \lambda^* \in C_{J_{\lambda^*}}$ and

$$\|\hat{\lambda} - \lambda^*\|_{\ell_1} \leq 2 \sum_{j \in J_{\lambda^*}} |\hat{\lambda}_j - \lambda_j^*| \leq 2\sqrt{d(\lambda^*)} \left( \sum_{j \in J_{\lambda^*}} |\hat{\lambda}_j - \lambda_j^*|^2 \right)^{1/2}$$

$$\leq 2\beta_2(J_{\lambda^*}; \Pi_n)\sqrt{d(\lambda^*)}\|f_{\hat{\lambda}} - f_{\lambda^*}\|_{L_2(\Pi_n)} = 0,$$

implying the result.                                                                    □

In particular, it means that as soon as the restricted isometry condition holds for the empirical distribution $\Pi_n$ for a sufficiently large $d$ with a sufficiently small $\delta_d$, the method (7.3) provides a solution of the sparse recovery problem for any target vector $\lambda^*$ such that $f_* = f_{\lambda^*}$ and $d(\lambda^*) \leq d$. To be more precise, it follows from the bounds of the previous section that the condition $\delta_{3d}(\Pi_n) \leq 1/8$ would suffice. Candes [38] gives sharper bounds. The restricted isometry condition for $\Pi_n$ (which can be also viewed as a condition on the design matrix $A$) has been also referred to as *the uniform uncertainty principle (UUP)* (see, e.g., Candes and Tao [44]). It is computationally hard to check UUP for a given large design matrix $A$. Moreover, it is hard to construct $n \times N$-matrices for which UUP holds. The main approach is based on using random matrices of special type and proving that for such matrices UUP holds for a sufficiently large $d$ with a high probability. We will discuss below a slightly different approach in which it is assumed that the design points $X_1, \ldots, X_n$ are i.i.d. with common distribution $\Pi$. It will be proved directly (without checking UUP for the random matrix $A$) that, under certain conditions, (7.3) does provide a solution of sparse recovery problem with a high probability.

Recall the definitions of $\psi_\alpha$-norms (see Appendix A.1) and, for $C > 0, A \geq 1$, define

$$\Lambda_S := \left\{ \lambda \in \mathbb{R}^N : C\beta(J_\lambda; \Pi) \max_{1 \leq k \leq N} \|h_k(X)\|_{\psi_1} \sqrt{\frac{A \log N}{n}} \leq 1/4 \right\}.$$

We will interpret $\Lambda_S$ as a set of "sparse" vectors. Note that in the case when the dictionary is $L_2(\Pi)$-orthonormal, $\beta(J; \Pi) \leq \sqrt{\operatorname{card}(J)}$, so, indeed, $\Lambda_S$ consists of vectors with a sufficiently small $d(\lambda)$ (that is, sparse).

In what follows we assume that $A \log N \leq n$.

Recall that

$$L = \left\{ \lambda \in \mathbb{R}^N : f_\lambda(X_j) = f_*(X_j), \ j = 1, \ldots, n \right\}.$$

**Theorem 7.2.** *Suppose* $f_* = f_{\lambda^*}$, $\lambda^* \in \mathbb{R}^N$. *Let* $A \geq 1$. *There exists a constant* $C$ *in the definition of the set* $\Lambda_S$ *such that with probability at least* $1 - N^{-A}$,

$$\text{either } L \cap \Lambda_S = \emptyset, \text{ or } L \cap \Lambda_S = \{\hat{\lambda}\}.$$

*In particular, if* $\lambda^* \in \Lambda_S$, *then with the same probability* $\hat{\lambda} = \lambda^*$.

*Proof.* The following lemma is used in the proof.

**Lemma 7.4.** *There exists a constant $C > 0$ such that for all $A \geq 1$ with probability at least $1 - N^{-A}$*

$$\sup_{\|u\|_{\ell_1} \leq 1} \left|(\Pi_n - \Pi)(|f_u|)\right| \leq C \max_{1 \leq k \leq N} \|h_k(X)\|_{\psi_1} \left(\sqrt{\frac{A \log N}{n}} \bigvee \frac{A \log N}{n}\right).$$

*Proof.* Let $R_n(f)$ be the Rademacher process. We will use symmetrization inequality and then contraction inequality for exponential moments (see Sects. 2.1, 2.2). For $t > 0$, we get

$$\mathbb{E} \exp\left\{t \sup_{\|u\|_{\ell_1} \leq 1} \left|(\Pi_n - \Pi)(|f_u|)\right|\right\} \leq \mathbb{E} \exp\left\{2t \sup_{\|u\|_{\ell_1} \leq 1} \left|R_n(|f_u|)\right|\right\}$$

$$\leq \mathbb{E} \exp\left\{4t \sup_{\|u\|_{\ell_1} \leq 1} \left|R_n(f_u)\right|\right\}.$$

Since the mapping $u \mapsto R_n(f_u)$ is linear, the supremum of $R_n(f_u)$ over the set $\{\|u\|_{\ell_1} \leq 1\}$ (which is a convex polytope) is attained at one of its vertices, and we get

$$\mathbb{E} \exp\left\{t \sup_{\|u\|_{\ell_1} \leq 1} \left|(\Pi_n - \Pi)(|f_u|)\right|\right\} \leq \mathbb{E} \exp\left\{4t \max_{1 \leq k \leq N} \left|R_n(h_k)\right|\right\}$$

$$= N \max_{1 \leq k \leq N} \mathbb{E}\left[\exp\left\{4t R_n(h_k)\right\} \bigvee \exp\left\{-4t R_n(h_k)\right\}\right]$$

$$\leq 2N \max_{1 \leq k \leq N} \mathbb{E} \exp\left\{4t R_n(h_k)\right\} \leq 2N \max_{1 \leq k \leq N} \left(\mathbb{E} \exp\left\{4\frac{t}{n}\varepsilon h_k(X)\right\}\right)^n.$$

To bound the last expectation and to complete the proof, follow the standard proof of Bernstein's inequality. $\square$

Assume that $L \cap \Lambda_S \neq \emptyset$ and let $\lambda \in L \cap \Lambda_S$. Arguing as in the proof of Proposition 7.2, we get that, for all $\lambda \in L$, $\hat{\lambda} - \lambda \in C_{J_\lambda}$ and $\|f_{\hat{\lambda}} - f_\lambda\|_{L_1(\Pi_n)} = 0$. Therefore,

$$\|\hat{\lambda} - \lambda\|_{\ell_1} \leq \sum_{j \notin J_\lambda} |\hat{\lambda}_j| + \sum_{j \in J_\lambda} |\lambda_j - \hat{\lambda}_j|$$

$$\leq 2 \sum_{j \in J_\lambda} |\lambda_j - \hat{\lambda}_j| \leq 2\beta(J_\lambda)\|f_{\hat{\lambda}} - f_\lambda\|_{L_1(\Pi)}. \tag{7.6}$$

We will now upper bound $\|f_{\hat{\lambda}} - f_\lambda\|_{L_1(\Pi)}$ in terms of $\|\hat{\lambda} - \lambda\|_{\ell_1}$, which will imply the result. First, note that

$$\|f_{\hat{\lambda}} - f_\lambda\|_{L_1(\Pi)} = \|f_{\hat{\lambda}} - f_\lambda\|_{L_1(\Pi_n)} + (\Pi - \Pi_n)(|f_{\hat{\lambda}} - f_\lambda|)$$

$$\leq \sup_{\|u\|_{\ell_1} \leq 1} \left| (\Pi_n - \Pi)(|f_u|) \right| \|\hat{\lambda} - \lambda\|_{\ell_1}. \tag{7.7}$$

By Lemma 7.4, with probability at least $1 - N^{-A}$ (under the assumption that $A \log N \leq n$)

$$\sup_{\|u\|_{\ell_1} \leq 1} \left| (\Pi_n - \Pi)(|f_u|) \right| \leq C \max_{1 \leq k \leq N} \|h_k\|_{\psi_1} \sqrt{\frac{A \log N}{n}}.$$

This yields the following bound that holds with probability at least $1 - N^{-A}$:

$$\|f_{\hat{\lambda}} - f_\lambda\|_{L_1(\Pi)} \leq C \max_{1 \leq k \leq N} \|h_k\|_{\psi_1} \sqrt{\frac{A \log N}{n}} \|\hat{\lambda} - \lambda\|_{\ell_1}. \tag{7.8}$$

Together with (7.6), this implies

$$\|\hat{\lambda} - \lambda\|_{\ell_1} \leq 2C\beta(J_\lambda) \max_{1 \leq k \leq N} \|h_k\|_{\psi_1} \sqrt{\frac{A \log N}{n}} \|\hat{\lambda} - \lambda\|_{\ell_1}.$$

It follows that, for $\lambda \in L \cap \Lambda_S$, with probability at least $1 - N^{-A}$,

$$\|\hat{\lambda} - \lambda\|_{\ell_1} \leq \frac{1}{2}\|\hat{\lambda} - \lambda\|_{\ell_1},$$

and, hence, $\hat{\lambda} = \lambda$.                                                                        $\square$

It is of interest to study the problem under the following condition on the dictionary and on the distribution $\Pi$: for all $\lambda \in C_J$

$$\left\| \sum_{j=1}^N \lambda_j h_j \right\|_{L_1(\Pi)} \leq \left\| \sum_{j=1}^N \lambda_j h_j \right\|_{L_2(\Pi)} \leq B(J) \left\| \sum_{j=1}^N \lambda_j h_j \right\|_{L_1(\Pi)} \tag{7.9}$$

with some constant $B(J) > 0$. This inequality always holds with some $B(J) > 0$ since any two norms on a finite dimensional space are equivalent. In fact, the first bound is just Cauchy–Schwarz inequality. However, in general, the constant $B(J)$ does depend on $J$ and we are interested in the situation when there is no such dependence (or, at least, $B(J)$ does not grow too fast as $\mathrm{card}(J) \to \infty$).

**Example.** • *Gaussian dictionary.* It will be said that $h_1, \ldots, h_N$ is a Gaussian dictionary with respect to $\Pi$ iff $(h_1(X), \ldots, h_N(X))$ has a normal distribution in $\mathbb{R}^N$, $X$ having distribution $\Pi$. In this case, condition (7.9) holds for all $\lambda \in \mathbb{R}^N$ with $B(J) = B$ that does not depend on the dimension $d(J)$. Moreover, all the

$L_p$ norms for $p \geq 1$ and even $\psi_1$- and $\psi_2$-norms of $\sum_{j=1}^{N} \lambda_j h_j$ are equivalent up to numerical constants.

- *Gaussian orthonormal dictionary.* In this special case of Gaussian dictionaries, $h_1(X), \ldots, h_N(X)$ are i.i.d. standard normal random variables.
- *Rademacher (Bernoulli) dictionary.* In this example $h_1(X), \ldots, h_N(X)$ are i.i.d. Rademacher random variables. Condition (7.9) holds for this dictionary with an absolute constant $B(J) = B$. Moreover, as in the case of Gaussian dictionaries, all the Orlicz norms between $L_1$ and $\psi_2$ are equivalent on the linear span of the dictionary up to numerical constants. This fact follows from the classical Khinchin inequality (see Bobkov and Houdré [27] for a discussion of Khinchin type inequalities and their connections with isoperimetric constants).
- $\psi_\alpha$-*dictionary.* Let $\alpha \geq 1$. It will be said that $h_1, \ldots, h_N$ is a $\psi_\alpha$-dictionary with respect to $\Pi$ iff

$$\|f_\lambda\|_{\psi_\alpha} \leq B \|f_\lambda\|_{L_1(\Pi)}, \ \lambda \in \mathbb{R}^N$$

with an absolute constant $B$. Condition (7.9) obviously holds for $\psi_\alpha$-dictionaries. In particular, $\psi_2$-dictionaries will be also called *subgaussian dictionaries*. Clearly, this includes the examples of Gaussian and Rademacher dictionaries.

- *Log-concave dictionary.* Recall that a probability measure $\mu$ in $\mathbb{R}^N$ is called *log-concave* iff

$$\mu(tA + (1-t)B) \geq (\mu(A))^t (\mu(B))^{1-t}$$

for all Borel sets $A, B \subset \mathbb{R}^N$ and all $t \in [0, 1]$. A log-concave measure $\mu$ is always supported in an affine subspace of $\mathbb{R}^N$ (that might coincide with the whole space). Moreover, it has a density on its support that is a log-concave function (i.e., its logarithm is concave). In particular, if $K \subset \mathbb{R}^N$ is a bounded convex set, then uniform distribution in $K$ is log-concave. It will be said that a dictionary $\{h_1, \ldots, h_N\}$ is *log-concave* with respect to $\Pi$ iff the random vector $(h_1(X), \ldots, h_N(X))$ has a log-concave distribution, $X$ having distribution $\Pi$. A well known result of Borell [28] (see also Ledoux [100], Proposition 2.14) implies that log-concave dictionaries satisfy the condition (7.9) with an absolute constant $B(J) = B$ (that does not depend on $J$). Moreover, the same result implies that for log-concave dictionaries

$$\|f_\lambda\|_{\psi_1} \leq B \|f_\lambda\|_{L_1(\Pi)}, \ \lambda \in \mathbb{R}^N$$

with an absolute constant $B$. Thus, logconcave dictionaries are examples of $\psi_1$-dictionary.

Under the condition (7.9),

$$\beta(J) \leq B(J)\beta_2(J)\sqrt{d(J)}. \tag{7.10}$$

If $\beta_2(J)$ is bounded (as in the case of orthonormal dictionaries), then $\beta(J)$ is "small" for sets $J$ of small cardinality $d(J)$. In this case, the definition of the set of "sparse vectors" $\Lambda_S$ can be rewritten in terms of $\beta_2$.

However, we will give below another version of this result slightly improving the logarithmic factor in the definition of the set of sparse vectors $\Lambda_S$ and providing bounds on the norms $\| \cdot \|_{L_2(\Pi)}$ and $\| \cdot \|_{\ell_2}$.

Denote

$$\beta_2(d) := \beta_2(d;\Pi) := \max\left\{\beta_2(J): \ J \subset \{1,\ldots,N\}, \ d(J) \le 2d\right\}.$$

Let

$$B(d) := \max\left\{B(J): \ J \subset \{1,\ldots,N\}, \ d(J) \le d\right\}.$$

Finally, denote $\bar{d}$ the largest $d$ satisfying the conditions $d \le \frac{N}{e} - 1$, $\frac{Ad\log(N/d)}{n} \le 1$, and

$$CB(d)\beta_2(d) \sup_{\|u\|_{\ell_2}\le 1, d(u)\le d} \|f_u\|_{\psi_1} \sqrt{\frac{Ad\log(N/d)}{n}} \le 1/4.$$

We will now use the following definition of the set of "sparse" vectors:

$$\Lambda_{S,2} := \{\lambda \in \mathbb{R}^N : d(\lambda) \le \bar{d}\}.$$

Recall the notation

$$\binom{n}{\le k} := \sum_{j=0}^{k} \binom{n}{j}.$$

Suppose $f_* = f_{\lambda^*}$, $\lambda^* \in \mathbb{R}^N$. Let $A \ge 1$. There exists a constant $C$ in the definition of the set $\Lambda_S$ such that with probability at least $1 - N^{-A}$,

**Theorem 7.3.** *Suppose that* $f_* = f_{\lambda^*}$, $\lambda^* \in \mathbb{R}^N$ *and that condition (7.9) holds. Let* $A \ge 1$. *There exists a constant* $C$ *in the definition of* $\Lambda_{S,2}$ *such that, with probability at least*

$$1 - 5^{-\bar{d}A}\binom{N}{\le \bar{d}}^{-A},$$

*either* $L \cap \Lambda_{S,2} = \emptyset$, *or* $L \cap \Lambda_S = \{\hat{\lambda}\}$.

*In particular, if* $\lambda^* \in \Lambda_{S,2}$, *then with the same probability* $\hat{\lambda} = \lambda^*$.

*Proof.* We will use the following lemma.

**Lemma 7.5.** *For* $J \subset \{1,\ldots,N\}$ *with* $d(J) \le d$, *let* $K_J := C_J \cap U_{\ell_2}$. *There exists a constant* $C > 0$ *such that, for all* $A \ge 1$ *with probability at least*

$$1 - 5^{-dA}\binom{N}{\le d}^{-A},$$

*the following bound holds:*

$$\sup_{u \in K_J} \left| (\Pi_n - \Pi)(|f_u|) \right|$$

$$\leq C \sup_{\|u\|_{\ell_2} \leq 1, d(u) \leq d} \|f_u\|_{\psi_1} \left( \sqrt{\frac{Ad \log(N/d)}{n}} \bigvee \frac{Ad \log(N/d)}{n} \right).$$

*Proof.* It follows from statement (ii) of Lemma 7.1 with $b = 1$ that

$$K_J \subset 6 \operatorname{conv}(\mathcal{M}_d),$$

where $\mathcal{M}_d$ is a set of vectors $u$ from the unit ball $\{u \in \mathbb{R}^N : \|u\|_{\ell_2} \leq 1\}$ such that $d(u) \leq d$ and

$$\operatorname{card}(\mathcal{M}_d) \leq 5^d \binom{N}{\leq d}.$$

Now, it is enough to repeat the proof of Lemma 7.4. In particular, we use symmetrization and contraction inequalities to reduce bounding the exponential moment of

$$\sup_{u \in K_J} \left| (\Pi_n - \Pi)(|f_u|) \right|$$

to bounding the exponential moment of $\sup_{u \in \mathcal{M}_d} |R_n(f_u)|$, $\operatorname{card}(\mathcal{M}_d)$ playing now the role of $N$. The bound on $\operatorname{card}(\mathcal{M}_d)$ implies that with some $c > 0$

$$\log(\operatorname{card}(\mathcal{M}_d)) \leq cd \log \frac{N}{d},$$

and it is easy to complete the proof.                                      □

We now follow the proof of Theorem 7.2 with straightforward modifications. Assume that $L \cap \Lambda_{S,2} \neq \emptyset$ and let $\lambda \in L \cap \Lambda_{S,2}$. Instead of (7.7), we use

$$\|f_{\hat{\lambda}} - f_\lambda\|_{L_1(\Pi)} = \|f_{\hat{\lambda}} - f_\lambda\|_{L_1(\Pi_n)} + (\Pi - \Pi_n)(|f_{\hat{\lambda}} - f_\lambda|)$$

$$\leq \sup_{\|u\|_{\ell_2} \leq 1, u \in C_{J_\lambda}} \left| (\Pi_n - \Pi)(|f_u|) \right| \|\hat{\lambda} - \lambda\|_{\ell_2}. \tag{7.11}$$

To bound $\|\hat{\lambda} - \lambda\|_{\ell_2}$ note that, as in the proof of Theorem 7.2, $\hat{\lambda} - \lambda \in C_{J_\lambda}$ and apply Lemma 7.1 to $u = \hat{\lambda} - \lambda$, $J = J_\lambda$:

$$\|\hat{\lambda} - \lambda\|_{\ell_2} \leq 2 \left( \sum_{j \in J_0 \cup J_1} |\hat{\lambda}_j - \lambda_j|^2 \right)^{1/2} \leq 2\beta_2(d(\lambda)) \|f_{\hat{\lambda}} - f_\lambda\|_{L_2(\Pi)}. \tag{7.12}$$

Use Lemma 7.5 to bound

$$\sup_{\|u\|_{\ell_2}\le 1, u\in C_{J_\lambda}} \left|(\varPi_n - \varPi)(|f_u|)\right|$$

$$\le C \sup_{\|u\|_{\ell_2}\le 1, d(u)\le \bar{d}} \|f_u\|_{\psi_1} \sqrt{\frac{A\bar{d}\log(N/\bar{d})}{n}}, \qquad (7.13)$$

which holds with probability at least $1 - 5^{-\bar{d}A}\binom{N}{\le \bar{d}}^{-A}$. It remains to substitute bounds (7.12) and (7.13) in (7.11), to use (7.9) and to solve the resulting inequality with respect to $\|f_{\hat{\lambda}} - f_\lambda\|_{L_2(\varPi)}$. It follows that the last norm is equal to 0. In view of (7.12), this implies that $\hat{\lambda} = \lambda$.                                                  □

**Remark.** Note that, in the case of $L_2(\varPi)$-orthonormal logconcave dictionary, Theorem 7.3 easily implies that $\hat{\lambda} = \lambda^*$ with a high probability provided that

$$\frac{Ad(\lambda^*)\log(N/d(\lambda^*))}{n} \le c$$

for a sufficiently small $c$. Recently, Adamczak et al. [2] obtained sharp bounds on empirical restricted isometry constants $\delta_d(\varPi_n)$ for such dictionaries that imply bounds on $d(\lambda^*)$ for which sparse recovery is possible with a little bit worse logarithmic factor than what follows from Theorem 7.3 (of course, in this theorem we are not providing any control of $\delta_d(\varPi_n)$).

## 7.4   The Dantzig Selector

We now turn to the case when the target function $f_*$ is observed in an additive noise. Moreover, it will not be assumed that $f_*$ belongs to the linear span of the dictionary, but rather that it can be well approximated in the linear span. Consider the following regression model with random design

$$Y_j = f_*(X_j) + \xi_j, \quad j = 1,\dots,n,$$

where $X, X_1, \dots, X_n$ are i.i.d. random variables in a measurable space $(S, \mathscr{A})$ with distribution $\varPi$ and $\xi, \xi_1, \dots, \xi_n$ are i.i.d. random variables with $\mathbb{E}\xi = 0$ independent of $(X_1, \dots, X_n)$. Candes and Tao [44] developed a method of sparse recovery based on linear programming suitable in this more general framework. They called it *the Dantzig selector*.

Given $\varepsilon > 0$, let

$$\hat{\Lambda}_\varepsilon := \left\{\lambda \in \mathbb{R}^N : \max_{1\le k\le N}\left|n^{-1}\sum_{j=1}^n (f_\lambda(X_j) - Y_j)h_k(X_j)\right| \le \varepsilon\right\}$$

and define the Dantzig selector as

$$\hat{\lambda} := \hat{\lambda}^{\varepsilon} \in \text{Argmin}_{\lambda \in \hat{\Lambda}_{\varepsilon}} \|\lambda\|_{\ell_1}.$$

It is easy to reduce the computation of $\hat{\lambda}^{\varepsilon}$ to a linear program. The Dantzig selector is closely related to the $\ell_1$-penalization method (called "LASSO" in statistical literature, see Tibshirani [141]) and defined as a solution of the following penalized empirical risk minimization problem:

$$n^{-1} \sum_{j=1}^{n} (f_{\lambda}(X_j) - Y_j)^2 + 2\varepsilon \|\lambda\|_{\ell_1} =: L_n(\lambda) + 2\varepsilon \|\lambda\|_{\ell_1} \longrightarrow \min. \qquad (7.14)$$

The set of constraints of the Dantzig selector can be written as

$$\hat{\Lambda}_{\varepsilon} = \left\{ \lambda : \left\| \nabla L_n(\lambda) \right\|_{\ell_{\infty}} \leq \varepsilon \right\}$$

and the condition $\lambda \in \hat{\Lambda}_{\varepsilon}$ is necessary for $\lambda$ to be a solution of (7.14).

In [44], Candes and Tao studied the performance of the Dantzig selector in the case of fixed design regression (nonrandom points $X_1, \ldots, X_n$) under the assumption that the design matrix $A = \left( h_j(X_i) \right)_{i=1,n; j=1,N}$ satisfies the uniform uncertainty principle (UUP). They stated that UUP holds with a high probability for some random design matrices such as the "Gaussian ensemble" and the "Bernoulli or Rademacher ensemble" (using the terminology of the previous section, Gaussian and Rademacher dictionaries).

We will prove several "sparsity oracle inequalities" for the Dantzig selector in spirit of recent results of Bunea et al. [36], van de Geer [63], Koltchinskii [84] in the case of $\ell_1$- or $\ell_p$-penalized empirical risk minimization. We follow the paper of Koltchinskii [85] that relies only on elementary empirical and Rademacher processes methods (symmetrization and contraction inequalities for Rademacher processes and Bernstein type exponential bounds), but does not use more advanced techniques, such as concentration of measure and generic chaining. It is also close to the approach of Sect. 7.3 and to recent papers by Rudelson and Vershynin [129] and Mendelson et al. [117]. As in Sect. 7.3, our proofs of oracle inequalities in the random design case are more direct, they are not based on a reduction to the fixed design case and checking UUP for random matrices. The results also cover broader families of design distributions. In particular, the assumption that the dictionary is $L_2(\Pi)$-orthonormal is replaced by the assumption that it satisfies the restricted isometry condition with respect to $\Pi$.

In what follows, the values of $\varepsilon > 0$, $A > 0$ and $C > 0$ will be fixed and it will be assumed that $\frac{A \log N}{n} \leq 1$. Consider the following set

$$\Lambda := \Lambda_\varepsilon(A) := \left\{ \lambda \in \mathbb{R}^N : \left| \langle f_\lambda - f_*, h_k \rangle_{L_2(\Pi)} \right| \right.$$

$$\left. + C \left( \|(f_\lambda - f_*)(X) h_k(X)\|_{\psi_1} + \|\xi h_k(X)\|_{\psi_1} \right) \sqrt{\frac{A \log N}{n}} \leq \varepsilon, \ k = 1, \ldots, N \right\},$$

consisting of vectors $\lambda$ ("oracles") such that $f_\lambda$ provides a good approximation of $f_*$. In fact, $\lambda \in \Lambda_\varepsilon(A)$ implies that

$$\max_{1 \leq k \leq N} \left| \langle f_\lambda - f_*, h_k \rangle_{L_2(\Pi)} \right| \leq \varepsilon. \tag{7.15}$$

This means that $f_\lambda - f_*$ is "almost orthogonal" to the linear span of the dictionary. Thus, $f_\lambda$ is close to the projection of $f_*$ on the linear span. Condition (7.15) is necessary for $\lambda$ to be a minimal point of

$$\lambda \mapsto \|f_\lambda - f_*\|^2_{L_2(\Pi)} + 2\varepsilon \|\lambda\|_{\ell_1},$$

and minimizing the last function is a "population version" of LASSO problem (7.14) ($\lambda \in \hat{\Lambda}_\varepsilon$ is a necessary condition for (7.14)). Of course, the condition

$$\varepsilon \geq \max_{1 \leq k \leq N} \|\xi h_k(X)\|_{\psi_1} \sqrt{\frac{A \log N}{n}}$$

is necessary for $\Lambda_\varepsilon(A) \neq \emptyset$. It will be clear from the proof of Theorem 7.4 below that $\lambda \in \Lambda_\varepsilon(A)$ implies $\lambda \in \hat{\Lambda}_\varepsilon$ with a high probability.

The next Theorems 7.4 and 7.5 show that if there exists a sufficiently sparse vector $\lambda$ in the set $\hat{\Lambda}_\varepsilon$ of constraints of the Dantzig selector, then, with a high probability, the Dantzig selector belongs to a small ball around $\lambda$ in such norms as $\|\cdot\|_{\ell_1}, \|\cdot\|_{\ell_2}$. At the same time, the function $f_{\hat{\lambda}}$ belongs to a small ball around $f_\lambda$ with respect to such norms as $\|\cdot\|_{L_1(\Pi)}$ or $\|\cdot\|_{L_2(\Pi)}$. The radius of this ball is determined by the degree of sparsity of $\lambda$ and by the properties of the dictionary characterized by such quantities as $\beta$ or $\beta_2$ (see Sect. 7.2). Essentially, the results show that the Dantzig selector is adaptive to unknown degree of sparsity of the problem, provided that the dictionary is not too far from being orthonormal in $L_2(\Pi)$.

Recall the definition of the set of "sparse" vectors $\Lambda_S$ from the previous section. Let

$$\tilde{\Lambda} = \tilde{\Lambda}_\varepsilon(A) := \Lambda_\varepsilon(A) \cap \Lambda_S.$$

**Theorem 7.4.** *There exists a constant $C$ in the definitions of $\Lambda_\varepsilon(A)$, $\Lambda_S$ such that, for $A \geq 1$ with probability at least $1 - N^{-A}$, the following bounds hold for all $\lambda \in \hat{\Lambda}_\varepsilon \cap \Lambda_S$:*

$$\|f_{\hat{\lambda}} - f_\lambda\|_{L_1(\Pi)} \leq 16\beta(J_\lambda)\varepsilon$$

*and*

$$\|\hat{\lambda} - \lambda\|_{\ell_1} \leq 32\beta^2(J_\lambda)\varepsilon.$$

*This implies that*

$$\|f_{\hat{\lambda}} - f_*\|_{L_1(\Pi)} \leq \inf_{\lambda \in \bar{\Lambda}_\varepsilon(A)} \left[ \|f_\lambda - f_*\|_{L_1(\Pi)} + 16\beta(J_\lambda)\varepsilon \right].$$

*If, in addition $f_* = f_{\lambda^*}, \lambda^* \in \mathbb{R}^N$, then also*

$$\|\hat{\lambda} - \lambda^*\|_{\ell_1} \leq \inf_{\lambda \in \bar{\Lambda}_\varepsilon(A)} \left[ \|\lambda - \lambda^*\|_{\ell_1} + 32\beta^2(J_\lambda)\varepsilon \right].$$

*Proof.* We use the following lemma based on Bernstein's inequality for $\psi_1$-random variables (see Sect. A.2).

**Lemma 7.6.** *Let $\eta^{(k)}, \eta_1^{(k)}, \ldots, \eta_n^{(k)}$ be i.i.d. random variables with $\mathbb{E}\eta^{(k)} = 0$ and $\|\eta^{(k)}\|_{\psi_1} < +\infty, k = 1, \ldots, N$. There exists a numerical constant $C > 0$ such that, for $A \geq 1$ with probability at least $1 - N^{-A}$ for all $k = 1, \ldots, N$,*

$$\left| n^{-1} \sum_{j=1}^n \eta_j^{(k)} \right| \leq C \|\eta^{(k)}\|_{\psi_1} \left( \sqrt{\frac{A \log N}{n}} \bigvee \frac{A \log N}{n} \right).$$

For $\lambda \in \hat{\Lambda}_\varepsilon \cap \Lambda_S$, we will upper bound the norms $\|\hat{\lambda} - \lambda\|_{\ell_1}, \|f_{\hat{\lambda}} - f_\lambda\|_{L_1(\Pi)}$ in terms of each other and solve the resulting inequalities, which will yield the first two bounds of the theorem. As in the proof of Proposition 7.2 and Theorems 7.2, 7.3, $\lambda \in \hat{\Lambda}_\varepsilon$ and the definition of $\hat{\lambda}$ imply that $\hat{\lambda} - \lambda \in C_{J_\lambda}$ and

$$\|\hat{\lambda} - \lambda\|_{\ell_1} \leq 2\beta(J_\lambda)\|f_{\hat{\lambda}} - f_\lambda\|_{L_1(\Pi)}. \tag{7.16}$$

It remains to upper bound $\|f_{\hat{\lambda}} - f_\lambda\|_{L_1(\Pi)}$ in terms of $\|\hat{\lambda} - \lambda\|_{\ell_1}$. To this end, note that

$$\|f_{\hat{\lambda}} - f_\lambda\|_{L_1(\Pi)} = \|f_{\hat{\lambda}} - f_\lambda\|_{L_1(\Pi_n)} + (\Pi - \Pi_n)(|f_{\hat{\lambda}} - f_\lambda|)$$

$$\leq \|f_{\hat{\lambda}} - f_\lambda\|_{L_1(\Pi_n)} + \sup_{\|u\|_{\ell_1} \leq 1} \left| (\Pi_n - \Pi)(|f_u|) \right| \|\hat{\lambda} - \lambda\|_{\ell_1}. \tag{7.17}$$

The first term in the right hand side can be bounded as follows

$$\|f_{\hat{\lambda}} - f_\lambda\|_{L_1(\Pi_n)}^2 \leq \|f_{\hat{\lambda}} - f_\lambda\|_{L_2(\Pi_n)}^2 = \langle f_{\hat{\lambda}} - f_\lambda, f_{\hat{\lambda}} - f_\lambda \rangle_{L_2(\Pi_n)}$$

$$= \sum_{k=1}^N (\hat{\lambda}_k - \lambda_k)\langle f_{\hat{\lambda}} - f_\lambda, h_k \rangle_{L_2(\Pi_n)} \leq \|\hat{\lambda} - \lambda\|_{\ell_1} \max_{1 \leq k \leq N} \left| \langle f_{\hat{\lambda}} - f_\lambda, h_k \rangle_{L_2(\Pi_n)} \right|.$$

Both $\hat{\lambda} \in \hat{\Lambda}$ and $\lambda \in \hat{\Lambda}$, implying that

$$\max_{1 \le k \le N} \left| \langle f_{\hat{\lambda}} - f_{\lambda}, h_k \rangle_{L_2(\Pi_n)} \right|$$

$$\le \max_{1 \le k \le N} \left| n^{-1} \sum_{j=1}^{n} (f_{\lambda}(X_j) - Y_j) h_k(X_j) \right|$$

$$+ \max_{1 \le k \le N} \left| n^{-1} \sum_{j=1}^{n} (f_{\hat{\lambda}}(X_j) - Y_j) h_k(X_j) \right| \le 2\varepsilon.$$

Therefore,

$$\| f_{\hat{\lambda}} - f_{\lambda} \|_{L_1(\Pi_n)} \le \sqrt{2\varepsilon \| \hat{\lambda} - \lambda \|_{\ell_1}}.$$

Now we bound the second term in the right hand side of (7.17). Under the assumption $A \log N \le n$, Lemma 7.4 implies that with probability at least $1 - N^{-A}$

$$\sup_{\|u\|_{\ell_1} \le 1} \left| (\Pi_n - \Pi)(|f_u|) \right| \le C \max_{1 \le k \le N} \|h_k\|_{\psi_1} \sqrt{\frac{A \log N}{n}}.$$

Hence, we conclude from (7.17) that

$$\| f_{\hat{\lambda}} - f_{\lambda} \|_{L_1(\Pi)} \le \sqrt{2\varepsilon \| \hat{\lambda} - \lambda \|_{\ell_1}} + C \max_{1 \le k \le N} \|h_k\|_{\psi_1} \sqrt{\frac{A \log N}{n}} \| \hat{\lambda} - \lambda \|_{\ell_1}. \quad (7.18)$$

Combining this with (7.16) yields

$$\| f_{\hat{\lambda}} - f_{\lambda} \|_{L_1(\Pi)} \le \sqrt{4\varepsilon \beta(J_{\lambda}) \| f_{\hat{\lambda}} - f_{\lambda} \|_{L_1(\Pi)}}$$

$$+ 2C \max_{1 \le k \le N} \|h_k\|_{\psi_1} \sqrt{\frac{A \log N}{n}} \beta(J_{\lambda}) \| f_{\hat{\lambda}} - f_{\lambda} \|_{L_1(\Pi)}.$$

By the definition of $\Lambda_S$,

$$2C \max_{1 \le k \le N} \|h_k\|_{\psi_1} \sqrt{\frac{A \log N}{n}} \beta(J_{\lambda}) \le 1/2,$$

so, we end up with

$$\| f_{\hat{\lambda}} - f_{\lambda} \|_{L_1(\Pi)} \le 2 \sqrt{4\varepsilon \beta(J_{\lambda}) \| f_{\hat{\lambda}} - f_{\lambda} \|_{L_1(\Pi)}},$$

which implies the first bound of the theorem. The second bound holds because of (7.16).

Observe that for all $\lambda \in \Lambda$,

$$\left| n^{-1} \sum_{j=1}^{n} (f_\lambda(X_j) - Y_j) h_k(X_j) \right| \leq \left| \langle f_\lambda - f_*, h_k \rangle_{L_2(\Pi)} \right|$$

$$+ \left| n^{-1} \sum_{j=1}^{n} \left[ (f_\lambda(X_j) - f_*(X_j)) h_k(X_j) - \mathbb{E}(f_\lambda(X) - f_*(X)) h_k(X) \right] \right|$$

$$+ \left| n^{-1} \sum_{j=1}^{n} \xi_j h_k(X_j) \right|.$$

Lemma 7.6 can be used to bound the second and the third terms: with probability at least $1 - 2N^{-A}$

$$\max_{1 \leq k \leq N} \left| n^{-1} \sum_{j=1}^{n} (f_\lambda(X_j) - Y_j) h_k(X_j) \right| \leq \max_{1 \leq k \leq N} \left[ \left| \langle f_\lambda - f_*, h_k \rangle_{L_2(\Pi)} \right| \right.$$

$$\left. + C \left( \|(f_\lambda - f_*)(X) h_k(X)\|_{\psi_1} + \|\xi h_k(X)\|_{\psi_1} \right) \sqrt{\frac{A \log N}{n}} \right] \leq \varepsilon.$$

This proves that for all $\lambda \in \Lambda$, with probability at least $1 - 2N^{-A}$, we also have $\lambda \in \hat{\Lambda}$.

For each of the remaining two bounds, let $\bar{\lambda}$ be the vector for which the infimum in the right hand side of the bound is attained. With probability at least $1 - 2N^{-A}$, $\bar{\lambda} \in \hat{\Lambda}_\varepsilon \cap \Lambda_S$. Hence, it is enough to use the first two bounds of the theorem and the triangle inequality to finish the proof.                                                         $\square$

We will give another result about the Dantzig selector in which the properties of the dictionary are characterized by the quantity $\beta_2$ instead of $\beta$. Recall the definition of the set of "sparse" vectors $\Lambda_{S,2}$ from the previous section and related notations $(\beta_2(d), B(d), \text{etc})$ and define

$$\tilde{\Lambda}^2 = \tilde{\Lambda}_\varepsilon^2(A) := \Lambda_\varepsilon(A) \cap \Lambda_{S,2}.$$

**Theorem 7.5.** *Suppose condition (7.9) holds. There exists a constant $C$ in the definitions of $\Lambda_\varepsilon(A), \Lambda_{S,2}$ such that, for $A \geq 1$ with probability at least*

$$1 - 5^{-\bar{d}A} \left( \frac{N}{\leq \bar{d}} \right)^{-A},$$

*the following bounds hold for all $\lambda \in \hat{\Lambda}_\varepsilon \cap \Lambda_{S,2}$ :*

$$\|f_{\hat{\lambda}} - f_\lambda\|_{L_2(\Pi)} \leq 16 B^2(d(\lambda)) \beta_2(d(\lambda)) \sqrt{d(\lambda)} \varepsilon$$

*and*

$$\|\hat{\lambda} - \lambda\|_{\ell_2} \leq 32 B^2(d(\lambda))\beta_2^2(d(\lambda)) \sqrt{d(\lambda)}\varepsilon.$$

*Also, with probability at least* $1 - N^{-A}$,

$$\|f_{\hat{\lambda}} - f_*\|_{L_2(\Pi)} \leq \inf_{\lambda \in \tilde{\Lambda}_\varepsilon^2(A)} \left[ \|f_\lambda - f_*\|_{L_2(\Pi)} + 16 B^2(d(\lambda))\beta_2(d(\lambda)) \sqrt{d(\lambda)}\varepsilon \right].$$

*If* $f_* = f_{\lambda^*}, \lambda^* \in \mathbb{R}^N$, *then*

$$\|\hat{\lambda} - \lambda^*\|_{\ell_2} \leq \inf_{\lambda \in \tilde{\Lambda}_\varepsilon^2(A)} \left[ \|\lambda - \lambda^*\|_{\ell_2} + 32 B^2(d(\lambda))\beta_2^2(d(\lambda)) \sqrt{d(\lambda)}\varepsilon \right].$$

*Proof.* We follow the proof of Theorem 7.4. For $\lambda \in \hat{\Lambda}_\varepsilon \cap \Lambda_{S,2}$, we use the following bound instead of (7.17):

$$\|f_{\hat{\lambda}} - f_\lambda\|_{L_1(\Pi)} = \|f_{\hat{\lambda}} - f_\lambda\|_{L_1(\Pi_n)} + (\Pi - \Pi_n)(|f_{\hat{\lambda}} - f_\lambda|)$$

$$\leq \|f_{\hat{\lambda}} - f_\lambda\|_{L_1(\Pi_n)} + \sup_{\|u\|_{\ell_2} \leq 1, u \in C_{J_\lambda}} \left| (\Pi_n - \Pi)(|f_u|) \right| \|\hat{\lambda} - \lambda\|_{\ell_2}. \quad (7.19)$$

Again, we have $\hat{\lambda} - \lambda \in C_{J_\lambda}$, and, using Lemma 7.1, we get for $u = \hat{\lambda} - \lambda$ and $J = J_\lambda$:

$$\|\hat{\lambda} - \lambda\|_{\ell_2} \leq 2 \left( \sum_{j \in J_0 \cup J_1} |\hat{\lambda}_j - \lambda_j|^2 \right)^{1/2} \leq 2\beta_2(d(\lambda)) \|f_{\hat{\lambda}} - f_\lambda\|_{L_2(\Pi)}. \quad (7.20)$$

Lemma 7.5 now yields

$$\sup_{\|u\|_{\ell_2} \leq 1, u \in C_{J_\lambda}} \left| (\Pi_n - \Pi)(|f_u|) \right|$$

$$\leq C \sup_{\|u\|_{\ell_2} \leq 1, d(u) \leq \bar{d}} \|f_u\|_{\psi_1} \sqrt{\frac{A\bar{d} \log(N/\bar{d})}{n}}, \quad (7.21)$$

which holds with probability at least

$$1 - 5^{-\bar{d}A} \binom{N}{\leq \bar{d}}^{-A}.$$

As in the proof of Theorem 7.4, we bound the first term in the right hand side of (7.19):

$$\|f_{\hat{\lambda}} - f_\lambda\|_{L_1(\Pi_n)} \leq \sqrt{2\varepsilon \|\hat{\lambda} - \lambda\|_{\ell_1}}. \quad (7.22)$$

In addition,

$$\|\hat{\lambda} - \lambda\|_{\ell_1} \le 2 \sum_{j \in J} |\hat{\lambda}_j - \lambda_j| \le 2\sqrt{d(\lambda)} \left( \sum_{j \in J \cup J_1} |\hat{\lambda}_j - \lambda_j|^2 \right)^{1/2}$$

$$\le 2\beta_2(d(\lambda))\sqrt{d(\lambda)}\|f_{\hat{\lambda}} - f_\lambda\|_{L_2(\Pi)}. \tag{7.23}$$

Substitute bounds (7.20)–(7.22) and (7.23) into (7.19), use (7.9) and solve the resulting inequality with respect to $\|f_{\hat{\lambda}} - f_\lambda\|_{L_2(\Pi)}$. This gives the first bound of the theorem.

The second bound follows from (7.20) and the remaining two bounds are proved exactly as in Theorem 7.4. □

In the fixed design case, the following result holds. Its proof is a simplified version of the proofs of Theorems 7.4, 7.5.

**Theorem 7.6.** *Suppose $X_1, \ldots, X_n$ are nonrandom design points in $S$ and let $\Pi_n$ be the empirical measure based on $X_1, \ldots, X_n$. Suppose also $f_* = f_{\lambda^*}, \lambda^* \in \mathbb{R}^N$. There exists a constant $C > 0$ such that for all $A \ge 1$ and for all*

$$\varepsilon \ge C\|\xi\|_{\psi_2} \max_{1 \le k \le N} \|h_k\|_{L_2(\Pi_n)} \sqrt{\frac{A \log N}{n}},$$

*with probability at least $1 - N^{-A}$ the following bounds hold:*

$$\tag{7.24}$$

$$\|f_{\hat{\lambda}} - f_{\lambda^*}\|_{L_2(\Pi_n)} \le 4\beta_2(J_{\lambda^*}, \Pi_n)\sqrt{d(\lambda^*)}\varepsilon,$$

$$\|\hat{\lambda} - \lambda^*\|_{\ell_1} \le 8\beta_2^2(J_{\lambda^*}, \Pi_n)d(\lambda^*)\varepsilon$$

*and*

$$\|\hat{\lambda} - \lambda^*\|_{\ell_2} \le 8\beta_2^2(d(\lambda^*), \Pi_n)\sqrt{d(\lambda^*)}\varepsilon.$$

*Proof.* As in the proof of Theorem 7.4,

$$\|f_{\hat{\lambda}} - f_{\lambda^*}\|_{L_2(\Pi_n)} \le \sqrt{2\varepsilon\|\hat{\lambda} - \lambda^*\|_{\ell_1}} \tag{7.25}$$

and

$$\|\hat{\lambda} - \lambda^*\|_{\ell_1} \le 2\beta_2(J_{\lambda^*}, \Pi_n)\sqrt{d(\lambda^*)}\|f_{\hat{\lambda}} - f_{\lambda^*}\|_{L_2(\Pi_n)}. \tag{7.26}$$

These bounds hold provided that $\lambda^* \in \hat{\Lambda}_\varepsilon$, or

$$\max_{1 \le k \le N} \left| n^{-1} \sum_{j=1}^n \xi_j h_k(X_j) \right| \le \varepsilon.$$

If $\|\xi\|_{\psi_2} < +\infty$ and

$$\varepsilon \geq C \|\xi\|_{\psi_2} \max_{1 \leq k \leq N} \|h_k\|_{L_2(\Pi_n)} \sqrt{\frac{A \log N}{n}},$$

then usual bounds for random variables in Orlicz spaces imply that $\lambda^* \in \hat{\Lambda}_\varepsilon$ with probability at least $1 - N^{-A}$.

Combining (7.25) and (7.26) shows that with probability at least $1 - N^{-A}$

$$\|f_{\hat{\lambda}} - f_{\lambda^*}\|_{L_2(\Pi_n)} \leq 4\beta_2(J_{\lambda^*}, \Pi_n)\sqrt{d(\lambda^*)}\varepsilon$$

and

$$\|\hat{\lambda} - \lambda^*\|_{\ell_1} \leq 8\beta_2^2(J_{\lambda^*}, \Pi_n)d(\lambda^*)\varepsilon.$$

Using Lemma 7.1 and arguing as in the proof of Theorem 7.5, we also get

$$\|\hat{\lambda} - \lambda^*\|_{\ell_2} \leq 8\beta_2^2(d(\lambda^*), \Pi_n)\sqrt{d(\lambda^*)}\varepsilon.$$

$\square$

Bounding $\beta_2(J, \Pi_n)$ in terms of restricted isometry constants (see Lemma 7.2), essentially, allows one to recover Theorem 1 of Candes and Tao [44] that was the first result about the Dantzig selector in the fixed design case. Instead of doing this, we turn again to the case of random design regression and conclude this section with the derivation of the results of Candes and Tao [44] in the random design case.

To simplify the matter, assume that the following conditions hold:

- The dictionary $\{h_1, \ldots, h_N\}$ is $L_2(\Pi)$-orthonormal and, for some numerical constant $B > 0$,

$$\frac{1}{B}\|\lambda\|_{\ell_2} \leq \left\|\sum_{j=1}^N \lambda_j h_j\right\|_{L_1(\Pi)} \leq B\|\lambda\|_{\ell_2}$$

and

$$\frac{1}{B}\|\lambda\|_{\ell_2} \leq \left\|\sum_{j=1}^N \lambda_j h_j\right\|_{L_{\psi_2}(\Pi)} \leq B\|\lambda\|_{\ell_2}, \quad \lambda \in \mathbb{R}^N.$$

This is the case, for instance, for Gaussian and Rademacher dictionaries.
- The noise $\{\xi_j\}$ is a sequence of i.i.d. normal random variables with mean 0 and variance $\sigma^2$.
- Finally, $f_* = f_{\lambda^*}$, $\lambda^* \in \mathbb{R}^N$.

The following corollary can be derived from the last bound of Theorem 7.5.

**Corollary 7.1.** *There exist constants* $C, D$ *with the following property. Let* $A \geq 1$ *and suppose that*

$$D\sqrt{\frac{Ad(\lambda^*) \log N}{n}} \leq 1.$$

*Then, for all* $\varepsilon$ *satisfying the condition*

$$\varepsilon \geq D\sigma \sqrt{\frac{A \log N}{n}},$$

*the following bound holds with probability at least* $1 - N^{-A}$:

$$\|\hat{\lambda} - \lambda^*\|_{\ell_2}^2 \leq C \sum_{j=1}^{N}(|\lambda_j^*|^2 \wedge \varepsilon^2) = C \inf_{J \subset \{1,\dots,N\}} \left[ \sum_{j \notin J} |\lambda_j^*|^2 + d(J)\varepsilon^2 \right]. \quad (7.27)$$

*In particular, this implies that*

$$\|\hat{\lambda} - \lambda^*\|_{\ell_2}^2 \leq Cd(\lambda^*)\varepsilon^2.$$

The proof of (7.27) is based on applying the last bound of Theorem 7.5 to the oracle $\lambda = \bar{\lambda}^*$ defined as follows:

$$\bar{\lambda}_j^* = \lambda_j^* I(|\lambda_j^*| \geq \varepsilon/3), \ j = 1, \dots, N.$$

## 7.5 Further Comments

Theoretical study of sparse recovery methods based on the $\ell_1$-norm minimization started with the work of Donoho [52–55] who understood the connections of these problems with convex geometry in high dimensional spaces (other important references include [42, 43, 56]). Rudelson and Vershynin [129] followed by Mendelson et al. [117] used ideas and methods of high dimensional probability and asymptotic geometric analysis (concentration of measure, generic chaining) in further development of the theory of sparse recovery.

Geometric properties of the dictionaries discussed in Sect. 7.2 have been used in many recent papers on sparse recovery as well as in other areas of analysis and probability.

The Dantzig selector was introduced by Candes and Tao [44] who proved sparsity oracle inequalities for this estimator. In the same paper, they also introduced the restricted isometry constants that have been frequently used to quantify the properties of the dictionary needed for sparse recovery.

Here we followed the approach to oracle inequalities for the Dantzig selector as well as to the analysis of noiseless sparse recovery problems developed in [85].

# Chapter 8
# Convex Penalization in Sparse Recovery

We will discuss the role of penalized empirical risk minimization with convex penalties in sparse recovery problems. This includes the $\ell_1$-norm (LASSO) penalty as well as strictly convex and smooth penalties, such as the negative entropy penalty for sparse recovery in convex hulls. The goal is to show that, when the target function can be well approximated by a "sparse" linear combination of functions from a given dictionary, then solutions of penalized empirical risk minimization problems with $\ell_1$ and some other convex penalties are "approximately sparse" and they approximate the target function with an error that depends on the "sparsity". As a result of this analysis, we derive sparsity oracle inequalities showing the dependence of the excess risk of the empirical solution on the underlying sparsity of the problem. These inequalities also involve various distribution dependent geometric characteristics of the dictionary (such as restricted isometry constants and alignment coefficients) and the error of sparse recovery crucially depends on the geometry of the dictionary.

## 8.1 General Aspects of Convex Penalization

In this chapter we study an approach to sparse recovery based on penalized empirical risk minimization of the following form:

$$\hat{\lambda}^\varepsilon := \operatorname{argmin}_{\lambda \in D}\left[ P_n(\ell \bullet f_\lambda) + \varepsilon \sum_{j=1}^N \psi(\lambda_j) \right]. \tag{8.1}$$

We use the notations of Chap. 1, in particular, we denote

$$f_\lambda := \sum_{j=1}^N \lambda_j h_j, \ \lambda \in \mathbb{R}^N,$$

V. Koltchinskii, *Oracle Inequalities in Empirical Risk Minimization and Sparse Recovery Problems*, Lecture Notes in Mathematics 2033, DOI 10.1007/978-3-642-22147-7_8,
© Springer-Verlag Berlin Heidelberg 2011

where $\mathcal{H} := \{h_1, \ldots, h_N\}$ is a given finite dictionary of measurable functions from $S$ into $[-1, 1]$. The cardinality of the dictionary is usually very large (often, larger than the sample size $n$). We will assume in what follows that $N \geq (\log n)^{\gamma}$ for some $\gamma > 0$ (this is needed only to avoid additional terms of the order $\frac{\log \log n}{n}$ in several inequalities).

We will also assume that $\psi$ is a convex even function and $\varepsilon \geq 0$ is a regularization parameter, and that $D \subset \mathbb{R}^N$ is a closed convex set.

The excess risk of $f$ is defined as

$$\mathcal{E}(f) := P(\ell \bullet f) - \inf_{g: S \mapsto \mathbb{R}} P(\ell \bullet g) = P(\ell \bullet f) - P(\ell \bullet f_*),$$

where the infimum is taken over all measurable functions and it is assumed, for simplicity, that it is attained at $f_* \in L_2(\Pi)$. Moreover, it will be assumed in what follows that $f_*$ is uniformly bounded by a constant $M$.

**Definition 8.1.** It will be said that $\ell : T \times \mathbb{R} \mapsto \mathbb{R}_+$ is a loss function of *quadratic type* iff the following assumptions are satisfied:

(i) For all $y \in T$, $\ell(y, \cdot)$ is convex.
(ii) For all $y \in T$, $\ell(y, \cdot)$ is twice differentiable, $\ell_u''$ is a uniformly bounded function in $T \times \mathbb{R}$ and

$$\sup_{y \in T} \ell(y; 0) < +\infty, \quad \sup_{y \in T} |\ell_u'(y; 0)| < +\infty.$$

(iii) Moreover, denote

$$\tau(R) := \frac{1}{2} \inf_{y \in T} \inf_{|u| \leq R} \ell_u''(y, u). \tag{8.2}$$

Then it is assumed that $\tau(R) > 0$, $R > 0$. Without loss of generality, it will be also assumed that $\tau(R) \leq 1$, $R > 0$ (otherwise, it can be replaced by a lower bound).

For losses of quadratic type, the following property is obvious:

$$\tau(\|f\|_{\infty} \vee M)\|f - f_*\|_{L_2(\Pi)}^2 \leq \mathcal{E}(f) \leq C\|f - f_*\|_{L_2(\Pi)}^2,$$

where $C := \frac{1}{2} \sup_{y \in T, u \in \mathbb{R}} \ell_u''(y, u)$.

There are many important examples of loss functions of quadratic type, most notably, the quadratic loss $\ell(y, u) := (y - u)^2$ in the case when $T \subset \mathbb{R}$ is a bounded set. In this case, we can choose $\tau = 1$. In regression problems with a bounded response variable, one can also consider more general loss functions of the form $\ell(y, u) := \phi(y - u)$, where $\phi$ is an even nonnegative convex twice continuously differentiable function with $\phi''$ uniformly bounded in $\mathbb{R}$, $\phi(0) = 0$ and $\phi''(u) > 0$, $u \in \mathbb{R}$. In binary classification setting (that is, when $T = \{-1, 1\}$), one can choose the loss $\ell(y, u) = \phi(yu)$ with $\phi$ being a nonnegative decreasing convex twice continuously differentiable function such that $\phi''$ is uniformly bounded in $\mathbb{R}$ and

$\phi''(u) > 0$, $u \in \mathbb{R}$. The loss function $\phi(u) = \log_2(1 + e^{-u})$ (often called the logit loss) is a typical example.

The condition that the second derivative $\ell''_u$ is uniformly bounded in $T \times \mathbb{R}$ can be often replaced by its uniform boundedness in $T \times [-a, a]$, where $[-a, a]$ is a suitable interval. This allows one to cover several other choices of the loss function, such as the exponential loss $\ell(y, u) := e^{-yu}$ in binary classification.

Clearly, the conditions that the loss $\ell$, the penalty function $\psi$ and the domain $D$ are convex make the optimization problem (8.1) convex and, at least in principle, computationally tractable; numerous methods of convex optimization can be used to solve it (see, e.g., Ben-Tal and Nemirovski [20]).

In the recent literature, there has been considerable attention to the problem of sparse recovery using LASSO type penalties, which is a special case of problem (8.1). In this case, $D = \mathbb{R}^N$, so this is a problem of sparse recovery in the linear span l.s.($\mathcal{H}$) of the dictionary, and $\psi(u) = |u|$, which means penalization with $\ell_1$-norm. It is also usually assumed that $\ell(y, u) = (y - u)^2$ (the case of regression with quadratic loss). In this setting, it has been shown that sparse recovery is possible under some geometric assumptions on the dictionary. They are often expressed in terms of the Gram matrix of the dictionary, which in the case of random design models is the matrix

$$H := \left( \langle h_i, h_j \rangle_{L_2(\Pi)} \right)_{i,j=1,N}.$$

They take form of various conditions on the entries of this matrix ("coherence coefficients"), or on its submatrices (in spirit of "uniform uncertainty principle" or "restricted isometry" conditions, see Sect. 7.2). The essence of these assumptions is to try to keep the dictionary not too far from being orthonormal in $L_2(\Pi)$ which, in some sense, is an ideal case for sparse recovery (see, e.g., Donoho [52–55], Candes and Tao [44], Rudelson and Vershynin [129], Mendelson et al. [117], Bunea et al. [36], van de Geer [63], Koltchinskii [82,84,85], Bickel et al. [22] among many other papers that study both the random design and the fixed design problems).

We will study several special cases of problem (8.1). LASSO or $\ell_1$-penalty is the most common choice when $D = \mathbb{R}^N$, but it can be used in some other cases, too, for instance, when $D = U_{\ell_1}$ (the unit ball of $\ell_1$). This leads to a problem of sparse recovery in the symmetric convex hull

$$\text{conv}_s(\mathcal{H}) := \left\{ f_\lambda : \lambda \in U_{\ell_1} \right\},$$

which can be viewed as a version of convex aggregation problem. Note that empirical risk minimization with no penalty does not allow one to achieve sparse recovery or even error rate faster than $n^{-1/2}$ in this case (see Lecue and Mendelson [99] for a counterexample). More generally, one can consider the case of $D = U_{\ell_p}$, the unit ball in the space $\ell_p$, with $p \geq 1$ and with $\psi(u) = |u|^p$ (that is, the penalty becomes $\|\lambda\|^p_{\ell_p}$); the same penalty can be also used when $D = \mathbb{R}^N$. It was shown by Koltchinskii [84] that sparse recovery is still possible if $p$ is close enough to 1 (say, of the order $1 + 1/\log N$). Another interesting example is

$$D = \Lambda := \left\{ \lambda \in \mathbb{R}^N : \lambda_j \geq 0, \sum_{j=1}^{N} \lambda_j = 1 \right\},$$

that is, $D$ is the simplex of all probability distributions in $\{1, \ldots, N\}$. This corresponds to the sparse recovery problem in the convex hull of the dictionary

$$\mathrm{conv}(\mathscr{H}) := \left\{ f_\lambda : \lambda \in \Lambda \right\}.$$

A possible choice of penalty in this case is

$$-H(\lambda) = \sum_{j=1}^{N} \lambda_j \log \lambda_j,$$

where $H(\lambda)$ is the entropy of probability distribution $\lambda$; this corresponds to the choice $\psi(u) = u \log u$. Such a problem was studied in Koltchinskii [86] and it will be also discussed below. We will also show in Sect. 9.4 that sparse recovery in convex hulls can be achieved by empirical risk minimization with no penalty (which is not possible in the case of symmetric convex hulls).

We will follow the approach of [84, 86]. This approach is based on the analysis of necessary conditions of extremum in problem (8.1). For simplicity, consider the case of $D = \mathbb{R}^N$. In this case, for $\hat{\lambda}^\varepsilon$ to be a solution of (8.1), it is necessary that $0 \in \partial L_{n,\varepsilon}(\hat{\lambda}^\varepsilon)$, where

$$L_{n,\varepsilon}(\lambda) := P_n(\ell \bullet f_\lambda) + \varepsilon \sum_{j=1}^{N} \psi(\lambda_j)$$

and $\partial$ denotes the subdifferential of convex functions. If $\psi$ is smooth, this leads to the equations

$$P_n(\ell' \bullet f_{\hat{\lambda}^\varepsilon}) h_j + \varepsilon \psi'(\hat{\lambda}_j^\varepsilon) = 0, \ j = 1, \ldots, N. \tag{8.3}$$

Define

$$L_\varepsilon(\lambda) := P(\ell \bullet f_\lambda) + \varepsilon \sum_{j=1}^{N} \psi(\lambda_j)$$

and

$$\nabla L_\varepsilon(\lambda) := \left( P(\ell' \bullet f_\lambda) h_j + \varepsilon \psi'(\lambda_j) \right)_{j=1,\ldots,N}.$$

The vector $\nabla L_\varepsilon(\lambda)$ is the gradient and the subgradient of the convex function $L_\varepsilon(\lambda)$ at point $\lambda$. It follows from (8.3) that

$$P_n(\ell' \bullet f_{\hat{\lambda}^\varepsilon})(f_{\hat{\lambda}^\varepsilon} - f_\lambda) + \varepsilon \sum_{j=1}^{N} \psi'(\hat{\lambda}_j^\varepsilon)(\hat{\lambda}_j^\varepsilon - \lambda_j) = 0$$

and we also have

$$P(\ell' \bullet f_\lambda)(f_{\hat{\lambda}^\varepsilon} - f_\lambda) + \varepsilon \sum_{j=1}^{N} \psi'(\lambda_j)(\hat{\lambda}_j^\varepsilon - \lambda_j) = \left\langle \nabla L_\varepsilon(\lambda), \hat{\lambda}^\varepsilon - \lambda \right\rangle_{\ell_2}.$$

Subtracting the second equation from the first one yields the relationship

$$P(\ell' \bullet f_{\hat{\lambda}^\varepsilon} - \ell' \bullet f_\lambda)(f_{\hat{\lambda}^\varepsilon} - f_\lambda) + \varepsilon \sum_{j=1}^{N} (\psi'(\hat{\lambda}_j^\varepsilon) - \psi'(\lambda_j))(\hat{\lambda}_j^\varepsilon - \lambda_j)$$

$$= \left\langle \nabla L_\varepsilon(\lambda), \lambda - \hat{\lambda}^\varepsilon \right\rangle_{\ell_2} + (P - P_n)(\ell' \bullet f_{\hat{\lambda}^\varepsilon})(f_{\hat{\lambda}^\varepsilon} - f_\lambda).$$

If $\ell$ is a loss of quadratic type and, in addition, $\tau(+\infty) > 0$, then

$$P(\ell' \bullet f_{\hat{\lambda}^\varepsilon} - \ell' \bullet f_\lambda)(f_{\hat{\lambda}^\varepsilon} - f_\lambda) \geq c \|f_{\hat{\lambda}^\varepsilon} - f_\lambda\|_{L_2(\Pi)}^2$$

with some constant $c > 0$ depending only on $\ell$ and the following inequality holds

$$c \|f_{\hat{\lambda}^\varepsilon} - f_\lambda\|_{L_2(\Pi)}^2 + \varepsilon \sum_{j=1}^{N} (\psi'(\hat{\lambda}_j^\varepsilon) - \psi'(\lambda_j))(\hat{\lambda}_j^\varepsilon - \lambda_j)$$

$$\leq \left\langle \nabla L_\varepsilon(\lambda), \lambda - \hat{\lambda}^\varepsilon \right\rangle_{\ell_2} + (P - P_n)(\ell' \bullet f_{\hat{\lambda}^\varepsilon})(f_{\hat{\lambda}^\varepsilon} - f_\lambda). \tag{8.4}$$

Inequality (8.4) provides some information about "sparsity" of $\hat{\lambda}^\varepsilon$ in terms of "sparsity" of the oracle $\lambda$ and it also provides tight bounds on $\|f_{\hat{\lambda}^\varepsilon} - f_\lambda\|_{L_2(\Pi)}$. Indeed, if $J = J_\lambda = \mathrm{supp}(\lambda)$ and $\psi'(0) = 0$ (which is the case, for instance, when $\psi(u) = u^p$ for some $p > 1$), then

$$\sum_{j=1}^{N} (\psi'(\hat{\lambda}_j^\varepsilon) - \psi'(\lambda_j))(\hat{\lambda}_j^\varepsilon - \lambda_j) \geq \sum_{j \notin J} \psi'(\hat{\lambda}_j^\varepsilon)\hat{\lambda}_j^\varepsilon = \sum_{j \notin J} |\psi'(\hat{\lambda}_j^\varepsilon)||\hat{\lambda}_j^\varepsilon|$$

(note that all the terms in the sum in the left hand side are nonnegative since $\psi$ is convex and $\psi'$ is nondecreasing). Thus, the following bound holds

$$c\|f_{\hat{\lambda}^{\varepsilon}} - f_{\lambda}\|_{L_2(\Pi)}^2 + \varepsilon \sum_{j \notin J} |\psi'(\hat{\lambda}_j^{\varepsilon})||\hat{\lambda}_j^{\varepsilon}|$$

$$\leq \left\langle \nabla L_{\varepsilon}(\lambda), \lambda - \hat{\lambda}^{\varepsilon}\right\rangle_{\ell_2} + (P - P_n)(\ell' \bullet f_{\hat{\lambda}^{\varepsilon}})(f_{\hat{\lambda}^{\varepsilon}} - f_{\lambda}), \qquad (8.5)$$

in which the left hand side measures the $L_2$-distance of $f_{\hat{\lambda}^{\varepsilon}}$ from the oracle $f_{\lambda}$ as well as the degree of sparsity of the empirical solution $\hat{\lambda}^{\varepsilon}$. This inequality will be applied to sparse vectors $\lambda$ ("oracles") such that the term $\left\langle \nabla L_{\varepsilon}(\lambda), \lambda - \hat{\lambda}^{\varepsilon}\right\rangle_{\ell_2}$ is either negative, or, if positive, then small enough. This is the case, for instance, when the subgradient $\nabla L_{\varepsilon}(\lambda)$ is small in certain sense. In such cases, the left hand side is controlled by the empirical process

$$(P - P_n)(\ell' \bullet f_{\hat{\lambda}^{\varepsilon}})(f_{\hat{\lambda}^{\varepsilon}} - f_{\lambda}).$$

It happens that its size, in turn, depends on the $L_2$-distance $\|f_{\hat{\lambda}^{\varepsilon}} - f_{\lambda}\|_{L_2(\Pi)}$ and on the measure of "sparsity" of $\hat{\lambda}^{\varepsilon}$, $\sum_{j \notin J} |\psi'(\hat{\lambda}_j^{\varepsilon})||\hat{\lambda}_j^{\varepsilon}|$, which are precisely the quantities involved in the left hand side of bound (8.5). Writing these bounds precisely yields an inequality on these two quantities which can be solved to derive the explicit bounds. In the case of strictly convex smooth penalty function $\psi$ (such as $\psi(u) = |u|^p$, $p > 1$ or $\psi(u) = u \log u$), the same approach can be used also in the case of "approximately sparse" oracles $\lambda$ (since the function $\psi'$ is strictly increasing and smooth). A natural choice of oracle is

$$\lambda^{\varepsilon} := \operatorname{argmin}_{\lambda \in D}\left[ P(\ell \bullet f_{\lambda}) + \varepsilon \sum_{j=1}^N \psi(\lambda_j)\right], \qquad (8.6)$$

for which in the smooth case $\left\langle \nabla L_{\varepsilon}(\lambda^{\varepsilon}), \lambda^{\varepsilon} - \hat{\lambda}^{\varepsilon}\right\rangle_{\ell_2} \leq 0$ (if $D = \mathbb{R}^N$, we even have $\nabla L_{\varepsilon}(\lambda^{\varepsilon}) = 0$). For this oracle, the bounds on $\|f_{\hat{\lambda}^{\varepsilon}} - f_{\lambda^{\varepsilon}}\|_{L_2(\Pi)}$ and on the degree of sparsity of $\hat{\lambda}^{\varepsilon}$ do not depend on the properties of the dictionary, but only on "approximate sparsity" of $\lambda^{\varepsilon}$. As a consequence, it is also possible to bound the "random error" $|\mathscr{E}(f_{\hat{\lambda}^{\varepsilon}}) - \mathscr{E}(f_{\lambda^{\varepsilon}})|$ in terms of "approximate sparsity" of $\lambda^{\varepsilon}$. It happens that bounding the "approximation error" $\mathscr{E}(f_{\lambda^{\varepsilon}})$ is a different problem with not entirely the same geometric parameters responsible for the size of the error. The approximation error is much more sensitive to the properties of the dictionary, in particular, of its Gram matrix $H$ that depends on the unknown design distribution $\Pi$.

The case of $\ell_1$-penalty is more complicated since the penalty is neither strictly convex, nor smooth. In this case there is no special advantage in using $\lambda^{\varepsilon}$ as an oracle since this vector is not necessarily sparse. It is rather approximately sparse, but bound (8.4) does not provide a way to control the random $L_2$-error $\|f_{\hat{\lambda}^{\varepsilon}} - f_{\lambda^{\varepsilon}}\|_{L_2(\Pi)}$ in terms of approximate sparsity of the oracle (note that in this case $\psi'(\lambda) = \operatorname{sign}(\lambda)$). A possible way to tackle the problem is to study a set of oracles $\lambda$ for which $\left\langle \nabla L_\varepsilon(\lambda), \lambda - \hat{\lambda}^\varepsilon \right\rangle_{\ell_2}$ is negative, or, if positive, then small enough. This can be expressed in terms of certain quantities that describe a way in which the subgradient $\nabla L_\varepsilon(\lambda)$ is *aligned* with the dictionary. Such quantities also emerge rather naturally in attempts to control the approximation error $\mathscr{E}(f_{\lambda^\varepsilon})$ in the case of smooth strictly convex penalties.

In this chapter, we concentrate on the case when the domain $D$ is bounded. In [84], for the $\ell_p$-penalization with $p$ close to 1, upper and lower bounds on $\|\hat{\lambda}^\varepsilon\|_{\ell_1}$ in terms of $\|\lambda^{c\varepsilon}\|_{\ell_1}$ for proper values of $c$ have been proved (when the domain $D$ is not necessarily bounded). Such bounds can be used to extend oracle inequalities of the following sections to the case of unbounded domain. We do not pursue this approach here, but in Chap. 9, we will obtain several results for sparse recovery in unbounded domains as corollaries of more general statement concerning low rank matrix recovery. This will be done when $\ell$ is the quadratic loss.

## 8.2   $\ell_1$-Penalization and Oracle Inequalities

The following penalized empirical risk minimization problem will be studied:

$$\hat{\lambda}^\varepsilon := \mathrm{argmin}_{\lambda \in U_{\ell_1}} \left[ P_n(\ell \bullet f_\lambda) + \varepsilon\|\lambda\|_{\ell_1} \right], \tag{8.7}$$

where $\varepsilon \geq 0$ is a regularization parameter. As always, we denote $\lambda^\varepsilon$ a solution of the "true" version of the problem:

$$\lambda^\varepsilon := \mathrm{argmin}_{\lambda \in U_{\ell_1}} \left[ P(\ell \bullet f_\lambda) + \varepsilon\|\lambda\|_{\ell_1} \right].$$

Let

$$L_\varepsilon(\lambda) := P(\ell \bullet f_\lambda) + \varepsilon\|\lambda\|_{\ell_1}.$$

For $\lambda \in \mathbb{R}^N$, let $\nabla L_\varepsilon(\lambda) \in \partial L_\varepsilon(\lambda)$ be the vector with components

$$P(\ell' \bullet f_\lambda)h_j + \varepsilon s_j(\lambda), \ j = 1, \ldots, N$$

where $s_j = s_j(\lambda) = \mathrm{sign}(\lambda_j)$ (assume that $\mathrm{sign}(0) = 0$). The vector $\nabla L_\varepsilon(\lambda)$ is a subgradient of the function $L_\varepsilon$ at point $\lambda$. Note that $\partial|u| = \{+1\}$ for $u > 0$, $\partial|u| = \{-1\}$ for $u < 0$ and $\partial|u| = [-1, 1]$ for $u = 0$.

In the case of $\ell_1$-penalization, we are going to compare the empirical solution $\hat{\lambda}^\varepsilon$ with an oracle $\lambda \in U_{\ell_1}$ that will be characterized by its "sparsity" as well as by a measure of "alignment" of the subgradient $\nabla L_\varepsilon(\lambda) \in \partial L_\varepsilon(\lambda)$ with the dictionary.

We will use the following versions of the alignment coefficient for vectors $\nabla L_\varepsilon(\lambda)$ and $s(\lambda)$:

$$\alpha_+(\varepsilon, \lambda) := a_H^{(\infty)}\left(U_{\ell_1}, \lambda, \nabla L_\varepsilon(\lambda)\right) \vee 0$$

and

$$\alpha(\lambda) := a_H^{(2)}\left(U_{\ell_1}, \lambda, s(\lambda)\right) \vee 0, \quad \alpha_+(\lambda) := a_H^{(\infty)}\left(U_{\ell_1}, \lambda, s(\lambda)\right) \vee 0.$$

Clearly, $\alpha(\lambda) \leq \alpha_+(\lambda)$ and it is easy to check that

$$\alpha_+(\varepsilon, \lambda) \leq \|P_{\mathscr{L}}(\ell' \bullet f_\lambda)\|_{L_2(P)} + \varepsilon\alpha_+(\lambda),$$

where $\mathscr{L}$ denotes the linear span of the dictionary $\{h_1, \ldots, h_N\}$ in the space $L_2(P)$ (with a minor abuse of notation, we view functions $h_j$ defined on $S$ as functions on $S \times T$) and $P_{\mathscr{L}}$ denotes the orthogonal projection on $\mathscr{L} \subset L_2(P)$. In the case of quadratic type losses, the first term in the right hand side can be upper bounded as follows:

$$\|P_{\mathscr{L}}(\ell' \bullet f_\lambda)\|_{L_2(P)} = \|P_{\mathscr{L}}(\ell' \bullet f_\lambda - \ell' \bullet f_*)\|_{L_2(P)} \leq C\|f_\lambda - f_*\|_{L_2(\Pi)},$$

where $C$ depends only on $\ell$. Thus, the quantity $\|P_{\mathscr{L}}(\ell' \bullet f_\lambda)\|_{L_2(P)}$ is upper bounded by the $L_2$-error of approximation of the target function $f_*$ in the linear span of the dictionary. The second term $\alpha_+(\lambda)$ is based on the alignment coefficient of vector $s(\lambda)$ with the dictionary. It depends on the sparsity of oracle $\lambda$ as well as on the geometry of the dictionary.

**Theorem 8.1.** *There exist constants $D > 0$ and $C > 0$ depending only on $\ell$ such that, for all $\bar{\lambda} \in U_{\ell_1}$, for $J = \mathrm{supp}(\bar{\lambda})$ and $d := d(J) = \mathrm{card}(J)$, for all $A \geq 1$ and for all*

$$\varepsilon \geq D\sqrt{\frac{d + A \log N}{n}}, \tag{8.8}$$

*the following bound holds with probability at least $1 - N^{-A}$:*

$$\|f_{\hat{\lambda}^\varepsilon} - f_{\bar{\lambda}}\|_{L_2(\Pi)}^2 + \varepsilon \sum_{j \notin J} |\hat{\lambda}_j^\varepsilon| \leq C\left[\frac{d + A \log N}{n} \bigvee \alpha_+^2(\varepsilon, \bar{\lambda})\right].$$

*Moreover, with the same probability*

$$\|f_{\hat{\lambda}^\varepsilon} - f_{\bar{\lambda}}\|_{L_2(\Pi)}^2 + \varepsilon \sum_{j \notin J} |\hat{\lambda}_j^\varepsilon| \leq C\left[\frac{d + A \log N}{n} \bigvee \left\|P_{\mathscr{L}}(\ell' \bullet f_{\bar{\lambda}})\right\|_{L_2(P)}^2 \bigvee \alpha^2(\bar{\lambda})\varepsilon^2\right].$$

Note that, if we formally pass to the limit as $n \to \infty$ in the bounds of the theorem, we get the following bounds for the true solution $\lambda^\varepsilon$ that hold for all $\varepsilon > 0$:

$$\|f_{\lambda^\varepsilon} - f_{\bar{\lambda}}\|_{L_2(\Pi)}^2 + \varepsilon \sum_{j \notin J} |\lambda_j^\varepsilon| \le C\alpha_+^2(\varepsilon, \bar{\lambda})$$

and

$$\|f_{\lambda^\varepsilon} - f_{\bar{\lambda}}\|_{L_2(\Pi)}^2 + \varepsilon \sum_{j \notin J} |\lambda_j^\varepsilon| \le C\left[\left\|P_{\mathscr{L}}(\ell' \bullet f_{\bar{\lambda}})\right\|_{L_2(P)}^2 \bigvee \alpha^2(\bar{\lambda})\varepsilon^2\right].$$

These bounds can be proved directly by modifying and simplifying the proofs in the empirical case given below. They show that the true penalized solution $\lambda^\varepsilon$ provides an approximation of "sparse" oracle vectors $\bar{\lambda} \in U_{\ell_1}$ that are, in some sense, well aligned with the dictionary. In particular, the second bound shows that $f_{\lambda^\varepsilon}$ is close in the space $L_2(\Pi)$ to "sparse" oracles $f_{\bar{\lambda}}$ such that the vector $s(\bar{\lambda})$ is well aligned with the dictionary and $f_{\bar{\lambda}}$ is close to the target function $f_*$ in $L_2(\Pi)$. Moreover, $\lambda^\varepsilon$ is "approximately sparse" in the sense that it is supported in $\text{supp}(\bar{\lambda})$ up to a small $\ell_1$-error. The same properties hold for the empirical solution $\hat{\lambda}^\varepsilon$ with an additional error term $\frac{d+A\log N}{n}$ that depends only on the degree of sparsity of $\bar{\lambda}$, but not on the geometry of the dictionary. In some sense, the meaning of this result is that the empirical solution $\hat{\lambda}^\varepsilon$ provides "sparse recovery" if and only if the true solution $\lambda^\varepsilon$ does (regardless of the properties of the dictionary). This is even more apparent in the versions of these results for strictly convex penalties discussed in the next section.

No condition on the dictionary is needed for the bounds of the theorem to be true (except uniform boundedness of functions $h_j$). On the other hand, the assumption on $\varepsilon$, $\varepsilon \ge D\sqrt{\frac{d+A\log N}{n}}$, essentially, relates the regularization parameter to the unknown sparsity of the problem. To get around this difficulty, we will prove another version of the theorem in which it is only assumed that $\varepsilon \ge D\sqrt{\frac{A\log N}{n}}$, but, on the other hand, there is more dependence of the error bounds on the geometry of the dictionary. At the same time, the error in this result is controlled not by $d = \text{card}(J)$, but rather by the dimension of a linear space $L$ providing a reasonably good approximation of the functions $\{h_j : j \in J\}$ (such a dimension could be much smaller than $\text{card}(J)$). To formulate the result, some further notation will be needed.

Given a linear subspace $L \subset L_2(\Pi)$, denote

$$U(L) := \sup_{f \in L, \|f\|_{L_2(\Pi)}=1} \|f\|_\infty + 1.$$

If $I_L : (L, \|\cdot\|_{L_2(\Pi)}) \mapsto (L, \|\cdot\|_\infty)$ is the identity operator, then $U(L) - 1$ is the norm of the operator $I_L$. We will use this quantity only for finite dimensional subspaces. In such case, for any $L_2(\Pi)$-orthonormal basis $\phi_1, \ldots, \phi_d$ of $L$,

$$U(L) \le \max_{1 \le j \le d} \|\phi_j\|_\infty \sqrt{d} + 1,$$

where $d := \dim(L)$. In what follows, let $P_L$ be the orthogonal projector onto $L$ and $L^{\perp}$ be the orthogonal complement of $L$. We are interested in subspaces $L$ such that

- $\dim(L)$ and $U(L)$ are not very large.
- Functions $\{h_j : j \in J\}$ in the "relevant" part of the dictionary can be approximated well by the functions from $L$ so that the quantity $\max_{j \in J} \|P_{L^{\perp}} h_j\|_{L_2(\Pi)}$ is small.

**Theorem 8.2.** *Suppose that*

$$\varepsilon \geq D \sqrt{\frac{A \log N}{n}} \tag{8.9}$$

*with a large enough constant $D > 0$ depending only on $\ell$. For all $\bar{\lambda} \in U_{\ell_1}$, for $J = \operatorname{supp}(\bar{\lambda})$, for all subspaces $L$ of $L_2(\Pi)$ with $d := \dim(L)$ and for all $A \geq 1$, the following bound holds with probability at least $1 - N^{-A}$ and with a constant $C > 0$ depending only on $\ell$:*

$$\|f_{\hat{\lambda}^{\varepsilon}} - f_{\bar{\lambda}}\|^2_{L_2(\Pi)} + \varepsilon \sum_{j \notin J} |\hat{\lambda}^{\varepsilon}_j| \tag{8.10}$$

$$\leq C \left[ \frac{d + A \log N}{n} \bigvee \max_{j \in J} \|P_{L^{\perp}} h_j\|_{L_2(\Pi)} \right.$$

$$\left. \sqrt{\frac{A \log N}{n}} \bigvee \frac{U(L) \log N}{n} \bigvee \alpha^2_+(\varepsilon; \bar{\lambda}) \right].$$

*Moreover, with the same probability*

$$\|f_{\hat{\lambda}^{\varepsilon}} - f_{\bar{\lambda}}\|^2_{L_2(\Pi)} + \varepsilon \sum_{j \notin J} |\hat{\lambda}^{\varepsilon}_j| \tag{8.11}$$

$$\leq C \left[ \frac{d + A \log N}{n} \bigvee \max_{j \in J} \|P_{L^{\perp}} h_j\|_{L_2(\Pi)} \sqrt{\frac{A \log N}{n}} \bigvee \frac{U(L) \log N}{n} \bigvee \right.$$

$$\left. \left\| P_{\mathscr{L}}(\ell' \bullet f_{\bar{\lambda}}) \right\|^2_{L_2(P)} \bigvee \alpha^2(\bar{\lambda}) \varepsilon^2 \right].$$

The next two corollaries provide bounds on $\|\hat{\lambda}^{\varepsilon} - \bar{\lambda}\|_{\ell_1}$ in terms of the quantity $\beta_{2,2}(\bar{\lambda}, \Pi)$ (see Sect. 7.2.3); they follow in a straightforward way from the proofs of the theorems.

**Corollary 8.1.** *Under the assumptions and notations of Theorem 8.1, the following bound holds with probability at least $1 - N^{-A}$:*

$$\|f_{\hat{\lambda}^\varepsilon} - f_{\bar{\lambda}}\|_{L_2(\Pi)}^2 + \varepsilon\|\hat{\lambda}^\varepsilon - \bar{\lambda}\|_{\ell_1}$$

$$\leq C\left[\frac{d + A\log N}{n} \bigvee \left\|P_{\mathscr{L}}(\ell' \bullet f_{\bar{\lambda}})\right\|_{L_2(P)}^2 \bigvee \beta_{2,2}^2(\bar{\lambda}, \Pi)d\varepsilon^2\right].$$

**Corollary 8.2.** *Under the assumptions and notations of Theorem 8.2, the following bound holds with probability at least* $1 - N^{-A}$:

$$\|f_{\hat{\lambda}^\varepsilon} - f_{\bar{\lambda}}\|_{L_2(\Pi)}^2 + \varepsilon\|\hat{\lambda}^\varepsilon - \bar{\lambda}\|_{\ell_1} \tag{8.12}$$

$$\leq C\left[\frac{d + A\log N}{n} \bigvee \max_{j\in J}\|P_{L^\perp}h_j\|_{L_2(\Pi)}\sqrt{\frac{A\log N}{n}} \bigvee \frac{U(L)\log N}{n} \bigvee\right.$$

$$\left.\left\|P_{\mathscr{L}}(\ell' \bullet f_{\bar{\lambda}})\right\|_{L_2(P)}^2 \bigvee \beta_{2,2}^2(\bar{\lambda}, \Pi)d(J_{\bar{\lambda}})\varepsilon^2\right].$$

We now turn to the proof of Theorem 8.2.

*Proof.* Note that subgradients of convex function

$$\lambda \mapsto P_n(\ell \bullet f_\lambda) + \varepsilon\|\lambda\|_{\ell_1} =: L_{n,\varepsilon}(\lambda)$$

are the vectors in $\mathbb{R}^N$ with components

$$P_n(\ell' \bullet f_\lambda)h_j + \varepsilon\sigma_j, \quad j = 1, \ldots, N$$

where $\sigma_j \in [-1, 1]$, $\sigma_j = \mathrm{sign}(\lambda_j)$ if $\lambda_j \neq 0$. It follows from necessary conditions of extremum in problem (8.7) that there exist numbers $\hat{s}_j \in [-1, 1]$ such that $\hat{s}_j = \mathrm{sign}(\hat{\lambda}_j^\varepsilon)$ when $\hat{\lambda}_j^\varepsilon \neq 0$ and, for all $u \in T_{U_{\ell_1}}(\hat{\lambda}^\varepsilon)$,

$$\sum_{j=1}^N \left(P_n(\ell' \bullet f_{\hat{\lambda}^\varepsilon})h_j u_j + \varepsilon\hat{s}_j u_j\right) \geq 0. \tag{8.13}$$

Indeed, since $\hat{\lambda}^\varepsilon$ is a minimal point of $L_{n,\varepsilon}$ in $U_{\ell_1}$, there exists $w \in \partial L_{n,\varepsilon}(\hat{\lambda}^\varepsilon)$ such that $-w$ belongs to the normal cone $N_{U_{\ell_1}}(\hat{\lambda}^\varepsilon)$ of the convex set $U_{\ell_1}$ at point $\hat{\lambda}^\varepsilon$ (see Aubin and Ekeland [9], Chap. 4, Sect. 2, Corollary 6). This immediately implies (8.13). Since $\bar{\lambda} \in U_{\ell_1}$, we have $\bar{\lambda} - \hat{\lambda}^\varepsilon \in T_{U_{\ell_1}}(\hat{\lambda}^\varepsilon)$, and the next inequality follows from (8.13).

$$P_n(\ell' \bullet f_{\hat{\lambda}^\varepsilon})(f_{\hat{\lambda}^\varepsilon} - f_{\bar{\lambda}}) + \varepsilon\sum_{j=1}^N \hat{s}_j(\hat{\lambda}_j - \bar{\lambda}_j) \leq 0. \tag{8.14}$$

Recalling the definition $s_j = s_j(\bar{\lambda}) = \mathrm{sign}(\bar{\lambda}_j)$ and

$$\nabla L_\varepsilon(\bar{\lambda}) = \left(P(\ell' \bullet f_{\bar{\lambda}})h_j + \varepsilon s_j\right)_{j=1,\dots,N},$$

we also have

$$P(\ell' \bullet f_{\bar{\lambda}})(f_{\hat{\lambda}^\varepsilon} - f_{\bar{\lambda}}) + \varepsilon \sum_{j=1}^{N} s_j(\hat{\lambda}_j - \bar{\lambda}_j) = \left\langle \nabla L_\varepsilon(\bar{\lambda}), \hat{\lambda}^\varepsilon - \bar{\lambda} \right\rangle_{\ell_2}. \tag{8.15}$$

Subtracting (8.15) from (8.14) yields by a simple algebra

$$P_n(\ell' \bullet f_{\hat{\lambda}^\varepsilon} - \ell' \bullet f_{\bar{\lambda}})(f_{\hat{\lambda}^\varepsilon} - f_{\bar{\lambda}}) + \varepsilon \sum_{j=1}^{N}(\hat{s}_j - s_j)(\hat{\lambda}_j - \bar{\lambda}_j)$$

$$\leq \left\langle \nabla L_\varepsilon(\bar{\lambda}), \bar{\lambda} - \hat{\lambda}^\varepsilon \right\rangle_{\ell_2} + (P - P_n)(\ell' \bullet f_{\bar{\lambda}})(f_{\hat{\lambda}^\varepsilon} - f_{\bar{\lambda}}) \tag{8.16}$$

and

$$P(\ell' \bullet f_{\hat{\lambda}^\varepsilon} - \ell' \bullet f_{\bar{\lambda}})(f_{\hat{\lambda}^\varepsilon} - f_{\bar{\lambda}}) + \varepsilon \sum_{j=1}^{N}(\hat{s}_j - s_j)(\hat{\lambda}_j - \bar{\lambda}_j)$$

$$\leq \left\langle \nabla L_\varepsilon(\bar{\lambda}), \bar{\lambda} - \hat{\lambda}^\varepsilon \right\rangle_{\ell_2} + (P - P_n)(\ell' \bullet f_{\hat{\lambda}^\varepsilon})(f_{\hat{\lambda}^\varepsilon} - f_{\bar{\lambda}}). \tag{8.17}$$

We use inequalities (8.16) and (8.17) to control the "approximate sparsity" of empirical solution $\hat{\lambda}^\varepsilon$ in terms of "sparsity" of the "oracle" $\bar{\lambda}$ and to obtain bounds on $\|f_{\hat{\lambda}^\varepsilon} - f_{\bar{\lambda}}\|_{L_2(\Pi)}$. As always, we use notations $J := J_{\bar{\lambda}} := \mathrm{supp}(\bar{\lambda})$. By the conditions on the loss (namely, the boundedness of its second derivative away from 0), we have

$$P(\ell' \bullet f_{\hat{\lambda}^\varepsilon} - \ell' \bullet f_{\bar{\lambda}})(f_{\hat{\lambda}^\varepsilon} - f_{\bar{\lambda}}) \geq c\|f_{\hat{\lambda}^\varepsilon} - f_{\bar{\lambda}}\|_{L_2(\Pi)}^2,$$

where $c = \tau(1)$ (note that $\|f_{\bar{\lambda}}\|_\infty \leq 1$ and $\|f_{\hat{\lambda}^\varepsilon}\|_\infty \leq 1$). Observe also that, for all $j$,

$$(\hat{s}_j - s_j)(\hat{\lambda}_j - \bar{\lambda}_j) \geq 0$$

(by monotonicity of subdifferential of convex function $u \mapsto |u|$). For $j \notin J$, we have $\bar{\lambda}_j = 0$ and $s_j = 0$. Therefore, (8.17) implies that

$$c\|f_{\hat{\lambda}^\varepsilon} - f_{\bar{\lambda}}\|_{L_2(\Pi)}^2 + \varepsilon \sum_{j \notin J} |\hat{\lambda}_j|$$

$$\leq \left\langle \nabla L_\varepsilon(\bar{\lambda}), \bar{\lambda} - \hat{\lambda}^\varepsilon \right\rangle_{\ell_2} + (P - P_n)(\ell' \bullet f_{\hat{\lambda}^\varepsilon})(f_{\hat{\lambda}^\varepsilon} - f_{\bar{\lambda}}). \tag{8.18}$$

Consider first the case when

$$\left\langle \nabla L_\varepsilon(\bar\lambda), \bar\lambda - \hat\lambda^\varepsilon \right\rangle_{\ell_2} \geq (P - P_n)(\ell' \bullet f_{\hat\lambda^\varepsilon})(f_{\hat\lambda^\varepsilon} - f_{\bar\lambda}). \tag{8.19}$$

In this case, (8.18) implies that

$$c\|f_{\hat\lambda^\varepsilon} - f_{\bar\lambda}\|^2_{L_2(\Pi)} + \varepsilon \sum_{j \notin J} |\hat\lambda_j| \leq 2\left\langle \nabla L_\varepsilon(\bar\lambda), \bar\lambda - \hat\lambda^\varepsilon \right\rangle_{\ell_2}, \tag{8.20}$$

which, in view of the definition of $\alpha_+(\varepsilon, \bar\lambda)$, yields

$$c\|f_{\hat\lambda^\varepsilon} - f_{\bar\lambda}\|^2_{L_2(\Pi)} + \varepsilon \sum_{j \notin J} |\hat\lambda_j| \leq 2\alpha_+(\varepsilon, \bar\lambda)\|f_{\hat\lambda^\varepsilon} - f_{\bar\lambda}\|_{L_2(\Pi)}. \tag{8.21}$$

Therefore,

$$\|f_{\hat\lambda^\varepsilon} - f_{\bar\lambda}\|_{L_2(\Pi)} \leq \frac{2}{c}\alpha_+(\varepsilon, \bar\lambda),$$

and, as a consequence, with some constant $C > 0$ depending only on $\ell$

$$\|f_{\hat\lambda^\varepsilon} - f_{\bar\lambda}\|^2_{L_2(\Pi)} + \varepsilon \sum_{j \notin J} |\hat\lambda_j| \leq C\alpha_+^2(\varepsilon, \bar\lambda). \tag{8.22}$$

If

$$\left\langle \nabla L_\varepsilon(\bar\lambda), \bar\lambda - \hat\lambda^\varepsilon \right\rangle_{\ell_2} < (P - P_n)(\ell' \bullet f_{\hat\lambda^\varepsilon})(f_{\hat\lambda^\varepsilon} - f_{\bar\lambda}), \tag{8.23}$$

then (8.18) implies that

$$c\|f_{\hat\lambda^\varepsilon} - f_{\bar\lambda}\|^2_{L_2(\Pi)} + \varepsilon \sum_{j \notin J} |\hat\lambda_j| \leq 2(P - P_n)(\ell' \bullet f_{\hat\lambda^\varepsilon})(f_{\hat\lambda^\varepsilon} - f_{\bar\lambda}). \tag{8.24}$$

Denote

$$\Lambda(\delta; \Delta) := \left\{ \lambda \in U_{\ell_1} : \|f_\lambda - f_{\bar\lambda}\|_{L_2(\Pi)} \leq \delta, \sum_{j \notin J} |\lambda_j| \leq \Delta \right\},$$

$$\alpha_n(\delta; \Delta) := \sup\left\{ |(P_n - P)((\ell' \bullet f_\lambda)(f_\lambda - f_{\bar\lambda}))| : \lambda \in \Lambda(\delta; \Delta) \right\}.$$

To bound $\alpha_n(\delta, \Delta)$, the following lemma will be used.

**Lemma 8.1.** *Under the assumptions of Theorem 8.2, there exists a constant $C$ that depends only on $\ell$ such that with probability at least $1 - N^{-A}$, for all*

$$n^{-1/2} \leq \delta \leq 1 \text{ and } n^{-1/2} \leq \Delta \leq 1. \tag{8.25}$$

*the following bound holds:*

$$\alpha_n(\delta; \Delta) \le \beta_n(\delta; \Delta) := C\left[\delta\sqrt{\frac{d + A\log N}{n}} \bigvee \Delta\sqrt{\frac{A\log N}{n}}\right.$$

$$\left.\bigvee \max_{j \in J}\|P_{L^{\perp}}h_j\|_{L_2(\Pi)}\sqrt{\frac{A\log N}{n}} \bigvee \frac{U(L)\log N}{n} \bigvee \frac{A\log N}{n}\right]. \quad (8.26)$$

Take
$$\delta = \|f_{\hat{\lambda}^{\varepsilon}} - f_{\lambda^{\varepsilon}}\|_{L_2(\Pi)} \quad \text{and} \quad \Delta = \sum_{j \notin J}\hat{\lambda}_j^{\varepsilon}. \quad (8.27)$$

If $\delta \ge n^{-1/2}, \Delta \ge n^{-1/2}$, then Lemma 8.1 and (8.24) imply the following bound:

$$c\delta^2 + \varepsilon\Delta \le 2\beta_n(\delta, \Delta). \quad (8.28)$$

If $\delta < n^{-1/2}$ or $\Delta < n^{-1/2}$, then $\delta$ and $\Delta$ should be replaced in the expression for $\beta_n(\delta, \Delta)$ by $n^{-1/2}$. With this change, bound (8.28) still holds and the proof goes through with some simplifications. Thus, we will consider only the main case when $\delta \ge n^{-1/2}, \Delta \ge n^{-1/2}$. In this case, the inequality (8.28) has to be solved to complete the proof. It follows from this inequality (with a proper change of constant $C$) that

$$\varepsilon\Delta \le C\Delta\sqrt{\frac{A\log N}{n}} + C\left[\delta\sqrt{\frac{d + A\log N}{n}} \bigvee\right.$$

$$\left.\max_{j \in J}\|P_{L^{\perp}}h_j\|_{L_2(\Pi)}\sqrt{\frac{A\log N}{n}} \bigvee \frac{U(L)\log N}{n} \bigvee \frac{A\log N}{n}\right].$$

As soon as $D$ in condition (8.9) is such that $D \ge 2C$, we can write

$$\varepsilon\Delta \le C\left[\delta\sqrt{\frac{d + A\log N}{n}} \bigvee\right.$$

$$\left.\max_{j \in J}\|P_{L^{\perp}}h_j\|_{L_2(\Pi)}\sqrt{\frac{A\log N}{n}} \bigvee \frac{U(L)\log N}{n} \bigvee \frac{A\log N}{n}\right]$$

(again the value of constant $C$ might have changed). Under the assumption (8.9) on $\varepsilon$ (assuming also that $D \ge 1$), it is easy to derive that

$$\Delta \le \Delta(\delta) := C\left[\frac{\delta}{\varepsilon}\sqrt{\frac{d + A\log N}{n}} \bigvee\right.$$

$$\left.\max_{j \in J}\|P_{L^{\perp}}h_j\|_{L_2(\Pi)} \bigvee \frac{U(L)\log N}{n\varepsilon} \bigvee \sqrt{\frac{A\log N}{n}}\right].$$

Note that $\beta_n(\delta, \Delta)$ is nondecreasing in $\Delta$ and replace $\Delta$ in (8.28) by $\Delta(\delta)$ to get the following bound:

$$\delta^2 \le C \left[ \delta \sqrt{\frac{d + A \log N}{n}} \bigvee \frac{\delta}{\varepsilon} \sqrt{\frac{d + A \log N}{n}} \sqrt{\frac{A \log N}{n}} \right.$$

$$\bigvee \frac{U(L) \log N}{n \varepsilon} \sqrt{\frac{A \log N}{n}} \bigvee$$

$$\left. \max_{j \in J} \| P_{L^\perp} h_j \|_{L_2(\Pi)} \sqrt{\frac{A \log N}{n}} \bigvee \frac{U(L) \log N}{n} \bigvee \frac{A \log N}{n} \right].$$

We skip the second term in the maximum and modify the third term because $\frac{1}{\varepsilon} \sqrt{\frac{A \log N}{n}} \le 1$. As a result, we get

$$\delta^2 \le C \left[ \delta \sqrt{\frac{d + A \log N}{n}} \bigvee \max_{j \in J} \| P_{L^\perp} h_j \|_{L_2(\Pi)} \sqrt{\frac{A \log N}{n}} \right.$$

$$\left. \bigvee \frac{U(L) \log N}{n} \bigvee \frac{A \log N}{n} \right].$$

Solving the last inequality with respect to $\delta$ yields the following bound on $\delta^2$:

$$\delta^2 \le C \left[ \frac{d + A \log N}{n} \bigvee \max_{j \in J} \| P_{L^\perp} h_j \|_{L_2(\Pi)} \sqrt{\frac{A \log N}{n}} \bigvee \frac{U(L) \log N}{n} \right].$$
$$(8.29)$$

We substitute the last bound back into the expression for $\Delta(\delta)$ to get:

$$\Delta \le C \left[ \frac{d + A \log N}{n \varepsilon} \bigvee \max_{j \in J} \| P_{L^\perp} h_j \|_{L_2(\Pi)}^{1/2} \frac{1}{\varepsilon} \left( \frac{A \log N}{n} \right)^{1/4} \sqrt{\frac{d + A \log N}{n}} \bigvee \right.$$

$$\sqrt{\frac{U(L) \log N}{n \varepsilon}} \sqrt{\frac{d + A \log N}{n \varepsilon}} \bigvee \max_{j \in J} \| P_{L^\perp} h_j \|_{L_2(\Pi)} \bigvee \frac{U(L) \log N}{n \varepsilon}$$

$$\left. \bigvee \sqrt{\frac{A \log N}{n}} \right].$$

Using the inequality $ab \le (a^2 + b^2)/2$ and the condition $\frac{1}{\varepsilon} \sqrt{\frac{A \log N}{n}} \le 1$, we can simplify the resulting bound as follows

$$\Delta \le C\left[\frac{d + A \log N}{n\varepsilon} \bigvee \max_{j \in J} \|P_{L^{\perp}} h_j\|_{L_2(\Pi)} \bigvee \frac{U(L) \log N}{n\varepsilon} \bigvee \sqrt{\frac{A \log N}{n}}\right]$$
(8.30)

with a proper change of $C$ that depends only on $\ell$. Finally, bounds (8.29) and (8.30) can be substituted in the expression for $\beta_n(\delta, \Delta)$. By a simple computation and in view of Lemma 8.1, we get the following bound on $\alpha_n(\delta, \Delta)$ that holds for $\delta, \Delta$ defined by (8.27) with probability at least $1 - N^{-A}$:

$$\alpha_n(\delta, \Delta) \le C\left[\frac{d + A \log N}{n} + \max_{j \in J} \|P_{L^{\perp}} h_j\|_{L_2(\Pi)} \sqrt{\frac{A \log N}{n}} + \frac{U(L) \log N}{n}\right].$$

Combining this with (8.24) yields

$$c\|f_{\hat{\lambda}^{\varepsilon}} - f_{\bar{\lambda}}\|^2_{L_2(\Pi)} + \varepsilon \sum_{j \notin J} |\hat{\lambda}^{\varepsilon}_j|$$

$$\le C\left[\frac{d + A \log N}{n} + \max_{j \in J} \|P_{L^{\perp}} h_j\|_{L_2(\Pi)} \sqrt{\frac{A \log N}{n}} + \frac{U(L) \log N}{n}\right], \quad (8.31)$$

which holds under condition (8.23).

Together with bound (8.22), that is true under the alternative condition (8.19), this gives (8.10).

To prove bound (8.11), we again use (8.18), but this time we control the term

$$\left\langle \nabla L_{\varepsilon}(\bar{\lambda}), \bar{\lambda} - \hat{\lambda}^{\varepsilon}\right\rangle$$

somewhat differently. First note that

$$\left\langle \nabla L_{\varepsilon}(\bar{\lambda}), \bar{\lambda} - \hat{\lambda}^{\varepsilon}\right\rangle_{\ell_2} = \left\langle \ell' \bullet f_{\bar{\lambda}}, f_{\bar{\lambda}} - f_{\hat{\lambda}^{\varepsilon}}\right\rangle_{L_2(P)} + \varepsilon \langle s(\bar{\lambda}), \bar{\lambda} - \hat{\lambda}^{\varepsilon}\rangle_{\ell_2}.$$

This implies that

$$\left\langle \nabla L_{\varepsilon}(\bar{\lambda}), \bar{\lambda} - \hat{\lambda}^{\varepsilon}\right\rangle_{\ell_2} \le \left\|P_{\mathscr{L}}(\ell' \bullet f_{\bar{\lambda}})\right\|_{L_2(P)} \|f_{\bar{\lambda}} - f_{\hat{\lambda}^{\varepsilon}}\|_{L_2(\Pi)} + \varepsilon \sum_{j \in J} s_j(\bar{\lambda}_j - \hat{\lambda}^{\varepsilon}_j)$$

$$\le \frac{1}{2c}\left\|P_{\mathscr{L}}(\ell' \bullet f_{\bar{\lambda}})\right\|^2_{L_2(P)} + \frac{c}{2}\|f_{\bar{\lambda}} - f_{\hat{\lambda}^{\varepsilon}}\|^2_{L_2(\Pi)} + \varepsilon \sum_{j \in J} s_j(\bar{\lambda}_j - \hat{\lambda}^{\varepsilon}_j).$$

Combining this with bound (8.18) yields the following inequality

$$\frac{c}{2}\|f_{\bar{\lambda}} - f_{\hat{\lambda}^\varepsilon}\|^2_{L_2(\Pi)} + \varepsilon \sum_{j \notin J} |\hat{\lambda}^\varepsilon_j|$$

$$\leq \varepsilon \sum_{j \in J} s_j (\bar{\lambda}_j - \hat{\lambda}^\varepsilon_j) + \frac{1}{2c}\left\|P_{\mathscr{L}}(\ell' \bullet f_{\bar{\lambda}})\right\|^2_{L_2(P)} + (P - P_n)(\ell' \bullet f_{\hat{\lambda}^\varepsilon})(f_{\hat{\lambda}^\varepsilon} - f_{\bar{\lambda}}).$$

If

$$\varepsilon \sum_{j \in J} s_j (\bar{\lambda}_j - \hat{\lambda}^\varepsilon_j) \geq \frac{1}{2c}\left\|P_{\mathscr{L}}(\ell' \bullet f_{\bar{\lambda}})\right\|^2_{L_2(P)} + (P - P_n)(\ell' \bullet f_{\hat{\lambda}^\varepsilon})(f_{\hat{\lambda}^\varepsilon} - f_{\bar{\lambda}}),$$

then

$$\frac{c}{2}\|f_{\bar{\lambda}} - f_{\hat{\lambda}^\varepsilon}\|^2_{L_2(\Pi)} + \varepsilon \sum_{j \notin J} |\hat{\lambda}^\varepsilon_j| \leq 2\varepsilon \sum_{j \in J} s_j (\bar{\lambda}_j - \hat{\lambda}^\varepsilon_j),$$

which implies

$$\sum_{j \notin J} |\hat{\lambda}^\varepsilon_j| \leq 2 \sum_{j \in J} |\bar{\lambda}_j - \hat{\lambda}^\varepsilon_j|,$$

or $\hat{\lambda}^\varepsilon - \bar{\lambda} \in C_{2,\bar{\lambda}}$. The definition of $\alpha(\bar{\lambda})$ then implies the bound

$$\frac{c}{2}\|f_{\bar{\lambda}} - f_{\hat{\lambda}^\varepsilon}\|^2_{L_2(\Pi)} + \varepsilon \sum_{j \notin J} |\hat{\lambda}^\varepsilon_j| \leq 2\varepsilon\alpha(\bar{\lambda})\|f_{\bar{\lambda}} - f_{\hat{\lambda}^\varepsilon}\|_{L_2(\Pi)}.$$

Solving this inequality with respect to $\|f_{\bar{\lambda}} - f_{\hat{\lambda}^\varepsilon}\|_{L_2(\Pi)}$ proves (8.11) in this case.
  If

$$\varepsilon \sum_{j \in J} s_j (\bar{\lambda}_j - \hat{\lambda}^\varepsilon_j) \leq \frac{1}{2c}\left\|P_{\mathscr{L}}(\ell' \bullet f_{\bar{\lambda}})\right\|^2_{L_2(P)} + (P - P_n)(\ell' \bullet f_{\hat{\lambda}^\varepsilon})(f_{\hat{\lambda}^\varepsilon} - f_{\bar{\lambda}})$$

and

$$\frac{1}{2c}\left\|P_{\mathscr{L}}(\ell' \bullet f_{\bar{\lambda}})\right\|^2_{L_2(P)} \geq (P - P_n)(\ell' \bullet f_{\hat{\lambda}^\varepsilon})(f_{\hat{\lambda}^\varepsilon} - f_{\bar{\lambda}}),$$

we get

$$\frac{c}{2}\|f_{\bar{\lambda}} - f_{\hat{\lambda}^\varepsilon}\|^2_{L_2(\Pi)} + \varepsilon \sum_{j \notin J} |\hat{\lambda}^\varepsilon_j| \leq \frac{2}{c}\left\|P_{\mathscr{L}}(\ell' \bullet f_{\bar{\lambda}})\right\|^2_{L_2(P)},$$

which also implies (8.11) with a proper choice of constant $C$ in the bound.
  Thus, it remains to consider the case when

$$\varepsilon \sum_{j \in J} s_j (\bar{\lambda}_j - \hat{\lambda}^\varepsilon_j) \leq \frac{1}{2c}\left\|P_{\mathscr{L}}(\ell' \bullet f_{\bar{\lambda}})\right\|^2_{L_2(P)} + (P - P_n)(\ell' \bullet f_{\hat{\lambda}^\varepsilon})(f_{\hat{\lambda}^\varepsilon} - f_{\bar{\lambda}})$$

and

$$\frac{1}{2c}\left\|P_{\mathcal{L}}(\ell' \bullet f_{\hat{\lambda}})\right\|^2_{L_2(P)} \le (P - P_n)(\ell' \bullet f_{\hat{\lambda}^\varepsilon})(f_{\hat{\lambda}^\varepsilon} - f_{\hat{\lambda}}),$$

which implies

$$\frac{c}{2}\|f_{\hat{\lambda}} - f_{\hat{\lambda}^\varepsilon}\|^2_{L_2(\Pi)} + \varepsilon \sum_{j \notin J} |\hat{\lambda}^\varepsilon_j| \le 4(P - P_n)(\ell' \bullet f_{\hat{\lambda}^\varepsilon})(f_{\hat{\lambda}^\varepsilon} - f_{\hat{\lambda}}).$$

In this case, we repeat the argument based on Lemma 8.1 to show that with probability at least $1 - N^{-A}$

$$\frac{c}{2}\|f_{\hat{\lambda}^\varepsilon} - f_{\hat{\lambda}}\|^2_{L_2(\Pi)} + \varepsilon \sum_{j \notin J} |\hat{\lambda}^\varepsilon_j|$$

$$\le C\left[\frac{d + A\log N}{n} + \max_{j \in J}\|P_{L^\perp}h_j\|_{L_2(\Pi)}\sqrt{\frac{A\log N}{n}} + \frac{U(L)\log N}{n}\right],$$

which again implies (8.11). This completes the proof.          □

We will now give the proof of Lemma 8.1.

*Proof.* First we use Talagrand's concentration inequality to get that with probability at least $1 - e^{-t}$

$$\alpha_n(\delta; \Delta) \le 2\left[\mathbb{E}\alpha_n(\delta; \Delta) + C\delta\sqrt{\frac{t}{n}} + \frac{Ct}{n}\right]. \tag{8.32}$$

Next, symmetrization inequality followed by contraction inequality for Rademacher sums yield:

$$\mathbb{E}\alpha_n(\delta; \Delta) \le 2\mathbb{E}\sup\left\{|R_n((\ell' \bullet f_\lambda)(f_\lambda - f_{\hat{\lambda}}))| : \lambda \in \Lambda(\delta; \Delta)\right\}$$

$$\le C\mathbb{E}\sup\left\{|R_n(f_\lambda - f_{\hat{\lambda}})| : \lambda \in \Lambda(\delta; \Delta)\right\} \tag{8.33}$$

with a constant $C$ depending only on $\ell$. In contraction inequality part, we write

$$\ell'(f_\lambda(\cdot))(f_\lambda(\cdot) - f_{\hat{\lambda}}(\cdot)) = \ell'(f_{\hat{\lambda}}(\cdot) + u)u\Big|_{u = f_\lambda(\cdot) - f_{\hat{\lambda}}(\cdot)}$$

and use the fact that the function

$$[-1, 1] \ni u \mapsto \ell'(f_{\hat{\lambda}}(\cdot) + u)u$$

satisfies the Lipschitz condition with a constant depending only on $\ell$.

The following representation is straightforward:

$$f_\lambda - f_{\bar\lambda} = P_L(f_\lambda - f_{\bar\lambda}) + \sum_{j \in J}(\lambda_j - \bar\lambda_j)P_{L\perp}h_j + \sum_{j \notin J}\lambda_j P_{L\perp}h_j. \qquad (8.34)$$

For all $\lambda \in \Lambda(\delta, \Delta)$,

$$\|P_L(f_\lambda - f_{\bar\lambda})\|_{L_2(\Pi)} \leq \|f_\lambda - f_{\bar\lambda}\|_{L_2(\Pi)} \leq \delta$$

and $P_L(f_\lambda - f_{\bar\lambda}) \in L$. Since $L$ is a $d$-dimensional subspace,

$$\mathbb{E}\sup\left\{|R_n(P_L(f_\lambda - f_{\bar\lambda}))| : \lambda \in \Lambda(\delta; \Delta)\right\} \leq C\delta\sqrt{\frac{d}{n}}$$

(see Proposition 3.2). On the other hand, $\lambda, \bar\lambda \in U_{\ell_1}$, so, we have $\sum_{j \in J}|\lambda_j - \bar\lambda_j| \leq 2$. Hence,

$$\mathbb{E}\sup\left\{\left|R_n\left(\sum_{j \in J}(\lambda_j - \bar\lambda_j)P_{L\perp}h_j\right)\right| : \lambda \in \Lambda(\delta; \Delta)\right\} \leq 2\mathbb{E}\max_{j \in J}|R_n(P_{L\perp}h_j)|.$$

Note also that

$$\|P_{L\perp}h_j\|_\infty \leq \|P_L h_j\|_\infty + \|h_j\|_\infty \leq (U(L) - 1)\|P_L h_j\|_{L_2(\Pi)} + 1$$

$$\leq (U(L) - 1)\|h_j\|_{L_2(\Pi)} + 1 \leq U(L),$$

and Theorem 3.5 yields

$$\mathbb{E}\max_{j \in J}|R_n(P_{L\perp}h_j)| \leq C\left[\max_{j \in J}\|P_{L\perp}h_j\|_{L_2(\Pi)}\sqrt{\frac{\log N}{n}} + U(L)\frac{\log N}{n}\right].$$

Similarly, for all $\lambda \in \Lambda(\delta, \Delta)$, $\sum_{j \notin J}|\lambda_j| \leq \Delta$ and

$$\mathbb{E}\sup\left\{\left|R_n\left(\sum_{j \notin J}\lambda_j P_{L\perp}h_j\right)\right| : \lambda \in \Lambda(\delta; \Delta)\right\} \leq \Delta\mathbb{E}\max_{j \notin J}|R_n(P_{L\perp}h_j)|.$$

Another application of Theorem 3.5, together with the fact that

$$\|P_{L\perp}h_j\|_{L_2(\Pi)} \leq \|h_j\|_{L_2(\Pi)} \leq 1,$$

results in the bound

$$\mathbb{E}\max_{j \notin J}|R_n(P_{L\perp}h_j)| \leq C\left[\sqrt{\frac{\log N}{n}} + U(L)\frac{\log N}{n}\right],$$

Now we use representation (8.34) and bound (8.33). It easily follows that

$$\mathbb{E}\alpha_n(\delta, \Delta) \le C\left[\delta\sqrt{\frac{d}{n}} \bigvee \Delta\sqrt{\frac{\log N}{n}} \bigvee\right.$$

$$\left.\max_{j\in J}\|P_{L^\perp}h_j\|_{L_2(\Pi)}\sqrt{\frac{\log N}{n}} \bigvee \frac{U(L)\log N}{n}\right]. \tag{8.35}$$

Substituting this bound into (8.32) shows that with probability $1 - e^{-t}$

$$\alpha_n(\delta, \Delta) \le \tilde{\beta}_n(\delta, \Delta, t) := C\left[\delta\sqrt{\frac{d}{n}} \bigvee \Delta\sqrt{\frac{\log N}{n}} \bigvee\right.$$

$$\left.\max_{j\in J}\|P_{L^\perp}h_j\|_{L_2(\Pi)}\sqrt{\frac{\log N}{n}} \bigvee \frac{U(L)\log N}{n} \bigvee \delta\sqrt{\frac{t}{n}} \bigvee \frac{t}{n}\right] \tag{8.36}$$

with a constant $C > 0$ depending only on $\ell$.

It remains to prove that, with a high probability, the above bounds hold uniformly in $\delta, \Delta$ satisfying (8.25). Let $\delta_j := 2^{-j}$ and $\Delta_j := 2^{-j}$. We will replace $t$ by $t + 2\log(j + 1) + 2\log(k + 1)$. By the union bound, with probability at least

$$1 - \sum_{j,k\ge0} \exp\{-t - 2\log(j + 1) - 2\log(k + 1)\}$$

$$= 1 - \left(\sum_{j\ge0}(j + 1)^{-2}\right)^2 \exp\{-t\} \ge 1 - 4e^{-t},$$

the following bound holds

$$\alpha_n(\delta; \Delta) \le \tilde{\beta}_n\left(\delta_j, \Delta_k, t + 2\log j + 2\log k\right),$$

for all $\delta$ and $\Delta$ satisfying (8.25) and for all $j, k$ such that

$$\delta \in (\delta_{j+1}, \delta_j] \text{ and } \Delta \in (\Delta_{k+1}, \Delta_k].$$

Using the fact that

$$2\log j \le 2\log\log_2\left(\frac{1}{\delta_j}\right) \le 2\log\log_2\left(\frac{2}{\delta}\right)$$

and

$$2\log k \le 2\log\log_2\left(\frac{2}{\Delta}\right),$$

we get

$$\tilde{\beta}_n\left(\delta_j, \Delta_k, t + 2\log j + 2\log k\right)$$

$$\leq \tilde{\beta}_n\left(2\delta, 2\Delta, t + 2\log\log_2\left(\frac{2}{\delta}\right) + 2\log\log_2\left(\frac{2}{\Delta}\right)\right) =: \bar{\beta}_n(\delta; \Delta; t).$$

As a result, with probability at least $1 - 4e^{-t}$, for all $\delta$ and $\Delta$ satisfying (8.25),

$$\alpha_n(\delta; \Delta) \leq \bar{\beta}_n(\delta; \Delta; t).$$

Take now $t = A \log N + \log 4$, so that $4e^{-t} = N^{-A}$. With some constant $C$ that depends only on $\ell$,

$$\bar{\beta}_n(\delta; \Delta; t) \leq C\left[\delta\sqrt{\frac{d}{n}} \bigvee \delta\sqrt{\frac{A\log N}{n}} \bigvee \delta\sqrt{\frac{2\log\log_2\left(\frac{2}{\delta}\right)}{n}} \bigvee \right.$$

$$\delta\sqrt{\frac{2\log\log_2\left(\frac{2}{\Delta}\right)}{n}} \bigvee \Delta\sqrt{\frac{\log N}{n}} \bigvee \max_{j\in J}\|P_{L^\perp}h_j\|_{L_2(\Pi)}\sqrt{\frac{\log N}{n}} \bigvee$$

$$\left. \frac{U(L)\log N}{n} \bigvee \frac{2\log\log_2\left(\frac{2}{\delta}\right)}{n} \bigvee \frac{2\log\log_2\left(\frac{2}{\Delta}\right)}{n} \bigvee \frac{A\log N}{n}\right].$$

For all $\delta$ and $\Delta$ satisfying (8.25),

$$\frac{2\log\log_2\left(\frac{2}{\delta}\right)}{n} \leq C\frac{\log\log n}{n} \quad\text{and}\quad \frac{2\log\log_2\left(\frac{2}{\Delta}\right)}{n} \leq C\frac{\log\log n}{n}.$$

Assumptions on $N, n$, imply that $A \log N \geq \gamma \log\log n$. Thus, for all $\delta$ and $\Delta$ satisfying (8.25),

$$\alpha_n(\delta, \Delta) \leq \bar{\beta}_n(\delta; \Delta; t) \leq C\left[\delta\sqrt{\frac{d}{n}} \bigvee \delta\sqrt{\frac{A\log N}{n}} \bigvee \Delta\sqrt{\frac{\log N}{n}} \bigvee \right.$$

$$\left. \max_{j\in J}\|P_{L^\perp}h_j\|_{L_2(\Pi)}\sqrt{\frac{\log N}{n}} \bigvee \frac{U(L)\log N}{n} \bigvee \frac{A\log N}{n}\right]. \tag{8.37}$$

The last bound holds with probability at least $1 - N^{-A}$ proving the lemma. □

The proof of Theorem 8.1 is quite similar. The following lemma is used instead of Lemma 8.1.

**Lemma 8.2.** *Under the assumptions of Theorem 8.1, there exists a constant $C$ that depends only on $\ell$ such that with probability at least $1 - N^{-A}$, for all*

$$n^{-1/2} \leq \delta \leq 1 \quad\text{and}\quad n^{-1/2} \leq \Delta \leq 1,$$

*the following bound holds:*

$$\alpha_n(\delta; \Delta) \le \beta_n(\delta; \Delta) :=$$

$$C\left[\delta\sqrt{\frac{d + A\log N}{n}} \bigvee \Delta\sqrt{\frac{d + A\log N}{n}} \bigvee \frac{A\log N}{n}\right]. \qquad (8.38)$$

In Theorems 8.1 and 8.2, we used a special version of subgradient $\nabla L_\varepsilon(\bar{\lambda})$. More generally, one can consider an arbitrary couple $(\bar{\lambda}, \nabla L_\varepsilon(\bar{\lambda}))$ where $\bar{\lambda} \in U_{\ell_1}$ and $\nabla L_\varepsilon(\bar{\lambda}) \in \partial L_\varepsilon(\bar{\lambda})$. This couple can be viewed as "an oracle" in our problem. As before,

$$\nabla L_\varepsilon(\bar{\lambda}) = \left((P(\ell' \bullet f_{\bar{\lambda}}))h_j + \varepsilon s_j\right)_{j=1,\dots,N},$$

but now $s_j = s_j(\bar{\lambda})$ are arbitrary numbers from $[-1, 1]$ satisfying the condition

$$s_j = \text{sign}(\bar{\lambda}_j), \quad \bar{\lambda}_j \ne 0.$$

The next results provide modifications of Theorems 8.1 and 8.2 for such more general "oracles".

Denote

$$\alpha^{(b)}(\lambda) := a_H^{(b)}\left(U_{\ell_1}, \lambda, s(\lambda)\right) \vee 0$$

for some fixed $b > 0$.

**Theorem 8.3.** *There exist constants $D > 0$ and $C > 0$ depending only on $\ell$ with the following property. Let $\bar{\lambda} \in U_{\ell_1}$ and*

$$\nabla L_\varepsilon(\bar{\lambda}) = \left((P(\ell' \bullet f_{\bar{\lambda}}))h_j + \varepsilon s_j\right)_{j=1,\dots,N} \in \partial L_\varepsilon(\bar{\lambda}).$$

*Let $J \subset \{1, \dots, N\}$, $J \supset \text{supp}(\bar{\lambda})$ with $d := d(J) = \text{card}(J)$. Suppose that, for some $\gamma \in (0, 1)$,*

$$|s_j| \le 1 - \gamma, \quad j \notin J.$$

*Then, for all $A \ge 1$ and for all*

$$\varepsilon \ge D\sqrt{\frac{d + A\log N}{n}}, \qquad (8.39)$$

*the following bound holds with probability at least $1 - N^{-A}$:*

$$\|f_{\hat{\lambda}^\varepsilon} - f_{\bar{\lambda}}\|^2_{L_2(\Pi)} + \varepsilon\gamma\sum_{j \notin J}|\hat{\lambda}_j^\varepsilon| \le C\left[\frac{d + A\log N}{n} \bigvee \alpha_+^2(\varepsilon, \bar{\lambda})\right].$$

*Moreover, with the same probability,*

$$\|f_{\hat{\lambda}^\varepsilon} - f_{\bar{\lambda}}\|^2_{L_2(\Pi)} + \varepsilon\gamma \sum_{j \notin J} |\hat{\lambda}^\varepsilon_j|$$

$$\leq C\left[\frac{d + A\log N}{n} \bigvee \left\|P_{\mathscr{L}}(\ell' \bullet f_{\bar{\lambda}})\right\|^2_{L_2(P)} \bigvee \left(\alpha^{(2/\gamma)}(\bar{\lambda})\right)^2 \varepsilon^2\right].$$

**Theorem 8.4.** *Suppose that*

$$\varepsilon \geq D\sqrt{\frac{A\log N}{n}} \tag{8.40}$$

*with a large enough constant $D > 0$ depending only on $\ell$. Let $\bar{\lambda} \in U_{\ell_1}$ and*

$$\nabla L_\varepsilon(\bar{\lambda}) = \left((P(\ell' \bullet f_{\bar{\lambda}}))h_j + \varepsilon s_j\right)_{j=1,\dots,N} \in \partial L_\varepsilon(\bar{\lambda}).$$

*Let $J \subset \{1,\dots,N\}$, $J \supset \text{supp}(\bar{\lambda})$. Suppose that, for some $\gamma \in (0,1)$,*

$$|s_j| \leq 1 - \gamma, \; j \notin J.$$

*Then, for all subspaces $L$ of $L_2(\Pi)$ with $d := \dim(L)$ and for all $A \geq 1$, the following bound holds with probability at least $1 - N^{-A}$ and with a constant $C > 0$ depending only on $\ell$:*

$$\|f_{\hat{\lambda}^\varepsilon} - f_{\bar{\lambda}}\|^2_{L_2(\Pi)} + \varepsilon\gamma \sum_{j \notin J} |\hat{\lambda}^\varepsilon_j| \tag{8.41}$$

$$\leq C\left[\frac{d + A\log N}{n} \bigvee \max_{j \in J} \|P_{L^\perp}h_j\|_{L_2(\Pi)} \sqrt{\frac{A\log N}{n}} \bigvee \right.$$

$$\left. \frac{U(L)\log N}{n} \bigvee \alpha^2_+(\varepsilon;\bar{\lambda})\right].$$

*Moreover, with the same probability*

$$\|f_{\hat{\lambda}^\varepsilon} - f_{\bar{\lambda}}\|^2_{L_2(\Pi)} + \varepsilon\gamma \sum_{j \notin J} |\hat{\lambda}^\varepsilon_j| \tag{8.42}$$

$$\leq C\left[\frac{d + A\log N}{n} \bigvee \max_{j \in J} \|P_{L^\perp}h_j\|_{L_2(\Pi)} \sqrt{\frac{A\log N}{n}} \bigvee \frac{U(L)\log N}{n}\right.$$

$$\left. \bigvee \left\|P_{\mathscr{L}}(\ell' \bullet f_{\bar{\lambda}})\right\|^2_{L_2(P)} \bigvee \left(\alpha^{(2/\gamma)}(\bar{\lambda})\right)^2 \varepsilon^2\right].$$

For some choices of vector $\bar{\lambda}$ and of subgradient $\nabla L_\varepsilon(\bar{\lambda})$, the alignment coefficient might be smaller than for the choice we used in Theorems 8.1 and 8.2 resulting in tighter bounds. An appealing choice would be $\bar{\lambda} = \lambda^\varepsilon$,

$$\lambda^\varepsilon = \mathrm{argmin}_{\lambda \in U_{\ell_1}}\left[ P(\ell \bullet f_\lambda) + \varepsilon \|\lambda\|_{\ell_1} \right],$$

since in this case it is possible to take $\nabla L_\varepsilon(\lambda^\varepsilon) \in \partial L_\varepsilon(\lambda^\varepsilon)$ such that

$$a_H^{(b)}(U_{\ell_1}, \lambda^\varepsilon, \nabla L_\varepsilon(\lambda^\varepsilon)) \leq 0$$

(this follows from the necessary conditions of extremum). Therefore, with this choice, we have $\alpha_+(\varepsilon, \lambda^\varepsilon) = 0$, which means that for the oracle vector $\lambda^\varepsilon$ there exists a version of subgradient that is "well aligned" with the dictionary.

We have the following corollaries in which both the $L_2$-error $\|f_{\hat{\lambda}^\varepsilon} - f_{\lambda^\varepsilon}\|_{L_2(\Pi)}$ and the degree of "approximate sparsity" of the empirical solution $\hat{\lambda}^\varepsilon$ are controlled by the "sparsity" of the "oracle" without any geometric assumptions on the dictionary.

**Corollary 8.3.** *There exist constants $D > 0$ and $C > 0$ depending only on $\ell$ with the following property. Let*

$$\nabla L_\varepsilon(\lambda^\varepsilon) = \left((P(\ell' \bullet f_{\lambda^\varepsilon}))h_j + \varepsilon s_j\right)_{j=1,\ldots,N} \in \partial L_\varepsilon(\lambda^\varepsilon)$$

*be such that, for all $u \in T_{U_{\ell_1}}(\lambda^\varepsilon)$,*

$$\langle \nabla L_\varepsilon(\lambda^\varepsilon), u\rangle_{\ell_2} \geq 0.$$

*Let $J \subset \{1,\ldots,N\}$, $J \supset \mathrm{supp}(\bar{\lambda})$ with $d := d(J) = \mathrm{card}(J)$. Suppose that, for some $\gamma \in (0,1)$,*

$$|s_j| \leq 1 - \gamma, \ j \notin J.$$

*Then, for all $A \geq 1$ and for all*

$$\varepsilon \geq D\sqrt{\frac{d + A\log N}{n}}, \tag{8.43}$$

*the following bound holds with probability at least $1 - N^{-A}$:*

$$\|f_{\hat{\lambda}^\varepsilon} - f_{\lambda^\varepsilon}\|_{L_2(\Pi)}^2 + \varepsilon\gamma\sum_{j \notin J}|\hat{\lambda}_j^\varepsilon| \leq C\frac{d + A\log N}{n}.$$

**Corollary 8.4.** *Suppose that*

$$\varepsilon \geq D\sqrt{\frac{A\log N}{n}} \tag{8.44}$$

*with a large enough constant $D > 0$ depending only on $\ell$. Let*

$$\nabla L_\varepsilon(\lambda^\varepsilon) = \left( (P(\ell' \bullet f_{\lambda^\varepsilon}))h_j + \varepsilon s_j \right)_{j=1,\dots,N} \in \partial L_\varepsilon(\lambda^\varepsilon)$$

*be such that, for all $u \in T_{U_{\ell_1}}(\lambda^\varepsilon)$,*

$$\langle \nabla L_\varepsilon(\lambda^\varepsilon), u \rangle_{\ell_2} \geq 0.$$

*Let $J \subset \{1,\dots,N\}$, $J \supset \mathrm{supp}(\bar{\lambda})$. Suppose that, for some $\gamma \in (0,1)$,*

$$|s_j| \leq 1 - \gamma, \ j \notin J.$$

*Then, for all subspaces $L$ of $L_2(\Pi)$ with $d := \dim(L)$ and for all $A \geq 1$, the following bound holds with probability at least $1 - N^{-A}$ and with a constant $C > 0$ depending only on $\ell$:*

$$\| f_{\hat{\lambda}^\varepsilon} - f_{\lambda^\varepsilon} \|^2_{L_2(\Pi)} + \varepsilon\gamma \sum_{j \notin J} |\hat{\lambda}^\varepsilon_j| \tag{8.45}$$

$$\leq C\left[ \frac{d + A \log N}{n} \bigvee \max_{j \in J} \| P_{L^\perp} h_j \|_{L_2(\Pi)} \sqrt{\frac{A \log N}{n}} \bigvee \frac{U(L) \log N}{n} \right].$$

## 8.3  Entropy Penalization and Sparse Recovery in Convex Hulls: Random Error Bounds

As before, it will be assumed that $\ell$ is a loss function of quadratic type (see Definition 8.1). Denote

$$\Lambda := \{(\lambda_1,\dots,\lambda_N) : \lambda_j \geq 0, \ j = 1,\dots,N, \ \sum_{j=1}^N \lambda_j = 1\}.$$

The following penalized empirical risk minimization problem will be studied:

$$\hat{\lambda}^\varepsilon := \mathrm{argmin}_{\lambda \in \Lambda} \left[ P_n(\ell \bullet f_\lambda) - \varepsilon H(\lambda) \right]$$

$$= \mathrm{argmin}_{\lambda \in \Lambda} \left[ P_n(\ell \bullet f_\lambda) + \varepsilon \sum_{j=1}^N \lambda_j \log \lambda_j \right], \tag{8.46}$$

where $\varepsilon \geq 0$ is a regularization parameter and

$$H(\lambda) = -\sum_{j=1}^{N} \lambda_j \log \lambda_j$$

is the entropy of $\lambda$. Since, for all $y$, $\ell(y, \cdot)$ is convex, the empirical risk $P_n(\ell \bullet f_\lambda)$ is a convex function of $\lambda$. Since also the set $\Lambda$ is convex and so is the function $\lambda \mapsto -H(\lambda)$, the problem (8.46) is a convex optimization problem.

It is natural to compare this problem with its distribution dependent version

$$\lambda^\varepsilon := \mathrm{argmin}_{\lambda \in \Lambda} \left[ P(\ell \bullet f_\lambda) - \varepsilon H(\lambda) \right]$$
$$= \mathrm{argmin}_{\lambda \in \Lambda} \left[ P(\ell \bullet f_\lambda) + \varepsilon \sum_{j=1}^{N} \lambda_j \log \lambda_j \right]. \tag{8.47}$$

Note that the minimum of the penalty $-H(\lambda)$ is attained at the uniform distribution $\lambda_j = N^{-1}, j = 1, \ldots, N$. Because of this, at the first glance, $-H(\lambda)$ penalizes for "sparsity" rather than for "non-sparsity". However, we will show that if $\lambda^\varepsilon$ is "approximately sparse", then $\hat{\lambda}^\varepsilon$ has a similar property with a high probability. Moreover, the approximate sparsity of $\lambda^\varepsilon$ will allow us to control $\|f_{\hat{\lambda}^\varepsilon} - f_{\lambda^\varepsilon}\|_{L_2(\Pi)}$ and $K(\hat{\lambda}^\varepsilon, \lambda^\varepsilon)$, where

$$K(\lambda, \nu) := K(\lambda|\nu) + K(\nu|\lambda)$$

is the symmetrized Kullback–Leibler distance between $\lambda$ and $\nu$,

$$K(\lambda|\nu) := \sum_{j=1}^{N} \lambda_j \log\left(\frac{\lambda_j}{\nu_j}\right)$$

being the Kullback–Leibler divergence between $\lambda, \nu$.

In particular, it will follow from our results that for any set $J \subset \{1, \ldots, N\}$ with $\mathrm{card}(J) = d$ and such that

$$\sum_{j \notin J} \lambda_j^\varepsilon \leq \sqrt{\frac{\log N}{n}},$$

with a high probability,

$$\|f_{\hat{\lambda}^\varepsilon} - f_{\lambda^\varepsilon}\|_{L_2(\Pi)}^2 + \varepsilon K(\hat{\lambda}^\varepsilon; \lambda^\varepsilon) \leq C \frac{d + \log N}{n}.$$

This easily implies upper bounds on "the random error" $|\mathscr{E}(f_{\hat{\lambda}^\varepsilon}) - \mathscr{E}(f_{\lambda^\varepsilon})|$ in terms of "approximate sparsity" of $\lambda^\varepsilon$.

Some further geometric parameters (such as "the alignment coefficient" introduced in Sect. 7.2.3) provide a way to control "the approximation error" $\mathscr{E}(f_{\lambda^\varepsilon})$. As a result, if there exists a "sparse" vector $\lambda \in \Lambda$ for which the excess risk $\mathscr{E}(f_\lambda)$ is small and $\lambda$ is properly "aligned" with the dictionary, then $\lambda^\varepsilon$ is approximately sparse and its excess risk $\mathscr{E}(f_{\lambda^\varepsilon})$ is controlled by sparsity of $\lambda$ and its "alignment" with the dictionary. Together with sparsity bounds on the random error this yields oracle inequalities on the excess risk $\mathscr{E}(f_{\hat{\lambda}^\varepsilon})$ showing that this estimation method provides certain degree of adaptation to the unknown "sparsity" of the problem.

The first result in this direction is the following theorem that provides the bounds on approximate sparsity of $\hat{\lambda}^\varepsilon$ in terms of approximate sparsity of $\lambda^\varepsilon$ as well as the bounds on the $L_2$-error of approximation of $f_{\lambda^\varepsilon}$ by $f_{\hat{\lambda}^\varepsilon}$ and the Kullback–Leibler error of approximation of $\lambda^\varepsilon$ by $\hat{\lambda}^\varepsilon$.

**Theorem 8.5.** *There exist constants $D > 0$ and $C > 0$ depending only on $\ell$ such that, for all $J \subset \{1, \dots, N\}$ with $d := d(J) = \mathrm{card}(J)$, for all $A \geq 1$ and for all*

$$\varepsilon \geq D\sqrt{\frac{d + A\log N}{n}}, \tag{8.48}$$

*the following bounds hold with probability at least $1 - N^{-A}$:*

$$\sum_{j \notin J} \hat{\lambda}_j^\varepsilon \leq C\left[\sum_{j \notin J} \lambda_j^\varepsilon + \sqrt{\frac{d + A\log N}{n}}\right],$$

$$\sum_{j \notin J} \lambda_j^\varepsilon \leq C\left[\sum_{j \notin J} \hat{\lambda}_j^\varepsilon + \sqrt{\frac{d + A\log N}{n}}\right]$$

*and*

$$\|f_{\hat{\lambda}^\varepsilon} - f_{\lambda^\varepsilon}\|_{L_2(\Pi)}^2 + \varepsilon K(\hat{\lambda}^\varepsilon, \lambda^\varepsilon) \leq C\left[\frac{d + A\log N}{n} \bigvee \sum_{j \notin J} \lambda_j^\varepsilon \sqrt{\frac{d + A\log N}{n}}\right].$$

Similarly to what was done in Sect. 8.2, we will also establish another version of these bounds that hold for smaller values of $\varepsilon$ (the quantity $U(L)$ introduced in Sect. 8.2 will be involved in these bounds).

**Theorem 8.6.** *Suppose that*

$$\varepsilon \geq D\sqrt{\frac{A\log N}{n}} \tag{8.49}$$

*with a large enough constant $D > 0$ depending only on $\ell$. For all $J \subset \{1, \ldots, N\}$,
for all subspaces $L$ of $L_2(\Pi)$ with $d := \dim(L)$ and for all $A \geq 1$, the following
bounds hold with probability at least $1 - N^{-A}$ and with a constant $C > 0$ depending
only on $\ell$:*

$$\sum_{j \notin J} \hat{\lambda}_j^\varepsilon \leq C \left[ \sum_{j \notin J} \lambda_j^\varepsilon + \frac{d + A \log N}{n\varepsilon} + \max_{j \in J} \| P_{L^\perp} h_j \|_{L_2(\Pi)} + \frac{U(L) \log N}{n\varepsilon} \right], \quad (8.50)$$

$$\sum_{j \notin J} \lambda_j^\varepsilon \leq C \left[ \sum_{j \notin J} \hat{\lambda}_j^\varepsilon + \frac{d + A \log N}{n\varepsilon} + \max_{j \in J} \| P_{L^\perp} h_j \|_{L_2(\Pi)} + \frac{U(L) \log N}{n\varepsilon} \right] \quad (8.51)$$

*and*

$$\| f_{\hat{\lambda}^\varepsilon} - f_{\lambda^\varepsilon} \|_{L_2(\Pi)}^2 + \varepsilon K(\hat{\lambda}^\varepsilon, \lambda^\varepsilon) \leq C \left[ \frac{d + A \log N}{n} \bigvee \sum_{j \notin J} \lambda_j^\varepsilon \sqrt{\frac{A \log N}{n}} \bigvee \right.$$

$$\left. \max_{j \in J} \| P_{L^\perp} h_j \|_{L_2(\Pi)} \sqrt{\frac{A \log N}{n}} \bigvee \frac{U(L) \log N}{n} \right]. \quad (8.52)$$

If, for some $J$,

$$\sum_{j \notin J} \lambda_j^\varepsilon \leq \sqrt{\frac{A \log N}{n}}$$

and, for some $L$ with $U(L) \leq d$, $h_j \in L$, $j \in J$, then bound (8.52) simplifies and
becomes

$$\| f_{\hat{\lambda}^\varepsilon} - f_{\lambda^\varepsilon} \|_{L_2(\Pi)}^2 + \varepsilon K(\hat{\lambda}^\varepsilon, \lambda^\varepsilon) \leq C \frac{Ad \log N}{n}.$$

In particular, it means that the sizes of the random errors $\| f_{\hat{\lambda}^\varepsilon} - f_{\lambda^\varepsilon} \|_{L_2(\Pi)}^2$ and
$K(\hat{\lambda}^\varepsilon, \lambda^\varepsilon)$ are controlled by the dimension $d$ of the linear span $L$ of the "relevant
part" of the dictionary $\{h_j : j \in J\}$. Note that $d$ can be much smaller than $\mathrm{card}(J)$
in the case when the functions in the dictionary are not linearly independent (so, the
lack of "orthogonality" of the dictionary might help to reduce the random error).

The proofs of Theorems 8.5 and 8.6 are quite similar. We give only the proof of
Theorem 8.6.

*Proof.* We use the method described in Sec. 8.1. In the current case, necessary
conditions of minima in minimization problems defining $\lambda^\varepsilon$ and $\hat{\lambda}^\varepsilon$ can be written
as follows:

$$P(\ell' \bullet f_{\lambda^\varepsilon})(f_{\hat{\lambda}^\varepsilon} - f_{\lambda^\varepsilon}) + \varepsilon \sum_{j=1}^N (\log \lambda_j^\varepsilon + 1)(\hat{\lambda}_j^\varepsilon - \lambda_j^\varepsilon) \geq 0 \quad (8.53)$$

and

$$P_n(\ell' \bullet f_{\hat{\lambda}^\varepsilon})(f_{\hat{\lambda}^\varepsilon} - f_{\lambda^\varepsilon}) + \varepsilon \sum_{j=1}^N (\log \hat{\lambda}_j^\varepsilon + 1)(\hat{\lambda}_j^\varepsilon - \lambda_j^\varepsilon) \leq 0. \quad (8.54)$$

The inequality (8.53) follows from the fact that the directional derivative of the penalized risk function (smooth and convex)

$$\Lambda \ni \lambda \mapsto P(\ell \bullet f_\lambda) + \varepsilon \sum_{j=1}^{N} \lambda_j \log \lambda_j$$

at the point of its minimum $\lambda^\varepsilon$ is nonnegative in the direction of any point of the convex set $\Lambda$, in particular, in the direction of $\hat{\lambda}^\varepsilon$. The same observation in the case of penalized empirical risk leads to inequality (8.54). Subtract (8.53) from (8.54) and replace $P$ by $P_n$ in (8.54) to get

$$P\Big(\big((\ell' \bullet f_{\hat{\lambda}^\varepsilon}) - (\ell' \bullet f_{\lambda^\varepsilon})\big)(f_{\hat{\lambda}^\varepsilon} - f_{\lambda^\varepsilon})\Big) + \varepsilon \sum_{j=1}^{N} \Big(\log \hat{\lambda}_j^\varepsilon - \log \lambda_j^\varepsilon\Big)(\hat{\lambda}_j^\varepsilon - \lambda_j^\varepsilon)$$

$$\leq (P - P_n)(\ell' \bullet f_{\hat{\lambda}^\varepsilon})(f_{\hat{\lambda}^\varepsilon} - f_{\lambda^\varepsilon}). \tag{8.55}$$

It is easy to see that

$$\sum_{j=1}^{N} \Big(\log \hat{\lambda}_j^\varepsilon - \log \lambda_j^\varepsilon\Big)(\hat{\lambda}_j^\varepsilon - \lambda_j^\varepsilon) = \sum_{j=1}^{N} \Big(\log \frac{\hat{\lambda}_j^\varepsilon}{\lambda_j^\varepsilon}\Big)(\hat{\lambda}_j^\varepsilon - \lambda_j^\varepsilon) = K(\hat{\lambda}^\varepsilon, \lambda^\varepsilon)$$

and rewrite bound (8.55) as

$$P\Big(\big((\ell' \bullet f_{\hat{\lambda}^\varepsilon}) - (\ell' \bullet f_{\lambda^\varepsilon})\big)(f_{\hat{\lambda}^\varepsilon} - f_{\lambda^\varepsilon})\Big) + \varepsilon K(\hat{\lambda}^\varepsilon; \lambda^\varepsilon)$$

$$\leq (P - P_n)(\ell' \bullet f_{\hat{\lambda}^\varepsilon})(f_{\hat{\lambda}^\varepsilon} - f_{\lambda^\varepsilon}). \tag{8.56}$$

We use the following simple inequality

$$K(\hat{\lambda}^\varepsilon, \lambda^\varepsilon) = \sum_{j=1}^{N} \Big(\log \frac{\hat{\lambda}_j^\varepsilon}{\lambda_j^\varepsilon}\Big)(\hat{\lambda}_j^\varepsilon - \lambda_j^\varepsilon)$$

$$\geq \frac{\log 2}{2} \sum_{j:\hat{\lambda}_j^\varepsilon \geq 2\lambda_j^\varepsilon} \hat{\lambda}_j^\varepsilon + \frac{\log 2}{2} \sum_{j:\lambda_j^\varepsilon \geq 2\hat{\lambda}_j^\varepsilon} \lambda_j^\varepsilon, \tag{8.57}$$

which implies that for all $J \subset \{1, \ldots, N\}$

$$\sum_{j \notin J} \hat{\lambda}_j^\varepsilon \leq 2 \sum_{j \notin J} \lambda_j^\varepsilon + \frac{2}{\log 2} K(\hat{\lambda}^\varepsilon, \lambda^\varepsilon) \tag{8.58}$$

and

$$\sum_{j \notin J} \lambda_j^\varepsilon \leq 2 \sum_{j \notin J} \hat{\lambda}_j^\varepsilon + \frac{2}{\log 2} K(\hat{\lambda}^\varepsilon, \lambda^\varepsilon). \tag{8.59}$$

If $K(\hat{\lambda}^\varepsilon, \lambda^\varepsilon)$ is small, the last bounds show that "sparsity patterns" of vectors $\hat{\lambda}^\varepsilon$ and $\lambda^\varepsilon$ are closely related. Then, it follows from (8.56) that

$$\varepsilon \sum_{j \notin J} \hat{\lambda}_j^\varepsilon \leq 2\varepsilon \sum_{j \notin J} \lambda_j^\varepsilon + \frac{2}{\log 2} (P - P_n)(\ell' \bullet f_{\hat{\lambda}^\varepsilon})(f_{\hat{\lambda}^\varepsilon} - f_{\lambda^\varepsilon}). \tag{8.60}$$

As in the previous section, for the loss functions of quadratic type, we have

$$P\Big((\ell' \bullet f_{\hat{\lambda}^\varepsilon}) - (\ell' \bullet f_{\lambda^\varepsilon})\Big)(f_{\hat{\lambda}^\varepsilon} - f_{\lambda^\varepsilon}) \geq c\|f_{\hat{\lambda}^\varepsilon} - f_{\lambda^\varepsilon}\|_{L_2(\Pi)}^2,$$

where $c = \tau(1)$. Note that $\|f_{\lambda^\varepsilon}\|_\infty \leq 1$ and $\|f_{\hat{\lambda}^\varepsilon}\|_\infty \leq 1$. Then, bound (8.56) yields

$$c\|f_{\hat{\lambda}^\varepsilon} - f_{\lambda^\varepsilon}\|^2 + \varepsilon K(\hat{\lambda}^\varepsilon, \lambda^\varepsilon) \leq (P - P_n)(\ell' \bullet f_{\hat{\lambda}^\varepsilon})(f_{\hat{\lambda}^\varepsilon} - f_{\lambda^\varepsilon}). \tag{8.61}$$

Following the methodology of Sect. 8.1, we have now to control the empirical process $(P - P_n)(\ell' \bullet f_{\hat{\lambda}^\varepsilon})(f_{\hat{\lambda}^\varepsilon} - f_{\lambda^\varepsilon})$. To this end, let

$$\Lambda(\delta; \Delta) := \Big\{\lambda \in \Lambda : \|f_\lambda - f_{\lambda^\varepsilon}\|_{L_2(\Pi)} \leq \delta, \ \sum_{j \notin J} \lambda_j \leq \Delta\Big\}$$

and

$$\alpha_n(\delta; \Delta) := \sup\Big\{|(P_n - P)((\ell' \bullet f_\lambda)(f_\lambda - f_{\lambda^\varepsilon}))| : \lambda \in \Lambda(\delta; \Delta)\Big\}.$$

The following two lemmas are similar to Lemmas 8.2 and 8.1 of the previous section. Their proofs are also similar and we skip them.

**Lemma 8.3.** *Under the assumptions of Theorem 8.5, there exists constant $C$ that depends only on $\ell$ such that with probability at least $1 - N^{-A}$, for all*

$$n^{-1/2} \leq \delta \leq 1 \text{ and } n^{-1/2} \leq \Delta \leq 1$$

*the following bound holds:*

$$\alpha_n(\delta; \Delta) \leq \beta_n(\delta; \Delta) := C\Bigg[\delta\sqrt{\frac{d + A\log N}{n}} \bigvee \Delta\sqrt{\frac{d + A\log N}{n}}$$

$$\bigvee \sum_{j \notin J} \lambda_j^\varepsilon \sqrt{\frac{d + A\log N}{n}} \bigvee \frac{A\log N}{n}\Bigg]. \tag{8.62}$$

**Lemma 8.4.** *Under the assumptions of Theorem 8.6, there exists constant $C$ that depends only on $\ell$ such that with probability at least $1 - N^{-A}$, for all*

$$n^{-1/2} \leq \delta \leq 1 \quad \text{and} \quad n^{-1/2} \leq \Delta \leq 1 \tag{8.63}$$

*the following bound holds:*

$$
\alpha_n(\delta; \Delta) \leq \beta_n(\delta; \Delta) := C\left[ \delta\sqrt{\frac{d + A\log N}{n}} \bigvee \Delta\sqrt{\frac{A\log N}{n}} \right.
$$

$$
\bigvee \sum_{j \notin J} \lambda_j^{\varepsilon} \sqrt{\frac{A\log N}{n}} \bigvee \max_{j \in J} \|P_{L^{\perp}}h_j\|_{L_2(\Pi)} \sqrt{\frac{A\log N}{n}} \bigvee
$$

$$
\left. \frac{U(L)\log N}{n} \bigvee \frac{A\log N}{n} \right]. \tag{8.64}
$$

We now proceed exactly as in the proof of Theorem 8.2. Let

$$\delta = \|f_{\hat{\lambda}^{\varepsilon}} - f_{\lambda^{\varepsilon}}\|_{L_2(\Pi)} \quad \text{and} \quad \Delta = \sum_{j \notin J} \hat{\lambda}_j^{\varepsilon}, \tag{8.65}$$

and suppose $\delta \geq n^{-1/2}, \Delta \geq n^{-1/2}$ (the case $\delta < n^{-1/2}$ or $\Delta < n^{-1/2}$ is even simpler). Then, by Lemma 8.4 and bounds (8.61), (8.60), the following inequalities hold with probability at least $1 - N^{-A}$:

$$c\delta^2 \leq \beta_n(\delta, \Delta) \tag{8.66}$$

and

$$\varepsilon\Delta \leq 2\varepsilon \sum_{j \notin J} \lambda_j^{\varepsilon} + \frac{2}{\log 2}\beta_n(\delta, \Delta), \tag{8.67}$$

where $\beta_n(\delta, \Delta)$ is defined in (8.64). Thus, it remains to solve the inequalities (8.66), (8.67) to complete the proof. First, rewrite (8.67) (with a possible change of constant $C$) as

$$
\varepsilon\Delta \leq C\Delta\sqrt{\frac{A\log N}{n}} + C\left[ \varepsilon\sum_{j \notin J}\lambda_j^{\varepsilon} \bigvee \delta\sqrt{\frac{d + A\log N}{n}} \bigvee \right.
$$

$$
\left. \sum_{j \notin J}\lambda_j^{\varepsilon}\sqrt{\frac{A\log N}{n}} \bigvee \max_{j \in J}\|P_{L^{\perp}}h_j\|_{L_2(\Pi)}\sqrt{\frac{A\log N}{n}} \bigvee \frac{U(L)\log N}{n} \bigvee \frac{A\log N}{n} \right].
$$

If the constant $D$ in condition (8.49) satisfies $D \geq 2C \vee 1$, then the term

$$\sum_{j \notin J}\lambda_j^{\varepsilon}\sqrt{\frac{A\log N}{n}}$$

in the maximum can be dropped since it is smaller than the first term $\varepsilon \sum_{j \notin J} \lambda_j^\varepsilon$, and the bound can be easily rewritten as follows:

$$\Delta \le \Delta(\delta) := C\left[\sum_{j \notin J} \lambda_j^\varepsilon \bigvee \frac{\delta}{\varepsilon}\sqrt{\frac{d + A\log N}{n}} \bigvee \max_{j \in J}\|P_{L^\perp}h_j\|_{L_2(\Pi)} \bigvee \right.$$

$$\left. \frac{U(L)\log N}{n\varepsilon} \bigvee \sqrt{\frac{A\log N}{n}}\right].$$

Using the fact that $\beta_n(\delta, \Delta)$ is nondecreasing in $\Delta$, substituting $\Delta(\delta)$ instead of $\Delta$ in (8.66) and dropping the smallest terms, we get

$$\delta^2 \le C\left[\delta\sqrt{\frac{d + A\log N}{n}} \bigvee \sum_{j \notin J}\lambda_j^\varepsilon\sqrt{\frac{A\log N}{n}} \bigvee \right.$$

$$\left. \max_{j \in J}\|P_{L^\perp}h_j\|_{L_2(\Pi)}\sqrt{\frac{A\log N}{n}} \bigvee \frac{U(L)\log N}{n} \bigvee \frac{A\log N}{n}\right].$$

Solving the inequality yields the following bound on $\delta^2$:

$$\delta^2 \le C\left[\frac{d + A\log N}{n} \bigvee \sum_{j \notin J}\lambda_j^\varepsilon\sqrt{\frac{A\log N}{n}} \bigvee \right. \tag{8.68}$$

$$\left. \max_{j \in J}\|P_{L^\perp}h_j\|_{L_2(\Pi)}\sqrt{\frac{A\log N}{n}} \bigvee \frac{U(L)\log N}{n}\right].$$

We substitute this into the expression for $\Delta(\delta)$ which results in the following bound on $\Delta$:

$$\Delta \le C\left[\sum_{j \notin J}\lambda_j^\varepsilon \bigvee \frac{d + A\log N}{n\varepsilon} \bigvee \left(\sum_{j \notin J}\lambda_j^\varepsilon\right)^{1/2}\frac{1}{\varepsilon}\left(\frac{A\log N}{n}\right)^{1/4}\right.$$

$$\sqrt{\frac{d + A\log N}{n}} \bigvee \sqrt{\frac{U(L)\log N}{n\varepsilon}}\sqrt{\frac{d + A\log N}{n}} \bigvee$$

$$\max_{j \in J}\|P_{L^\perp}h_j\|_{L_2(\Pi)}^{1/2}\frac{1}{\varepsilon}\left(\frac{A\log N}{n}\right)^{1/4}\sqrt{\frac{d + A\log N}{n}} \bigvee$$

$$\left. \max_{j \in J}\|P_{L^\perp}h_j\|_{L_2(\Pi)} \bigvee \frac{U(L)\log N}{n\varepsilon} \bigvee \sqrt{\frac{A\log N}{n}}\right],$$

The inequality $ab \le (a^2 + b^2)/2$ and the condition $\frac{1}{\varepsilon}\sqrt{\frac{A\log N}{n}} \le 1$, allows us to simplify the last bound and to get

$$\Delta \leq C \left[ \sum_{j \notin J} \lambda_j^\varepsilon \bigvee \frac{d + A \log N}{n\varepsilon} \bigvee \max_{j \in J} \| P_{L^\perp} h_j \|_{L_2(\Pi)} \bigvee \right.$$

$$\left. \frac{U(L) \log N}{n\varepsilon} \bigvee \sqrt{\frac{A \log N}{n}} \right] \tag{8.69}$$

with a constant $C$ depending only on $\ell$. Substitute bounds (8.68) and (8.69) in the expression for $\beta_n(\delta, \Delta)$. With a little further work and using Lemma 8.4, we get the following bound on $\alpha_n(\delta, \Delta)$ that holds for $\delta, \Delta$ defined by (8.65) with probability at least $1 - N^{-A}$:

$$\alpha_n(\delta, \Delta) \leq C \left[ \frac{d + A \log N}{n} + \sum_{j \notin J} \lambda_j^\varepsilon \sqrt{\frac{A \log N}{n}} \bigvee \right.$$

$$\left. \max_{j \in J} \| P_{L^\perp} h_j \|_{L_2(\Pi)} \sqrt{\frac{A \log N}{n}} \bigvee \frac{U(L) \log N}{n} \right].$$

This bound and (8.61) imply that

$$c \| f_{\hat{\lambda}^\varepsilon} - f_{\lambda^\varepsilon} \|_{L_2(\Pi)}^2 + \varepsilon K(\hat{\lambda}^\varepsilon, \lambda^\varepsilon) \leq C \left[ \frac{d + A \log N}{n} + \sum_{j \notin J} \lambda_j^\varepsilon \sqrt{\frac{A \log N}{n}} \bigvee \right.$$

$$\left. \max_{j \in J} \| P_{L^\perp} h_j \|_{L_2(\Pi)} \sqrt{\frac{A \log N}{n}} \bigvee \frac{U(L) \log N}{n} \right], \tag{8.70}$$

and (8.52) follows. Bound (8.50) is an immediate consequence of (8.69); bound (8.51) follows from (8.59) and (8.70). □

From Theorems 8.5, 8.6 and the properties of the loss function, we will easily deduce the next result.

As in Sect. 8.2, let $\mathscr{L}$ be the linear span of the dictionary $\{h_1, \ldots, h_N\}$ in the space $L_2(P)$ and let $P_{\mathscr{L}}$ be the orthogonal projector on $\mathscr{L} \subset L_2(P)$. Define

$$g_\varepsilon := P_{\mathscr{L}}(\ell' \bullet f_{\lambda^\varepsilon}).$$

**Theorem 8.7.** *Under the conditions of Theorem 8.5, the following bound holds with probability at least $1 - N^{-A}$, with a constant $C > 0$ depending only on $\ell$ and with $d = \mathrm{card}(J)$:*

$$\left| P(\ell \bullet f_{\hat{\lambda}^\varepsilon}) - P(\ell \bullet f_{\lambda^\varepsilon}) \right| \le C\left[\frac{d + A \log N}{n} \bigvee \sum_{j \notin J} \lambda_j^\varepsilon \sqrt{\frac{d + A \log N}{n}}\right] \bigvee$$

$$C^{1/2}\|g_\varepsilon\|_{L_2(\Pi)}\left[\frac{d + A \log N}{n} \bigvee \sum_{j \notin J} \lambda_j^\varepsilon \sqrt{\frac{d + A \log N}{n}}\right]^{1/2}. \tag{8.71}$$

*Similarly, under the conditions of Theorem 8.6, with probability at least* $1 - N^{-A}$ *and with* $d = \dim(L)$

$$\left| P(\ell \bullet f_{\hat{\lambda}^\varepsilon}) - P(\ell \bullet f_{\lambda^\varepsilon}) \right|$$

$$\le C\left[\frac{d + A \log N}{n} \bigvee \left(\sum_{j \notin J} \lambda_j^\varepsilon \bigvee \max_{j \in J} \|P_{L^\perp} h_j\|_{L_2(\Pi)}\right)\right.$$

$$\left.\sqrt{\frac{A \log N}{n}} \bigvee \frac{U(L) \log N}{n}\right] \bigvee C^{1/2}\|g_\varepsilon\|_{L_2(\Pi)}\left[\frac{d + A \log N}{n} \bigvee\right.$$

$$\left.\left(\sum_{j \notin J} \lambda_j^\varepsilon \bigvee \max_{j \in J} \|P_{L^\perp} h_j\|_{L_2(\Pi)}\right)\sqrt{\frac{A \log N}{n}} \bigvee \frac{U(L) \log N}{n}\right]^{1/2}. \tag{8.72}$$

*Proof.* For the losses of quadratic type,

$$(\ell \bullet f_{\hat{\lambda}^\varepsilon})(x, y) - (\ell \bullet f_{\lambda^\varepsilon})(x, y) = (\ell' \bullet f_{\lambda^\varepsilon})(x, y)(f_{\hat{\lambda}^\varepsilon} - f_{\lambda^\varepsilon})(x) + R(x, y),$$

where

$$|R(x, y)| \le C(f_{\hat{\lambda}^\varepsilon} - f_{\lambda^\varepsilon})^2(x).$$

Integrate with respect to $P$ and get

$$\left| P(\ell \bullet f_{\hat{\lambda}^\varepsilon}) - P(\ell \bullet f_{\lambda^\varepsilon}) - P(\ell' \bullet f_{\lambda^\varepsilon})(f_{\hat{\lambda}^\varepsilon} - f_{\lambda^\varepsilon}) \right| \le C\|f_{\hat{\lambda}^\varepsilon} - f_{\lambda^\varepsilon}\|^2_{L_2(\Pi)}.$$

Since

$$\left| P(\ell' \bullet f_{\lambda^\varepsilon})(f_{\hat{\lambda}^\varepsilon} - f_{\lambda^\varepsilon}) \right| = \left| \left\langle \ell' \bullet f_{\lambda^\varepsilon}, f_{\hat{\lambda}^\varepsilon} - f_{\lambda^\varepsilon} \right\rangle_{L_2(P)} \right|$$

$$= \left| \left\langle P_{\mathscr{L}}(\ell' \bullet f_{\lambda^\varepsilon}), f_{\hat{\lambda}^\varepsilon} - f_{\lambda^\varepsilon} \right\rangle_{L_2(P)} \right| \le \|g_\varepsilon\|_{L_2(P)}\|f_{\hat{\lambda}^\varepsilon} - f_{\lambda^\varepsilon}\|_{L_2(\Pi)}$$

Theorems 8.5 and 8.6 imply the result.    □

Recall that $f_*$ is a function that minimizes the risk $P(\ell \bullet f)$ and that $f_*$ is uniformly bounded by a constant $M$. It follows from necessary conditions of minimum that

$$P(\ell' \bullet f_*)h_j = 0, \quad j = 1, \ldots, N,$$

or $\ell' \bullet f_* \in \mathcal{L}^{\perp}$. For any function $\bar{f}$ uniformly bounded by $M$ and such that $\ell' \bullet \bar{f} \in \mathcal{L}^{\perp}$ (for instance, for $f_*$), the following bounds hold

$$\|g_\varepsilon\|_{L_2(\Pi)} = \|P_{\mathcal{L}}(\ell' \bullet f_{\lambda^\varepsilon})\|_{L_2(P)} = \|P_{\mathcal{L}}(\ell' \bullet f_{\lambda^\varepsilon} - \ell' \bullet \bar{f})\|_{L_2(P)}$$

$$\leq \|(\ell' \bullet f_{\lambda^\varepsilon} - \ell' \bullet \bar{f})\|_{L_2(P)} \leq C\|f_{\lambda^\varepsilon} - \bar{f}\|_{L_2(\Pi)}$$

since $\ell'$ is Lipschitz with respect to the second variable.

Since $\ell$ is the loss of quadratic type, we have, for all $\lambda \in \Lambda$,

$$\mathcal{E}(f_\lambda) \geq \frac{1}{2}\tau(\|f_*\|_\infty \vee 1)\|f_\lambda - f_*\|^2_{L_2(\Pi)} =: \tau\|f_\lambda - f_*\|^2_{L_2(\Pi)}. \tag{8.73}$$

Note that

$$|\mathcal{E}(f_{\hat{\lambda}^\varepsilon}) - \mathcal{E}(f_{\lambda^\varepsilon})| = |P(\ell \bullet f_{\hat{\lambda}^\varepsilon}) - P(\ell \bullet f_{\lambda^\varepsilon})|.$$

Thus, Theorem 8.7 implies the following bound on the random error $|\mathcal{E}(f_{\hat{\lambda}^\varepsilon}) - \mathcal{E}(f_{\lambda^\varepsilon})|$: under the conditions of Theorem 8.5, with probability at least $1 - N^{-A}$

$$\left|\mathcal{E}(f_{\hat{\lambda}^\varepsilon}) - \mathcal{E}(f_{\lambda^\varepsilon})\right| \leq C\left[\frac{d + A\log N}{n} \vee \sum_{j \notin J}\lambda_j^\varepsilon \sqrt{\frac{d + A\log N}{n}}\right] \vee$$

$$C^{1/2}\sqrt{\frac{\mathcal{E}(f_{\lambda^\varepsilon})}{\tau}}\left[\frac{d + A\log N}{n} \vee \sum_{j \notin J}\lambda_j^\varepsilon \sqrt{\frac{d + A\log N}{n}}\right]^{1/2}, \tag{8.74}$$

where $d = d(J)$, and under the conditions of Theorem 8.6, with probability at least $1 - N^{-A}$

$$\left|\mathcal{E}(f_{\hat{\lambda}^\varepsilon}) - \mathcal{E}(f_{\lambda^\varepsilon})\right|$$

$$\leq C\left[\frac{d + A\log N}{n} \vee \left(\sum_{j \notin J}\lambda_j^\varepsilon \vee \max_{j \in J}\|P_{L^{\perp}}h_j\|_{L_2(\Pi)}\right)\right.$$

$$\left.\sqrt{\frac{A\log N}{n}} \vee \frac{U(L)\log N}{n}\right] \vee C^{1/2}\sqrt{\frac{\mathcal{E}(f_{\lambda^\varepsilon})}{\tau}}\left[\frac{d + A\log N}{n} \vee\right.$$

$$\left.\left(\sum_{j \notin J}\lambda_j^\varepsilon \vee \max_{j \in J}\|P_{L^{\perp}}h_j\|_{L_2(\Pi)}\right)\sqrt{\frac{A\log N}{n}} \vee \frac{U(L)\log N}{n}\right]^{1/2}, \tag{8.75}$$

where $d = \dim(L)$.

## 8.4   Approximation Error Bounds, Alignment and Oracle Inequalities

To consider the approximation error, we will use the definitions of alignment coefficients from Sect. 7.2.3. For $\lambda \in \mathbb{R}^N$, let $s_j^N(\lambda) := \log(eN^2\lambda_j)$, $j \in \text{supp}(\lambda)$ and $s_j^N(\lambda) := 0$, $j \notin \text{supp}(\lambda)$. Note that $\log \lambda_j + 1$ is the derivative of the function $\lambda \log \lambda$ involved in the definition of the penalty and, for $j \in \text{supp}(\lambda)$, $s_j^N(\lambda) = \log \lambda_j + 1 + 2 \log N$. Introduce the following vector

$$s^N(\lambda) := (s_1^N(\lambda), \dots, s_N^N(\lambda)).$$

We will show that both the approximation error $\mathscr{E}(f_{\lambda^\varepsilon})$ and the "approximate sparsity" of $\lambda^\varepsilon$ can be controlled in terms of the alignment coefficient of the vector $s^N(\lambda)$ for an arbitrary "oracle" vector $\lambda \in \Lambda$. We will use the following version of the alignment coefficient:

$$\alpha_N(\lambda) := a_H^{(b)}(\Lambda, \lambda, s^N(\lambda)) \vee 0,$$

where

$$b := b(\lambda) := 2\|s^N(\lambda)\|_{\ell_\infty}.$$

**Theorem 8.8.** *There exists a constant $C > 0$ that depends only on $\ell$ and on the constant $M$ such that $\|f_*\|_\infty \leq M$ with the following property. For all $\varepsilon > 0$ and all $\lambda \in \Lambda$,*

$$\mathscr{E}(f_{\lambda^\varepsilon}) + \varepsilon \sum_{j \notin \text{supp}(\lambda)} \lambda_j^\varepsilon \leq 2\mathscr{E}(f_\lambda) + C\left(\alpha_N^2(\lambda)\varepsilon^2 + \frac{\varepsilon}{N}\right). \tag{8.76}$$

*Proof.* The definition of $\lambda^\varepsilon$ implies that, for all $\lambda \in \Lambda$,

$$\mathscr{E}(f_{\lambda^\varepsilon}) + \varepsilon \sum_{j=1}^N \lambda_j^\varepsilon \log(N^2\lambda_j^\varepsilon) \leq \mathscr{E}(f_\lambda) + \varepsilon \sum_{j=1}^N \lambda_j \log(N^2\lambda_j)$$

By convexity of the function $u \mapsto u \log(N^2u)$ and the fact that its derivative is $\log(eN^2u)$,

$$\mathscr{E}(f_{\lambda^\varepsilon}) + \varepsilon \sum_{j \notin J_\lambda} \lambda_j^\varepsilon \log(N^2\lambda_j^\varepsilon)$$

$$\leq \mathscr{E}(f_\lambda) + \varepsilon \sum_{j \in J_\lambda} \left(\lambda_j \log(N^2\lambda_j) - \lambda_j^\varepsilon \log(N^2\lambda_j^\varepsilon)\right)$$

$$\leq \mathscr{E}(f_\lambda) + \varepsilon \sum_{j \in J_\lambda} \log(eN^2\lambda_j)(\lambda_j - \lambda_j^\varepsilon). \tag{8.77}$$

Note that

$$\varepsilon \sum_{j \notin J_\lambda} \lambda_j^\varepsilon = \varepsilon \sum_{j \notin J_\lambda} \lambda_j^\varepsilon \log(N^2 \lambda_j^\varepsilon)$$

$$+ \varepsilon \sum_{j \notin J_\lambda, \lambda_j^\varepsilon \leq eN^{-2}} \lambda_j^\varepsilon \left(1 - \log(N^2 \lambda_j^\varepsilon)\right) + \varepsilon \sum_{j \notin J_\lambda, \lambda_j^\varepsilon > eN^{-2}} \lambda_j^\varepsilon \left(1 - \log(N^2 \lambda_j^\varepsilon)\right).$$

We have

$$\varepsilon \sum_{j \notin J_\lambda, \lambda_j^\varepsilon > eN^{-2}} \lambda_j^\varepsilon \left(1 - \log(N^2 \lambda_j^\varepsilon)\right) \leq 0.$$

Moreover, the function

$$(0, eN^{-2}] \ni x \mapsto x(1 - \log(N^2 x))$$

is nonnegative, its maximum is attained at $x = N^{-2}$ and this maximum is equal to $N^{-2}$. Therefore, we have

$$\varepsilon \sum_{j \notin J_\lambda, \lambda_j^\varepsilon \leq eN^{-2}} \lambda_j^\varepsilon \left(1 - \log(N^2 \lambda_j^\varepsilon)\right) \leq \varepsilon \sum_{j \notin J_\lambda, \lambda_j^\varepsilon \leq eN^{-2}} N^{-2} \leq \varepsilon N^{-1}.$$

It follows that

$$\varepsilon \sum_{j \notin J_\lambda} \lambda_j^\varepsilon \leq \varepsilon \sum_{j \notin J_\lambda} \lambda_j^\varepsilon \log(N^2 \lambda_j^\varepsilon) + \varepsilon N^{-1}.$$

Recalling (8.77), we get

$$\mathscr{E}(f_{\lambda^\varepsilon}) + \varepsilon \sum_{j \notin J_\lambda} \lambda_j^\varepsilon \leq \mathscr{E}(f_\lambda) + \varepsilon \sum_{j \in J_\lambda} \log(eN^2 \lambda_j)(\lambda_j - \lambda_j^\varepsilon) + \varepsilon N^{-1}.$$

If

$$\mathscr{E}(f_\lambda) + \varepsilon N^{-1} \geq \varepsilon \sum_{j \in J_\lambda} \log(eN^2 \lambda_j)(\lambda_j - \lambda_j^\varepsilon),$$

then

$$\mathscr{E}(f_{\lambda^\varepsilon}) + \varepsilon \sum_{j \notin J_\lambda} \lambda_j^\varepsilon \leq 2\mathscr{E}(f_\lambda) + 2\varepsilon N^{-1},$$

and the bound of the theorem follows. Otherwise, we have

$$\mathscr{E}(f_{\lambda^\varepsilon}) + \varepsilon \sum_{j \notin J_\lambda} \lambda_j^\varepsilon \leq 2\varepsilon \sum_{j \in J_\lambda} \log(eN^2 \lambda_j)(\lambda_j - \lambda_j^\varepsilon),$$

which, in particular, implies that

$$\sum_{j \notin J_\lambda} \lambda_j^\varepsilon \le 2\|s_N(\lambda)\|_{\ell_\infty} \sum_{j \in J_\lambda} |\lambda_j - \lambda_j^\varepsilon|.$$

This means that $\lambda - \lambda^\varepsilon \in C_{b,\lambda}$. The definition of $\alpha_N(\lambda)$ implies in this case that

$$\mathcal{E}(f_{\lambda^\varepsilon}) + \varepsilon \sum_{j \notin J_\lambda} \lambda_j^\varepsilon \le 2\varepsilon \sum_{j \in J_\lambda} \log(eN^2\lambda_j)(\lambda_j - \lambda_j^\varepsilon) \le 2\varepsilon\alpha_N(\lambda)\|f_\lambda - f_{\lambda^\varepsilon}\|_{L_2(\Pi)}.$$

Since $\ell$ is a loss of quadratic type, we have

$$\|f_\lambda - f_{\lambda^\varepsilon}\|_{L_2(\Pi)} \le \|f_\lambda - f_*\|_{L_2(\Pi)} + \|f_{\lambda^\varepsilon} - f_*\|_{L_2(\Pi)} \le \sqrt{\frac{\mathcal{E}(f_\lambda)}{\tau}} + \sqrt{\frac{\mathcal{E}(f_{\lambda^\varepsilon})}{\tau}}$$

(see (8.73)). This yields

$$\mathcal{E}(f_{\lambda^\varepsilon}) + \varepsilon \sum_{j \notin J_\lambda} \lambda_j^\varepsilon \le 2\varepsilon\alpha_N(\lambda)\left(\sqrt{\frac{\mathcal{E}(f_\lambda)}{\tau}} + \sqrt{\frac{\mathcal{E}(f_{\lambda^\varepsilon})}{\tau}}\right).$$

Using the fact that

$$2\varepsilon\alpha_N(\lambda)\sqrt{\frac{\mathcal{E}(f_{\lambda^\varepsilon})}{\tau}} \le 2\frac{\alpha_N^2(\lambda)\varepsilon^2}{\tau} + \frac{1}{2}\mathcal{E}(f_{\lambda^\varepsilon})$$

and

$$2\varepsilon\alpha_N(\lambda)\sqrt{\frac{\mathcal{E}(f_\lambda)}{\tau}} \le 2\frac{\alpha_N^2(\lambda)\varepsilon^2}{\tau} + \frac{1}{2}\mathcal{E}(f_\lambda),$$

we get

$$\frac{1}{2}\mathcal{E}(f_{\lambda^\varepsilon}) + \varepsilon \sum_{j \notin J_\lambda} \lambda_j^\varepsilon \le \frac{1}{2}\mathcal{E}(f_\lambda) + 4\frac{\alpha_N^2(\lambda)\varepsilon^2}{\tau},$$

which completes the proof.                                                        $\square$

Theorem 8.8 and random error bounds (8.74), (8.75) imply oracle inequalities for the excess risk $\mathcal{E}(f_{\hat\lambda^\varepsilon})$. The next corollary is based on (8.75).

**Corollary 8.5.** *Under the conditions and the notations of Theorems 8.6, 8.8, for all* $\lambda \in \Lambda$ *with* $J = \mathrm{supp}(\lambda)$ *and for all subspaces* $L$ *of* $L_2(\Pi)$ *with* $d := \dim(L)$, *the following bound holds with probability at least* $1 - N^{-A}$ *and with a constant* $C$ *depending on* $\ell$ *and on* $M$:

$$\mathscr{E}(f_{\hat{\lambda}^\varepsilon}) \le 4\mathscr{E}(f_\lambda) + C\left(\frac{d + A\log N}{n} + \max_{j\in J} \|P_{L^\perp} h_j\|_{L_2(\Pi)} \sqrt{\frac{A\log N}{n}} + \right.$$

$$\left. \frac{U(L)\log N}{n} + \alpha_N^2(\lambda)\varepsilon^2 + \frac{\varepsilon}{N}\right).$$

**Remark.** Note that the constants in front of $\mathscr{E}(f_\lambda)$ in the bounds of Theorem 8.8 and Corollary 8.5 can be replaced by $1 + \delta, \delta > 0$ at a price of $C$ being dependent on $\delta$.

## 8.5 Further Comments

$\ell_1$-penalization in linear regression problems is often called LASSO, the term introduced by Tibshirani [141].

Sparsity oracle inequalities for this method have been studied by many authors, in particular, Bickel et al. [22], Bunea et al. [36], van de Geer [63], Koltchinskii [84]. In these papers, some form of restricted isometry property or its generalizations have been used (which means strong geometric assumptions on the dictionary viewed either as a subset of $L_2(\Pi_n)$ in the fixed design case, or as a subset of $L_2(\Pi)$ in the random design case). The version of sparsity oracle inequalities presented here is close to what was considered in [84]. Candes and Plan [39] study the problem of sparse recovery under weaker geometric assumptions in the case when the target vector is random.

Other type of risk bounds for LASSO (under very mild assumptions on the dictionary, but with "slow" error rates) were obtained by Bartlett et al. [18], Rigollet and Tsybakov [126], and Massart and Meynet [109].

Extensions of LASSO and related methods of complexity penalization to sparse recovery problems in high-dimensional additive modeling and multiple kernel learning can be found in Koltchinskii and Yuan [96, 97], Meier et al. [111].

There has been a considerable amount of work on entropy penalization in information theory and statistics, for instance, in problems of aggregation of statistical estimators using exponential weighting and in PAC-Bayesian methods of learning theory (see, e.g., McAllester [110], Catoni [46], Audibert [10], Zhang [154, 156, 157] and references therein). Dalalyan and Tsybakov [48] studied PAC-Bayesian method with special priors in sparse recovery problems.

The approach to sparse recovery in convex hulls based on entropy penalization was suggested by Koltchinskii [86] and it was followed in this chapter. In [86], this method was also used in density estimation problems (see [37] for another approach to sparse density estimation). Earlier, Koltchinskii [84] suggested to use $\|\cdot\|_{\ell_p}^p$ as complexity penalty, which is also a strictly convex function for $p > 1$. It was shown that, when $p = 1 + \frac{c}{\log N}$, the estimator based on penalized empirical risk minimization with such a penalty satisfies random error bounds and sparsity oracle inequalities of the same type as for entropy penalty. Koltchinskii and Minsker [91] studied extensions of the entropy penalization method to sparse recovery in infinite dictionaries.

# Chapter 9
# Low Rank Matrix Recovery: Nuclear Norm Penalization

In this chapter, we discuss a problem of estimation of a large target matrix based on a finite number of noisy measurements of linear functionals (often, random) of this matrix. The underlying assumption is that the target matrix is of small rank and the goal is to determine how the estimation error depends on the rank as well as on other important parameters of the problem such as the number of measurements and the variance of the noise. This problem can be viewed as a natural noncommutative extension of sparse recovery problems discussed in the previous chapters. As a matter of fact, low rank recovery is equivalent to sparse recovery when all the matrices in question are diagonal. There are several important instances of such problems, in particular, matrix completion [41, 45, 70, 124], matrix regression [40, 90, 127] and the problem of density matrix estimation in quantum state tomography [70, 71, 88]. We will study some of these problems using general empirical processes techniques developed in the first several chapters. Noncommutative Bernstein type inequalities established in Sect. 2.4 will play a very special role in our analysis. The main results will be obtained for Hermitian matrices. So called "Paulsen dilation" (see Sect. 2.4) can be then used to tackle the case of rectangular matrices. Throughout the chapter, we use the notations introduced in Sect. A.4.

## 9.1  Geometric Parameters of Low Rank Recovery and Other Preliminaries

In the results that follow, we will need matrix extensions of some of the geometric parameters introduced in Sect. 7.2.

Given a subspace $L \subset \mathbb{C}^m$, $P_L$ denotes the orthogonal projection onto $L$. We will need the following linear mappings $\mathscr{P}_L : \mathbb{H}_m(\mathbb{C}) \mapsto \mathbb{H}_m(\mathbb{C})$ and $\mathscr{P}_L^\perp : \mathbb{H}_m(\mathbb{C}) \mapsto \mathbb{H}_m(\mathbb{C})$ :

$$\mathscr{P}_L(B) := B - P_{L^\perp} B P_{L^\perp}, \quad \mathscr{P}_L^\perp(B) = P_{L^\perp} B P_{L^\perp}.$$

V. Koltchinskii, *Oracle Inequalities in Empirical Risk Minimization and Sparse Recovery Problems*, Lecture Notes in Mathematics 2033, DOI 10.1007/978-3-642-22147-7_9, © Springer-Verlag Berlin Heidelberg 2011

Note that, for all Hermitian matrices $B$, $\text{rank}(\mathscr{P}_L(B)) \leq 2\dim(L)$.

Given $b \in [0, +\infty]$, a subspace $L \subset \mathbb{C}^m$ and a closed convex subset $\mathbb{D} \subset \mathbb{H}_m(\mathbb{C})$, consider the following cone in the space $\mathbb{H}_m(\mathbb{C})$ :

$$\mathscr{K}(\mathbb{D}; L; b) := \left\{ B \in \text{l.s.}(\mathbb{D}) : \|\mathscr{P}_L^{\perp}(B)\|_1 \leq b\|\mathscr{P}_L(B)\|_1 \right\}.$$

Roughly, in the case when $\dim(L)$ is small, the cone $\mathscr{K}(\mathbb{D}; L; b)$ consists of matrices $B$ for which "the low rank part" $\mathscr{P}_L(B)$ is dominant and "the high rank part" $\mathscr{P}_L^{\perp}(B)$ is "small". Note that, for $b = 0$, $\mathscr{K}(\mathbb{D}; L; 0)$ is a subspace of matrices of low rank and, for $b = +\infty$, $\mathscr{K}(\mathbb{D}; L; +\infty)$ coincides with the whole linear span of $\mathbb{D}$.

Given a probability distribution $\Pi$ in $\mathbb{H}_m(\mathbb{C})$, define

$$\beta_2^{(b)}(\mathbb{D}; L; \Pi) := \inf\left\{ \beta > 0 : \|\mathscr{P}_L(B)\|_2 \leq \beta\|B\|_{L_2(\Pi)}, \ B \in \mathscr{K}(\mathbb{D}; L; b) \right\}.$$

Clearly, $\mathbb{D}_1 \subset \mathbb{D}_2$ implies that $\beta_2^{(b)}(\mathbb{D}_1; L; \Pi) \leq \beta_2^{(b)}(\mathbb{D}_2; L; \Pi)$. We will write

$$\beta_2^{(b)}(L; \Pi) := \beta_2^{(b)}(\mathbb{H}_m(\mathbb{C}); L; \Pi).$$

As in Sect. 7.2, we will also introduce a matrix version of restricted isometry constants. Namely, given $r \leq m$, define

$$\delta_r := \delta_r(\Pi) :=$$

$$\inf\left\{ \delta > 0 : (1 - \delta)\|B\|_2 \leq \|B\|_{L_2(\Pi)} \leq (1 + \delta)\|B\|_2, \ B \in \mathbb{H}_m(\mathbb{C}), \text{rank}(B) \leq r \right\}.$$

The quantity $\delta_r(\Pi)$ will be called *the matrix restricted isometry constant* of rank $r$ with respect to the distribution $\Pi$. A *matrix restricted isometry condition* holds for $\Pi$ if $\delta_r(\Pi)$ is "sufficiently small" for a certain value of $r$ (in low rank recovery problems, it usually depends on the rank of the target matrix).

Define also the following measure of "correlation" between two orthogonal (in the Hilbert–Schmidt sense) matrices of small rank:

$$\rho_r := \rho_r(\Pi) := \sup\left\{ \left| \frac{\langle B_1, B_2 \rangle_{L_2(\Pi)}}{\|B_1\|_{L_2(\Pi)}\|B_2\|_{L_2(\Pi)}} \right| : B_1, B_2 \in \mathbb{H}_m(\mathbb{C}), \right.$$

$$\left. \text{rank}(B_1) \leq 3r, \text{rank}(B_2) \leq r, \langle B_1, B_2 \rangle = 0 \right\}.$$

Finally, define

$$m_r := m_r(\Pi) := \inf\left\{ \|B\|_{L_2(\Pi)} : B \in \mathbb{H}_m(\mathbb{C}), \|B\|_2 = 1, \text{rank}(B) \leq r \right\}$$

and

$$M_r := M_r(\Pi) := \sup\left\{\|B\|_{L_2(\Pi)} : B \in \mathbb{H}_m(\mathbb{C}), \|B\|_2 = 1, \mathrm{rank}(B) \leq r\right\}.$$

If $m_r \leq 1 \leq M_r \leq 2$, the matrix restricted isometry constant can be written as

$$\delta_r = (M_r - 1) \vee (1 - m_r).$$

Also, a simple geometric argument shows that

$$\rho_r \leq \frac{1}{2}\left[\left(\frac{1+\delta_{4r}}{1-\delta_{3r}}\right)^2 + \left(\frac{1+\delta_{4r}}{1-\delta_r}\right)^2 - 2\right] \vee \frac{1}{2}\left[2 - \left(\frac{1-\delta_{4r}}{1+\delta_{3r}}\right)^2 - \left(\frac{1-\delta_{4r}}{1+\delta_r}\right)^2\right].$$

The next statement is a matrix version of Lemma 7.2 and its proof is a rather straightforward modification of the proof in the vector case.

**Lemma 9.1.** *Let $L \subset \mathbb{C}^m$ be a subspace with $\dim(L) = r$. Suppose that $\rho_r < \frac{m_{3r}}{b\sqrt{2}M_r}$. Then, for all $B \in \mathscr{K}(\mathbb{H}_m(\mathbb{C}); L; b)$,*

$$\|\mathscr{P}_L(B)\|_2 \leq \frac{1}{m_{3r} - b\sqrt{2}\rho_r M_r}\|B\|_{L_2(\Pi)},$$

*and, as a consequence,*

$$\beta_2^{(b)}(L;\Pi) \leq \frac{1}{m_{3r} - b\sqrt{2}\rho_r M_r}.$$

*Also, for all $B \in \mathscr{K}(\mathbb{H}_m(\mathbb{C}); L; b)$,*

$$\|B\|_2 \leq \frac{(2b^2+1)^{1/2}}{m_{3r} - b\sqrt{2}\rho_r M_r}\|B\|_{L_2(\Pi)}.$$

It follows from Lemma 9.1 that as soon as $\delta_{4r} \leq c$ for a sufficiently small $c > 0$, $\beta_2^{(b)}(L;\Pi)$ is bounded from above by a constant $C$ (depending on $c$) provided that $\dim(L) \leq r$.

To control the quantities $m_r$ and $M_r$, it is convenient to discretize the infimum and the supremum in their definitions, that is, to consider

$$m_r^\varepsilon := m_r^\varepsilon(\Pi) := \inf\left\{\|B\|_{L_2(\Pi)} : B \in \mathscr{S}_r^\varepsilon\right\}$$

and

$$M_r^\varepsilon := M_r^\varepsilon(\Pi) := \sup\left\{\|B\|_{L_2(\Pi)} : B \in \mathscr{S}_r^\varepsilon\right\},$$

where $\mathscr{S}_r^\varepsilon$ is a minimal proper $\varepsilon$-net for the set

$$\mathscr{S}_r := \left\{ B \in \mathbb{H}_m(\mathbb{C}) : \|B\|_2 = 1, \operatorname{rank}(B) \le r \right\}$$

(that is, a set of points of $\mathscr{S}_r$ of the smallest possible cardinality such that any $S \in \mathscr{S}_r$ is within distance $\varepsilon$ from the set).

**Lemma 9.2.** *For all $\varepsilon < 2^{-1/2}$, the following bounds hold:*

$$M_r(\Pi) \le \frac{M_r^\varepsilon(\Pi)}{1 - \sqrt{2}\varepsilon} \tag{9.1}$$

*and*

$$m_r(\Pi) \ge m_r^\varepsilon(\Pi) - \frac{\sqrt{2}M_r^\varepsilon(\Pi)\varepsilon}{1 - \sqrt{2}\varepsilon}. \tag{9.2}$$

*Proof.* Note that, for all $B_1, B_2 \in \mathscr{S}_r$,

$$\|B_1 - B_2\|_{L_2(\Pi)} \le \sqrt{2}M_r(\Pi)\|B_1 - B_2\|_2. \tag{9.3}$$

Indeed, since $\operatorname{rank}(B_1 - B_2) \le 2r$, this matrix can be represented as $B_1 - B_2 = A_1 + A_2$, where $A_1, A_2 \in \mathbb{H}_m(\mathbb{C})$, $\operatorname{rank}(A_1) \le r, \operatorname{rank}(A_2) \le r$ and $A_1 \perp A_2$ with respect to the Hilbert–Schmidt inner product (to obtain such a representation it is enough to write down the spectral decomposition of $B_1 - B_2$ and to split it into two orthogonal parts of rank at most $r$). Therefore,

$$\|B_1 - B_2\|_{L_2(\Pi)} \le \|A_1\|_{L_2(\Pi)} + \|A_2\|_{L_2(\Pi)} \le M_r(\Pi)(\|A_1\|_2 + \|A_2\|_2)$$

$$\le M_r(\Pi)\sqrt{2}(\|A_1\|_2^2 + \|A_2\|_2^2)^{1/2} = M_r(\Pi)\sqrt{2}\|A_1 + A_2\|_2 = \sqrt{2}M_r(\Pi)\|B_1 - B_2\|_2.$$

It immediately follows from (9.3) that

$$M_r(\Pi) \le M_r^\varepsilon(\Pi) + \sup_{B \in \mathscr{S}_r, B' \in \mathscr{S}_r^\varepsilon, \|B - B'\|_2 \le \varepsilon} \|B - B'\|_{L_2(\Pi)} \le M_r^\varepsilon(\Pi) + \sqrt{2}M_r(\Pi)\varepsilon,$$

which implies

$$M_r(\Pi) \le \frac{M_r^\varepsilon(\Pi)}{1 - \sqrt{2}\varepsilon}.$$

Similarly,

$$m_r(\Pi) \ge m_r^\varepsilon(\Pi) - \sqrt{2}M_r(\Pi)\varepsilon \ge m_r^\varepsilon(\Pi) - \frac{\sqrt{2}M_r^\varepsilon(\Pi)\varepsilon}{1 - \sqrt{2}\varepsilon}. \qquad \square$$

Clearly, as soon as

$$\sup_{B \in \mathscr{S}_r^\varepsilon} \left| \|B\|_{L_2(\Pi)}^2 - 1 \right| \le \lambda,$$

we have $M_r^\varepsilon \le \sqrt{1 + \lambda}$ and $m_r^\varepsilon \ge \sqrt{1 - \lambda}$, and it follows from Lemma 9.2 that

$$M_r(\Pi) \leq \frac{\sqrt{1+\lambda}}{1-\sqrt{2}\varepsilon} \tag{9.4}$$

and

$$m_r(\Pi) \geq \sqrt{1-\lambda} - \frac{\sqrt{2(1+\lambda)}\varepsilon}{1-\sqrt{2}\varepsilon}. \tag{9.5}$$

When both $\lambda$ and $\varepsilon$ are small enough, this guarantees that $M_r(\Pi)$ and $m_r(\Pi)$ are close to 1 and $\delta_r(\Pi)$ is small.

Lemma 9.2 is usually combined with the following bound on the covering numbers of the set $\mathscr{S}_r$ of all matrices of rank $r$ and of unit Hilbert–Schmidt norm (see also Candes and Plan [40]).

**Lemma 9.3.** *The following bound holds:*

$$\mathrm{card}(\mathscr{S}_r^\varepsilon) \leq \left(\frac{18}{\varepsilon}\right)^{(2m+1)r}.$$

*Proof.* Given two Hermitian matrices $B, \bar{B} \in \mathscr{S}_r$ with spectral representations

$$B = \sum_{j=1}^{r} \lambda_j (e_j \otimes e_j), \quad \bar{B} = \sum_{j=1}^{r} \bar{\lambda}_j (\bar{e}_j \otimes \bar{e}_j),$$

we have

$$\|B - \bar{B}\|_2 \leq \|\lambda - \bar{\lambda}\|_{\ell_2} + 2 \max_{1 \leq j \leq r} |e_j - \bar{e}_j|, \tag{9.6}$$

where $\lambda, \bar{\lambda} \in \mathbb{R}^r$ are the vectors of the eigenvalues of $B, \bar{B}$, respectively. Indeed, we have

$$\left\|\sum_{j=1}^{r} \lambda_j (e_j \otimes e_j) - \sum_{j=1}^{r} \bar{\lambda}_j (\bar{e}_j \otimes \bar{e}_j)\right\|_2 \leq \left\|\sum_{j=1}^{r} (\lambda_j - \bar{\lambda}_j)(e_j \otimes e_j)\right\|_2$$

$$+ \left\|\sum_{j=1}^{r} \bar{\lambda}_j ((e_j - \bar{e}_j) \otimes e_j)\right\|_2 + \left\|\sum_{j=1}^{r} \bar{\lambda}_j (\bar{e}_j \otimes (e_j - \bar{e}_j))\right\|_2,$$

and it is easy to see that the first term in the right hand side is equal to $\|\lambda - \bar{\lambda}\|_{\ell_2}$ and the two remaining terms are both bounded by $\max_{1 \leq j \leq r} |e_j - \bar{e}_j|$. For instance, we have

$$\left\|\sum_{j=1}^{r} \bar{\lambda}_j ((e_j - \bar{e}_j) \otimes e_j)\right\|_2^2 = \sum_{j=1}^{r} \bar{\lambda}_j^2 \|(e_j - \bar{e}_j) \otimes e_j\|_2^2$$

$$= \sum_{j=1}^{r} \bar{\lambda}_j^2 |e_j - \bar{e}_j|^2 |e_j|^2 \leq \sum_{j=1}^{r} \bar{\lambda}_j^2 \max_{1 \leq j \leq r} |e_j - \bar{e}_j|^2 \leq \max_{1 \leq j \leq r} |e_j - \bar{e}_j|^2,$$

where we used the facts that the matrices $(e_j - \bar{e}_j) \otimes e_j$ are orthogonal and, for $\bar{B} \in \mathscr{S}_r$, $\sum_{j=1}^{r} \bar{\lambda}_j^2 = 1$.

It remains to observe that there exists an $\varepsilon/3$-covering of the unit ball in $\mathbb{R}^r$ of cardinality at most $(\frac{9}{\varepsilon})^r$. On the other hand, there exists a proper $\varepsilon/6$-covering of the set

$$U := \left\{ (u_1, \ldots, u_r) : u_j \in \mathbb{C}^m, |u_j| = 1 \right\}$$

with respect to the metric

$$d((u_1, \ldots, u_r), (v_1, \ldots, v_r)) = \max_{1 \le j \le r} |u_j - v_j|$$

that has cardinality at most $(\frac{18}{\varepsilon})^{2mr}$. This also implies the existence of a proper $\varepsilon/3$-covering of a subset $V \subset U$,

$$V := \left\{ (e_1, \ldots, e_r) : e_1, \ldots, e_r \text{ orthonormal in } \mathbb{C}^m \right\}.$$

In view of (9.6), this implies the existence of an $\varepsilon$-covering of $\mathscr{S}_r$ of the desired cardinality.                                                                             $\square$

## 9.2  Matrix Regression with Fixed Design

In this section, we study the following regression problem

$$Y_j = \langle A, X_j \rangle + \xi_j, \quad j = 1, \ldots, n, \tag{9.7}$$

where $X_j \in \mathbb{H}_m(\mathbb{C})$, $j = 1, \ldots, n$ are nonrandom Hermitian $m \times m$ matrices, $\xi, \xi_j, j = 1, \ldots, n$ are i.i.d. mean zero random variables with $\sigma_\xi^2 := \mathbb{E}\xi^2 < +\infty$ (i.i.d. random noise) and $A$ is an unknown Hermitian target matrix to be estimated based on the observations $(X_1, Y_1), \ldots, (X_n, Y_n)$. Assume that $A \in \mathbb{D} \subset \mathbb{H}_m(\mathbb{C})$, where $\mathbb{D}$ is a given closed convex set of Hermitian matrices and consider the following nuclear norm penalized least squares estimator:

$$\hat{A}^\varepsilon := \operatorname{argmin}_{S \in \mathbb{D}} \left[ n^{-1} \sum_{j=1}^{n} (Y_j - \langle S, X_j \rangle)^2 + \varepsilon \|S\|_1 \right], \tag{9.8}$$

where $\varepsilon > 0$ is a regularization parameter. Our goal is to develop upper bounds on the prediction error $\|\hat{A}^\varepsilon - A\|_{L_2(\Pi_n)}^2$, where $\Pi_n$ is the empirical distribution based on $(X_1, \ldots, X_n)$.

We will use the quantity $\beta^{(b)}(\mathbb{D}; L; \Pi_n)$ with $b = 5$ and with $L := \operatorname{supp}(S)$, where $S \in \mathbb{H}_m(\mathbb{C})$. For simplicity, denote

$$\beta_n(S) := \beta^{(5)}(\mathbb{D}; \operatorname{supp}(S); \Pi_n).$$

We will also use the following characteristics of the noise $\xi_j, j = 1, \ldots, n$ and of the design matrices $X_j, j = 1, \ldots, n$ :

$$\sigma_\xi^2 := \mathbb{E}\xi^2, \; U_\xi^{(\alpha)} := \|\xi\|_{\psi_\alpha} \vee (2\sigma_\xi), \; \alpha \geq 1$$

and

$$\sigma_X^2 := \sigma_{X,n}^2 := \left\|n^{-1}\sum_{j=1}^n X_j^2\right\|, \; U_X := U_{X,n} := \max_{1 \leq j \leq n} \|X_j\|.$$

The next theorem is the main result of this section.

**Theorem 9.1.** *Let* $\alpha \geq 1, t > 0$ *and suppose that*

$$\varepsilon \geq D\left[\sigma_\xi \sigma_X \sqrt{\frac{t + \log(2m)}{n}} \vee U_\xi^{(\alpha)} U_X \log^{1/\alpha}\left(\frac{U_\xi^{(\alpha)} U_X}{\sigma_\xi \sigma_X}\right)\frac{t + \log(2m)}{n}\right].$$

*There exists a constant* $D > 0$ *in the above condition on* $\varepsilon$ *such that with probability at least* $1 - e^{-t}$

$$\|\hat{A}^\varepsilon - A\|_{L_2(\Pi_n)}^2 \leq \inf_{S \in \mathcal{D}}\left[\|S - A\|_{L_2(\Pi_n)}^2 + 2\varepsilon\|S\|_1\right] \tag{9.9}$$

*and*

$$\|\hat{A}^\varepsilon - A\|_{L_2(\Pi_n)}^2 \leq \inf_{S \in \mathcal{D}}\left[\|S - A\|_{L_2(\Pi_n)}^2 + \varepsilon^2\beta_n(S)\mathrm{rank}(S)\right]. \tag{9.10}$$

It immediately follows from the bounds of the theorem that

$$\|\hat{A}^\varepsilon - A\|_{L_2(\Pi_n)}^2 \leq \varepsilon^2\beta_n(A)\mathrm{rank}(A) \wedge 2\varepsilon\|A\|_1.$$

If, for $r = \mathrm{rank}(A)$, $\delta_r(\Pi_n)$ is sufficiently small (that is, $\Pi_n$ satisfies a "matrix restricted isometry" condition), then $\beta_n(A)$ is bounded by a constant and the bound becomes

$$\|\hat{A}^\varepsilon - A\|_{L_2(\Pi_n)}^2 \leq C\varepsilon^2\mathrm{rank}(A) \wedge 2\varepsilon\|A\|_1.$$

*Proof.* The definition of the estimator $\hat{A}^\varepsilon$ implies that, for all $S \in \mathbb{H}_m(\mathbb{C})$,

$$\|\hat{A}^\varepsilon\|_{L_2(\Pi_n)}^2 - \left\langle\frac{2}{n}\sum_{j=1}^n Y_j X_j, \hat{A}^\varepsilon\right\rangle + \varepsilon\|\hat{A}^\varepsilon\|_1$$

$$\leq \|S\|_{L_2(\Pi_n)}^2 - \left\langle\frac{2}{n}\sum_{j=1}^n Y_j X_j, S\right\rangle + \varepsilon\|S\|_1.$$

Also, note that $\mathbb{E}(Y_j X_j) = \langle A, X_j \rangle X_j$. This implies that

$$\frac{1}{n} \sum_{j=1}^{n} (Y_j X_j - \mathbb{E}(Y_j X_j)) = \frac{1}{n} \sum_{j=1}^{n} \xi_j X_j =: \Xi.$$

Therefore, we have

$$\|\hat{A}^\varepsilon\|_{L_2(\Pi_n)}^2 - 2\langle \hat{A}^\varepsilon, A \rangle_{L_2(\Pi_n)} \le \|S\|_{L_2(\Pi_n)}^2 - 2\langle S, A \rangle_{L_2(\Pi_n)}$$
$$+\langle 2\Xi, \hat{A}^\varepsilon - S \rangle + \varepsilon(\|S\|_1 - \|\hat{A}^\varepsilon\|_1),$$

which implies

$$\|\hat{A}^\varepsilon - A\|_{L_2(\Pi_n)}^2 \le \|S - A\|_{L_2(\Pi_n)}^2 + 2\Delta\|\hat{A}^\varepsilon - S\|_1 + \varepsilon(\|S\|_1 - \|\hat{A}^\varepsilon\|_1), \quad (9.11)$$

where $\Delta := \|\Xi\|$. Under the assumption $\varepsilon \ge 2\Delta$, this yields

$$\|\hat{A}^\varepsilon - A\|_{L_2(\Pi_n)}^2 \le \|S - A\|_{L_2(\Pi_n)}^2 + \varepsilon(\|\hat{A}^\varepsilon - S\|_1 + \|S\|_1 - \|\hat{A}^\varepsilon\|_1)$$
$$\le \|S - A\|_{L_2(\Pi_n)}^2 + 2\varepsilon\|S\|_1. \quad (9.12)$$

It follows from Theorem 2.7 that, for some constant $C > 0$, with probability at least $1 - e^{-t}$

$$\Delta \le C\left[\sigma_\xi \sigma_X \sqrt{\frac{t + \log(2m)}{n}} \bigvee U_\xi^{(\alpha)} U_X \log^{1/\alpha}\left(\frac{U_\xi^{(\alpha)} U_X}{\sigma_\xi \sigma_X}\right) \frac{t + \log(2m)}{n}\right]. \quad (9.13)$$

Thus, bound (9.9) follows from (9.12) provided that $D \ge 2C$.

To prove the second bound, we use a necessary condition of extremum in problem (9.8): there exists $\hat{V} \in \partial\|\hat{A}^\varepsilon\|_1$ such that, for all $S \in \mathbb{D}$,

$$2\langle \hat{A}^\varepsilon, \hat{A}^\varepsilon - S \rangle_{L_2(\Pi_n)} - \left\langle \frac{2}{n} \sum_{j=1}^{n} Y_j X_j, \hat{A}^\varepsilon - S \right\rangle + \varepsilon\langle \hat{V}, \hat{A}^\varepsilon - S \rangle \le 0. \quad (9.14)$$

To see this, note that since $\hat{A}^\varepsilon$ is a minimizer of the functional

$$L_n(S) := n^{-1} \sum_{j=1}^{n} (Y_j - \langle S, X_j \rangle)^2 + \varepsilon\|S\|_1,$$

there exists $B \in \partial L_n(\hat{A}^\varepsilon)$ such that $-B$ belongs to the normal cone of convex set $\mathbb{D}$ at the point $\hat{A}^\varepsilon$ (see, e.g., Aubin and Ekeland [9], Chap. 4, Sect. 2, Corollary 6). A simple computation of subdifferential of $L_n$ shows that such a $B$ has the following representation:

$$B := 2 \int_{\mathbb{H}_m(\mathbb{C})} \langle \hat{A}^\varepsilon, H \rangle H \, \Pi_n(dH) - \frac{2}{n} \sum_{j=1}^{n} Y_j X_j + \varepsilon \hat{V}$$

for some $\hat{V} \in \partial \| \hat{A}^\varepsilon \|_1$. Since $-B$ belongs to the normal cone of $\mathbb{D}$ at $\hat{A}^\varepsilon$,

$$\langle B, \hat{A}^\varepsilon - S \rangle \le 0,$$

and (9.14) holds. Consider an arbitrary $S \in \mathbb{D}$ of rank $r$ with spectral representation $S = \sum_{j=1}^{r} \lambda_j (e_j \otimes e_j)$ and with support $L$. Then, (9.14) easily implies that, for an arbitrary $V \in \partial \| S \|_1$,

$$2 \langle \hat{A}^\varepsilon - A, \hat{A}^\varepsilon - S \rangle_{L_2(\Pi_n)} + \varepsilon \langle \hat{V} - V, \hat{A}^\varepsilon - S \rangle \le -\varepsilon \langle V, \hat{A}^\varepsilon - S \rangle + \langle 2\Xi, \hat{A}^\varepsilon - S \rangle. \quad (9.15)$$

It follows from monotonicity of subdifferential of convex function $\| \cdot \|_1$ that

$$\langle \hat{V} - V, \hat{A}^\varepsilon - S \rangle \ge 0.$$

On the other hand, a well known computation of subdifferential of the nuclear norm (see Sect. A.4) implies that

$$V := \sum_{j=1}^{r} \text{sign}(\lambda_j)(e_j \otimes e_j) + P_{L^\perp} W P_{L^\perp} = \text{sign}(S) + P_{L^\perp} W P_{L^\perp},$$

where $W \in \mathbb{H}_m(\mathbb{C})$ and $\| W \| \le 1$. Since $L$ is the support of $S$, it follows from the duality between nuclear and operator norms that there exists a matrix $W$ such that $\| W \| \le 1$ and

$$\langle P_{L^\perp} W P_{L^\perp}, \hat{A}^\varepsilon - S \rangle = \langle P_{L^\perp} W P_{L^\perp}, \hat{A}^\varepsilon \rangle = \langle W, P_{L^\perp} \hat{A}^\varepsilon P_{L^\perp} \rangle = \| P_{L^\perp} \hat{A}^\varepsilon P_{L^\perp} \|_1.$$

For such a choice of $W$, (9.15) implies that

$$2 \langle \hat{A}^\varepsilon - A, \hat{A}^\varepsilon - S \rangle_{L_2(\Pi_n)} + \varepsilon \| P_{L^\perp} \hat{A}^\varepsilon P_{L^\perp} \|_1$$
$$\le -\varepsilon \langle \text{sign}(S), \hat{A}^\varepsilon - S \rangle + \langle 2\Xi, \hat{A}^\varepsilon - S \rangle. \quad (9.16)$$

We will also use the following simple identity:

$$2 \langle \hat{A}^\varepsilon - A, \hat{A}^\varepsilon - S \rangle_{L_2(\Pi_n)} = \| \hat{A}^\varepsilon - A \|_{L_2(\Pi_n)}^2 + \| \hat{A}^\varepsilon - S \|_{L_2(\Pi_n)}^2 - \| S - A \|_{L_2(\Pi_n)}^2. \quad (9.17)$$

Note that if $\langle \hat{A}^\varepsilon - A, \hat{A}^\varepsilon - S \rangle_{L_2(\Pi_n)} \le 0$, then (9.17) implies that $\| \hat{A}^\varepsilon - A \|_{L_2(\Pi_n)}^2 \le \| S - A \|_{L_2(\Pi_n)}^2$, and (9.10) trivially holds. On the other hand, if

$$\langle \hat{A}^\varepsilon - A, \hat{A}^\varepsilon - S \rangle_{L_2(\Pi_n)} \ge 0,$$

then it easily follows from (9.16) that

$$\varepsilon\|\mathscr{P}_L^\perp(\hat{A}^\varepsilon - S)\|_1 \le \varepsilon\|\mathscr{P}_L(\hat{A}^\varepsilon - S)\|_1 + 2\Delta(\|\mathscr{P}_L(\hat{A}^\varepsilon - S)\|_1 + \|\mathscr{P}_L^\perp(\hat{A}^\varepsilon - S)\|_1).$$

As a result, under the condition $\varepsilon \ge 3\Delta$, we get that

$$\|\mathscr{P}_L^\perp(\hat{A}^\varepsilon - S)\|_1 \le 5\|\mathscr{P}_L(\hat{A}^\varepsilon - S)\|_1, \tag{9.18}$$

or $\hat{A}^\varepsilon - S \in \mathscr{K}(\mathbb{D}, L, 5)$. Therefore, recalling the definition of $\beta_n(S)$, we also have

$$\|\mathscr{P}_L(\hat{A}^\varepsilon - S)\|_2 \le \beta_n(S)\|\hat{A}^\varepsilon - S\|_{L_2(\Pi_n)}. \tag{9.19}$$

Now, using the fact that

$$|\langle \text{sign}(S), \hat{A}^\varepsilon - S \rangle| = |\langle \text{sign}(S), \mathscr{P}_L(\hat{A}^\varepsilon - S) \rangle|$$
$$\le \|\text{sign}(S)\|_2 \|\mathscr{P}_L(\hat{A}^\varepsilon - S)\|_2 = \sqrt{\text{rank}(S)}\|\mathscr{P}_L(\hat{A}^\varepsilon - S)\|_2,$$

we can deduce from (9.16), (9.17) and (9.19) that

$$\|\hat{A}^\varepsilon - A\|_{L_2(\Pi_n)}^2 + \|\hat{A}^\varepsilon - S\|_{L_2(\Pi_n)}^2 + \varepsilon\|P_{L\perp}\hat{A}^\varepsilon P_{L\perp}\|_1 \tag{9.20}$$
$$\le \|S - A\|_{L_2(\Pi_n)}^2 + \varepsilon\sqrt{\text{rank}(S)}\beta_n(S)\|\hat{A}^\varepsilon - S\|_{L_2(\Pi_n)} + \langle 2\Xi, \hat{A}^\varepsilon - S \rangle.$$

Finally, we have

$$\langle \Xi, \hat{A}^\varepsilon - S \rangle = \langle \mathscr{P}_L(\Xi), \hat{A}^\varepsilon - S \rangle + \langle \mathscr{P}_L^\perp(\Xi), \hat{A}^\varepsilon - S \rangle \le \tag{9.21}$$
$$\le \Lambda\|\mathscr{P}_L(\hat{A}^\varepsilon - S)\|_2 + \Gamma\|\mathscr{P}_L^\perp(A^\varepsilon - S)\|_1,$$

where

$$\Lambda := \|\mathscr{P}_L(\Xi)\|_2, \quad \Gamma := \|\mathscr{P}_L^\perp(\Xi)\|.$$

Note that $\Gamma \le \|\Xi\| = 2\Delta$ and

$$\Lambda^2 = \|P_L\Xi\|_2^2 + \|P_L^\perp\Xi P_L\|_2^2$$
$$\le \text{rank}(S)\|P_L\Xi\|^2 + \text{rank}(S)\|P_L^\perp\Xi P_L\|^2 \le 2\text{rank}(S)\Delta^2,$$

where we used the facts that $\text{rank}(P_L\Xi) \le \text{rank}(S), \text{rank}(P_{L\perp}\Xi P_L) \le \text{rank}(S)$ and $\|P_L\Xi\| \le \|\Xi\|, \|P_{L\perp}\Xi P_L\| \le \|\Xi\|$. As a consequence,

$$\Lambda \le \sqrt{2\text{rank}(S)}\Delta.$$

Now, we can deduce from (9.20) and (9.21) that

$$\|\hat{A}^{\varepsilon} - A\|^2_{L_2(\Pi_n)} + \|\hat{A}^{\varepsilon} - S\|^2_{L_2(\Pi_n)} + \varepsilon\|P_{L^{\perp}}\hat{A}^{\varepsilon}P_{L^{\perp}}\|_1$$

$$\leq \|S - A\|^2_{L_2(\Pi_n)} + \varepsilon\sqrt{\mathrm{rank}(S)}\beta_n(S)\|\hat{A}^{\varepsilon} - S\|_{L_2(\Pi_n)}$$

$$+2\Delta\|P_{L^{\perp}}\hat{A}^{\varepsilon}P_{L^{\perp}}\|_1 + 2\sqrt{2\mathrm{rank}(S)}\Delta\beta_n(S)\|\hat{A}^{\varepsilon} - S\|_{L_2(\Pi_n)}. \qquad (9.22)$$

If $\varepsilon \geq 3\Delta$ (which, in view of (9.13), holds provided that $D \geq 3C$), then (9.22) implies that

$$\|\hat{A}^{\varepsilon} - A\|^2_{L_2(\Pi_n)} \leq \|S - A\|^2_{L_2(\Pi_n)} + \varepsilon^2\beta_n^2(S)\mathrm{rank}(S),$$

and (9.10) follows.                                                                    $\square$

It is worth mentioning that Theorem 9.1 implies sparsity oracle inequalities in the vector recovery problems discussed in the previous chapters. It is enough to use this theorem in the case when $\mathbb{D}$ is the space of all diagonal $m \times m$ matrices with real entries and the design matrices $X_j$ also belong to $\mathbb{D}$. In this case, for all $S \in \mathbb{D}$, $\mathrm{rank}(S)$ is equal to the number of nonzero diagonal entries. Also, if $e_1, \ldots, e_m$ denotes the canonical basis, $L = L_J$ is the subspace spanned on $\{e_j : j \in J\}$ and $\Pi$ is a probability distribution in $\mathbb{D}$, then the quantity $\beta_2^{(b)}(\mathbb{D}; L; \Pi)$ coincides with $\beta_2^{(b)}(J; \Pi)$ defined in Sect. 7.2. Note also that, for $S \in \mathbb{D}$, $\|S\|_1$ coincides with the $\ell_1$-norm of the vector of diagonal entries of $S$ and the operator norm $\|S\|$ coincides with the $\ell_{\infty}$-norm of the same vector. The quantities $\sigma_X^2$ and $U_X$ become

$$\sigma_X^2 = \left\|n^{-1}\sum_{j=1}^{n}X_j^2\right\|_{\ell_{\infty}}, \quad U_X := \max_{1 \leq j \leq n}\|X_j\|_{\ell_{\infty}}$$

(with an obvious interpretation of diagonal matrices $X_j$ as vectors of their diagonal entries).

Under the notations of Chaps. 7–8, it is easy to deduce from Theorem 9.1 a corollary for the LASSO-estimator

$$\hat{\lambda}^{\varepsilon} := \mathrm{argmin}_{\lambda \in \mathbb{R}^m}\left[n^{-1}\sum_{j=1}^{n}(Y_j - f_{\lambda}(X_j))^2 + \varepsilon\|\lambda\|_1\right] \qquad (9.23)$$

of parameter $\lambda_* \in \mathbb{R}^m$ in the following regression model with fixed design:

$$Y_j = f_{\lambda_*}(X_j) + \xi_j, \quad j = 1, \ldots, n.$$

Here

$$f_{\lambda} := \sum_{j=1}^{m}\lambda_j h_j, \quad \lambda = (\lambda_1, \ldots, \lambda_m) \in \mathbb{R}^m,$$

$h_1, \ldots, h_m : S \mapsto \mathbb{R}$ is a dictionary and $X_1, \ldots, X_n \in S$ are nonrandom design points.

In this case, denote

$$\sigma_X^2 := \max_{1 \le k \le m} \left| n^{-1} \sum_{j=1}^{n} h_k^2(X_j) \right|, \quad U_X := \max_{1 \le k \le m} \max_{1 \le j \le n} |h_k(X_j)|.$$

and

$$\beta_n(\lambda) := \beta_2^{(5)}(J_\lambda; \Pi_n),$$

where $J_\lambda := \mathrm{supp}(\lambda)$.

**Corollary 9.1.** *Let* $\alpha \ge 1, t > 0$ *and suppose that*

$$\varepsilon \ge D \left[ \sigma_\xi \sigma_X \sqrt{\frac{t + \log(2m)}{n}} \bigvee U_\xi^{(\alpha)} U_X \log^{1/\alpha} \left( \frac{U_\xi^{(\alpha)} U_X}{\sigma_\xi \sigma_X} \right) \frac{t + \log(2m)}{n} \right].$$

*There exists a constant* $D > 0$ *in the above condition on* $\varepsilon$ *such that with probability at least* $1 - e^{-t}$

$$\| f_{\hat{\lambda}^\varepsilon} - f_{\lambda_*} \|_{L_2(\Pi_n)}^2 \le \inf_{\lambda \in \mathbb{R}^m} \left[ \| f_\lambda - f_{\lambda_*} \|_{L_2(\Pi_n)}^2 + 2\varepsilon \|\lambda\|_{\ell_1} \right] \qquad (9.24)$$

*and*

$$\| f_{\hat{\lambda}^\varepsilon} - f_{\lambda_*} \|_{L_2(\Pi_n)}^2 \le \inf_{\lambda \in \mathbb{R}^m} \left[ \| \lambda - \lambda_* \|_{L_2(\Pi_n)}^2 + \varepsilon^2 \beta_n(\lambda) \mathrm{card}(J_\lambda) \right]. \qquad (9.25)$$

The case of matrix regression with rectangular $m_1 \times m_2$ matrices from $\mathbb{M}_{m_1,m_2}(\mathbb{R})$ can be easily reduced to the Hermitian case using so called *Paulsen dilation* already discussed in Sect. 2.4. In this case, we still deal with the regression model (9.7) with fixed design matrices $X_1, \ldots, X_n \in \mathbb{M}_{m_1,m_2}(\mathbb{R})$ and we are interested in the nuclear norm penalized least squares estimator

$$\hat{A}^\varepsilon := \mathrm{argmin}_{S \in \mathbb{M}_{m_1,m_2}(\mathbb{R})} \left[ n^{-1} \sum_{j=1}^{n} (Y_j - \langle S, X_j \rangle)^2 + \varepsilon \|S\|_1 \right]. \qquad (9.26)$$

Recall that $J : \mathbb{M}_{m_1,m_2}(\mathbb{R}) \mapsto \mathbb{H}_{m_1+m_2}(\mathbb{C})$ is defined as follows (see also Sect. 2.4)

$$JS := \begin{pmatrix} O & S \\ S^* & O \end{pmatrix}$$

and let $\bar{J} := \frac{1}{\sqrt{2}} J$. Also observe that

$$\langle \bar{J} S_1, \bar{J} S_2 \rangle = \langle S_1, S_2 \rangle, S_1, S_2 \in \mathbb{M}_{m_1,m_2}(\mathbb{R})$$

and, for a random matrix $X$ in $\mathbb{M}_{m_1,m_2}(\mathbb{R})$ with distribution $\Pi$,

$$\|A\|^2_{L_2(\Pi)} = \mathbb{E}\langle A, X\rangle^2 = \mathbb{E}\langle \bar{J}A, \bar{J}X\rangle^2 = \|\bar{J}A\|^2_{L_2(\Pi\circ\bar{J}^{-1})}.$$

Moreover, $\|\bar{J}S\|_1 = \sqrt{2}\|S\|_1$, $S \in \mathbb{M}_{m_1,m_2}(\mathbb{R})$.

Consider a linear subspace $\mathbb{D} := \bar{J}\mathbb{M}_{m_1,m_2}(\mathbb{R}) \subset \mathbb{H}_{m_1+m_2}(\mathbb{C})$. Then, it is straightforward to see that

$$\bar{J}\hat{A}^\varepsilon = \mathrm{argmin}_{S\in\mathbb{D}}\left[n^{-1}\sum_{j=1}^n (Y_j - \langle S, \bar{J}X_j\rangle)^2 + \frac{\varepsilon}{\sqrt{2}}\|S\|_1\right],$$

and, applying the bounds of Theorem 9.1 to $\bar{J}\hat{A}^\varepsilon$, one can derive similar bounds for $\hat{A}^\varepsilon$ in the rectangular matrix case. We leave further details to the reader.

## 9.3  Matrix Regression with Subgaussian Design

We will study a matrix regression problem

$$Y_j = f_*(X_j) + \xi_j, \; j = 1,\ldots,n, \tag{9.27}$$

where $\{X_j\}$ are i.i.d. subgaussian Hermitian $m \times m$ matrices, $\{\xi_j\}$ are i.i.d. mean zero random variables, $\{X_j\}$ and $\{\xi_j\}$ are independent. The goal is to estimate the regression function $f_* : \mathbb{H}_m(\mathbb{C}) \mapsto \mathbb{R}$. We are especially interested in the case when $f_*(\cdot)$ can be well approximated by a linear oracle $\langle S, \cdot\rangle$, where $S \in \mathbb{H}_m(\mathbb{C})$ is a Hermitian matrix of a small rank. We consider the following estimator based on penalized empirical risk minimization with quadratic loss and with nuclear norm penalty:

$$\hat{A}^\varepsilon := \mathrm{argmin}_{S\in\mathbb{D}}\left[n^{-1}\sum_{j=1}^n (Y_j - \langle S, X_j\rangle)^2 + \varepsilon\|S\|_1\right], \tag{9.28}$$

where $\mathbb{D} \subset \mathbb{H}_m(\mathbb{C})$ is a closed convex set that supposedly contains reasonably good oracles and that can coincide with the whole space $\mathbb{H}_m(\mathbb{C})$, and $\varepsilon > 0$ is a regularization parameter.

To be more specific, let $X$ be a Hermitian random matrix with distribution $\Pi$ such that, for some constant $\tau > 0$ and for all Hermitian matrices $A \in \mathbb{H}_m(\mathbb{C})$, $\langle A, X\rangle$ is a subgaussian random variable with parameter $\tau^2\|A\|^2_{L_2(\Pi)}$. This property implies that $\mathbb{E}X = 0$ and, for some constant $\tau_1 > 0$,

$$\left\|\langle A, X\rangle\right\|_{\psi_2} \le \tau_1\|A\|_{L_2(\Pi)}, \; A \in \mathbb{M}_m(\mathbb{C}). \tag{9.29}$$

We will also assume that, for some constant $\tau_2 > 0$ and for all $u, v \in \mathbb{C}^m$ with $|u| = |v| = 1$,

$$\mathbb{E}|\langle Xu, v\rangle|^2 = \|v \otimes u\|^2_{L_2(\Pi)} \leq \tau_2. \tag{9.30}$$

A Hermitian random matrix $X$ satisfying these conditions will be called a *subgaussian matrix*. If, in addition, $X$ satisfies the following assumption (that is stronger than (9.30))

$$\|A\|^2_{L_2(\Pi)} = \mathbb{E}|\langle A, X\rangle|^2 = \|A\|^2_2, \ A \in \mathbb{M}_m(\mathbb{C}), \tag{9.31}$$

then $X$ will be called an *isotropic subgaussian matrix*. As it was pointed out in Sect. 1.7, this includes the following important examples:

- *Gaussian matrices:* $X$ is a symmetric random matrix with real entries such that $\{X_{ij} : 1 \leq i \leq j \leq m\}$ are independent centered normal random variables with $\mathbb{E}X^2_{ii} = 1$, $i = 1, \ldots, m$ and $\mathbb{E}X^2_{ij} = \frac{1}{2}$, $i < j$;
- *Rademacher matrices:* $X_{ii} = \varepsilon_{ii}$, $i = 1, \ldots, m$ and $X_{ij} = \frac{1}{\sqrt{2}}\varepsilon_{ij}$, $i < j$, $\{\varepsilon_{ij} : 1 \leq i \leq j \leq m\}$ being i.i.d. Rademacher random variables.

Simple properties of Orlicz norms (see Sect. A.1) imply that for subgaussian matrices

$$\|A\|_{L_p(\Pi)} = \mathbb{E}^{1/p}\left|\langle A, X\rangle\right|^p \leq c_p \tau_1 \tau_2 \|A\|_{L_2(\Pi)}$$

and

$$\|A\|_{\psi_1} := \left\|\langle A, X\rangle\right\|_{\psi_1} \leq c\tau_1\tau_2\|A\|_{L_2(\Pi)}, \ A \in \mathbb{M}_m(\mathbb{C}), p \geq 1,$$

with some numerical constants $c_p > 0$ and $c > 0$.

The following fact is well known (see, e.g., [130], Proposition 2.4).

**Proposition 9.1.** *Let $X$ be a subgaussian $m \times m$ matrix. There exists a constant $B > 0$ such that*

$$\left\|\|X\|\right\|_{\psi_2} \leq B\sqrt{m}.$$

*Proof.* Consider an $\varepsilon$-net $M \subset S^{m-1} := \{u \in \mathbb{C}^m : |u| = 1\}$ of the smallest cardinality. Then, $\text{card}(M) \leq (1 + 2/\varepsilon)^{2m}$ and it is easy to check that

$$\|X\| = \sup_{u,v \in S^{m-1}} |\langle Xu, v\rangle| \leq (1 - 2\varepsilon)^{-2} \max_{u,v \in M} |\langle Xu, v\rangle|.$$

Let $\varepsilon = 1/4$. We will use standard bounds for Orlicz norms of a maximum (see Sect. A.1) to get that, for some constants $C_1, C_2, B > 0$,

$$\left\|\|X\|\right\|_{\psi_2} \leq 4\left\|\max_{u,v \in M}\langle Xu, v\rangle\right\|_{\psi_2} \leq C_1\psi_2^{-1}(\text{card}^2(M)) \max_{u,v \in M}\left\|\langle Xu, v\rangle\right\|_{\psi_2}$$

$$\leq C_2\sqrt{\log \text{card}(M)} \max_{u,v \in M}\|v \otimes u\|_{L_2(\Pi)} \leq B\sqrt{m}. \qquad \square$$

Given $S \in \mathbb{H}_m(\mathbb{C})$, denote $f_S$ the linear functional $f_S(\cdot) := \langle S, \cdot \rangle$. Our goal is to obtain oracle inequalities on the $L_2(\Pi)$ prediction error $\|f_{\hat{A}^\varepsilon} - f_*\|_{L_2(\Pi)}^2$ in terms of the $L_2(\Pi)$-approximation error $\|f_S - f_*\|_{L_2(\Pi)}^2$ of $f_*$ by low rank oracles $S \in \mathbb{D}$.

One possible approach to this problem is to show that, in the case of i.i.d. subgaussian design, the matrix restricted isometry property holds for the empirical distribution $\Pi_n$ with a high probability and then to use the oracle inequalities for fixed design matrix regression proved in Sect. 9.2 (see Candes and Plan [40], where a similar program was implemented). Below we develop a version of this approach in the case when $f_*$ is a linear functional, $f_*(\cdot) = \langle A, \cdot \rangle$, so, Theorem 9.1 can be applied directly. We also do it only in the case of *subgaussian isotropic* design. We do not derive an oracle inequality, just a bound on the Hilbert–Schmidt error $\|\hat{A}^\varepsilon - A\|_2^2$ in terms of the rank of $A$. Later in this section, we develop a more direct approach to oracle inequalities for the random design regression.

As in Sect. 9.2, denote $\sigma_\xi^2 := \mathbb{E}\xi^2$, $U_\xi^{(\alpha)} := \|\xi\|_{\psi_\alpha} \vee (2\sigma_\xi), \alpha \geq 1$. We will also use the following notations:

$$\sigma_{X,n}^2 = \left\| n^{-1} \sum_{j=1}^n X_j^2 \right\|, \quad U_{X,n} := \max_{1 \leq j \leq n} \|X_j\|.$$

**Theorem 9.2.** *Suppose that $X$ is an isotropic subgaussian matrix and $X_1, \ldots, X_n$ are its i.i.d. copies. Let $\alpha \geq 1, t > 0$ and suppose that*

$$\varepsilon \geq D\left[ \sigma_\xi \sigma_{X,n} \sqrt{\frac{t + \log(2m)}{n}} \bigvee \right.$$
$$\left. U_\xi^{(\alpha)} U_{X,n} \log^{1/\alpha}\left( \frac{U_\xi^{(\alpha)} U_{X,n}}{\sigma_\xi \sigma_{X,n}} \right) \frac{t + \log(2m)}{n} \right]. \tag{9.32}$$

*There exists a constant $D > 0$ in the above condition on $\varepsilon$ and constants $C > 0, \nu > 0, \beta > 0, \delta_0 > 0$ with the following property. For all $\delta \in (0, \delta_0)$ and for all*

$$n \geq \nu \delta^{-2} \log(1/\delta) m \, \mathrm{rank}(A),$$

*with probability at least $1 - e^{-t} - e^{-\beta n \delta^2}$,*

$$\|\hat{A}^\varepsilon - A\|_2^2 \leq C \min\left( \varepsilon \|A\|_1, \varepsilon^2 \, \mathrm{rank}(A) \right). \tag{9.33}$$

*Moreover, there exists a constant $D_1 > 0$ such that, for all $t \in (0, n)$, the condition (9.32) on $\varepsilon$ holds with probability at least $1 - 2e^{-t}$ provided that*

$$\varepsilon \geq D_1 \left[ \sigma_\xi \sqrt{\frac{m(t + \log(2m))}{n}} \bigvee \right.$$

$$\left. U_\xi^{(\alpha)} \log^{1/\alpha} \left( \frac{U_\xi^{(\alpha)}}{\sigma_\xi} \right) \frac{\sqrt{m}(t + \log n)(t + \log(2m))}{n} \right]. \tag{9.34}$$

*Proof.* We will use the bounds of Theorem 9.1 applying them to $S = A$. Denote $r := \mathrm{rank}(A)$. To bound $\beta_n(A)$, we apply Lemma 9.1 for the empirical measure $\Pi_n$. The quantities $M_r(\Pi_n)$, $m_{3r}(\Pi_n)$ and $\rho_r(\Pi_n)$ will be bounded in terms of the restricted isometry constants $\delta_{4r}(\Pi_n)$. It is enough to show that $\delta_{4r}(\Pi_n)$ is small enough (smaller than some $\delta_0 \in (0,1)$) to guarantee that $\beta_n(A)$ is bounded by a constant. Moreover, it follows from the proof of Theorem 9.1 (see (9.18)) that $\hat{A}^\varepsilon - A \in \mathcal{K}(\mathbb{D}, L, 5)$ with $L = \mathrm{supp}(A)$, and, in this case, the last bound of Lemma 9.1 implies that $\|\hat{A}^\varepsilon - A\|_2^2 \leq C_1 \|\hat{A} - A\|_{L_2(\Pi_n)}^2$ for some constant $C_1$. Thus, to prove (9.33), it is enough to control $\delta_{4r}(\Pi_n)$.

We use Lemmas 9.2 and 9.3 to derive the following result.

**Lemma 9.4.** *Suppose that $X$ is a subgaussian isotropic random matrix. There exist constants $\nu > 0, \gamma > 0$ such that for all $1 \leq r \leq m$ and all $\delta \in (0, 1/2)$*

$$\mathbb{P}\left\{ \delta_r(\Pi_n) \geq \delta \right\} \leq \exp\{-\beta n \delta^2\},$$

*provided that $n \geq \nu \delta^{-2} \log(1/\delta) m r$.*

*Proof.* Since $X$ is isotropic subgaussian, for all $B \in \mathbb{H}_m(\mathbb{C})$ with $\|B\|_2 = 1$, $\|\langle B, X \rangle^2\|_{\psi_1} \leq c$, for some constant $c$. Let $\varepsilon \in (0, 1/2)$ and $\lambda \in (0, 1)$. Using a version of Bernstein's inequality for random variables with bounded $\psi_1$-norms, the union bound and Lemma 9.3, we get that with some constant $c_1 > 0$

$$\mathbb{P}\left\{ \sup_{B \in \mathscr{S}_r^\varepsilon} \left| \|B\|_{L_2(\Pi_n)}^2 - 1 \right| \geq \lambda \right\}$$

$$\leq \mathrm{card}(\mathscr{S}_r^\varepsilon) \sup_{B \in \mathscr{S}_r^\varepsilon} \mathbb{P}\left\{ \left| n^{-1} \sum_{j=1}^n \langle B, X_j \rangle^2 - \mathbb{E}\langle B, X \rangle^2 \right| \geq \lambda \right\}$$

$$\leq 2\mathrm{card}(\mathscr{S}_r^\varepsilon) \exp\left\{ -c_1(\lambda^2 \wedge \lambda)n \right\} \leq 2\left( \frac{18}{\varepsilon} \right)^{(2m+1)r} \exp\left\{ -c_1(\lambda^2 \wedge \lambda)n \right\}.$$

Therefore, if

$$n \geq \frac{6 \log(18/\varepsilon)}{c_1 \lambda^2} m r,$$

then, with probability at least $1 - e^{-(c_1/2)\lambda^2 n}$,

$$\sup_{B \in \mathscr{S}_r^\varepsilon} \left| \|B\|_{L_2(\Pi_n)}^2 - 1 \right| \leq \lambda.$$

We can choose $\lambda, \varepsilon$ in such a way that

$$\frac{\sqrt{1+\lambda}}{1-\sqrt{2\varepsilon}} \leq 1+\delta \quad \text{and} \quad \sqrt{1-\lambda} - \frac{\sqrt{2(1+\lambda)}\varepsilon}{1-\sqrt{2\varepsilon}} \geq 1-\delta$$

(note that, to satisfy these bounds, it is enough to choose $\lambda$ and $\varepsilon$ proportional to $\delta$ with some numerical constants). Then, bounds (9.4) and (9.5) (that followed from Lemma 9.2) imply that $M_r(\Pi_n) \leq 1+\delta$, $m_r(\Pi_n) \geq 1-\delta$ and $\delta_r(\Pi_n) \leq \delta$, which holds with probability at least $1 - e^{-\beta n \delta^2}$ for some constant $\beta > 0$.                           □

It remains to prove the last statement of the Theorem. To this end, note that

$$\sigma_{X,n}^2 = \left\| n^{-1} \sum_{j=1}^{n} X_j^2 \right\| \leq \mathbb{E}\|X\|^2 + n^{-1} \sum_{j=1}^{n} (\|X_j\|^2 - \mathbb{E}\|X\|^2).$$

For a subgaussian isotropic matrix $X$, we have

$$\mathbb{E}\|X\|^2 \leq c_1 m, \quad \left\| \|X\|^2 \right\|_{\psi_1} \leq c_2 m,$$

for some constants $c_1, c_2 > 0$. It easily follows from a version of Bernstein's inequality for random variables with bounded $\psi_1$-norms (see Sect. A.2) that, for some constant $c_3 > 0$ and with probability at least $1 - e^{-t}$,

$$\left| n^{-1} \sum_{j=1}^{n} (\|X_j\|^2 - \mathbb{E}\|X\|^2) \right| \leq c_3 m \left( \sqrt{\frac{t}{n}} \bigvee \frac{t}{n} \right) \leq c_3 m$$

provided that $t \leq n$. Therefore, we also have that, for some $c_4 > 0$, $\sigma_{X,n}^2 \leq c_4 m$. On the other hand, it easily follows from the properties of Orlicz norms and the union bound that for some $c_5 > 0$

$$\mathbb{P}\left\{ U_{X,n} \geq c_5 \sqrt{m}(t+\log n) \right\} \leq n\mathbb{P}\left\{ \|X\| \geq c_5 \sqrt{m}(t+\log n) \right\} \leq n e^{-(t+\log n)} = e^{-t}.$$

Thus, with probability at least $1 - e^{-t}$,

$$U_{X,n} \leq c_5 \sqrt{m}(t + \log n).$$

Since in the bound (9.32) $\sigma_{X,n}, U_{X,n}$ can be replaced by upper bounds, it is easy to complete the proof.                           □

Theorem 9.2 essentially shows that, in the case of subgaussian isotropic design, the nuclear norm penalized least squares estimator $\hat{A}^\varepsilon$ recovers the target matrix with the Hilbert–Schmidt error $\|\hat{A}^\varepsilon - A\|_2^2$ of the order $C \frac{\sigma_\xi^2 m \, \mathrm{rank}(A)}{n}$ (with a proper choice of regularization parameter $\varepsilon$).

We now turn to a somewhat different approach to bounding the error of (9.28) without a reduction to the fixed design case. We will obtain a general oracle inequality for a subgaussian (not necessarily isotropic) design.

Assume that $\xi \in L_{\psi_2}(\mathbb{P})$ and denote

$$U_\xi^{(2)} := \|\xi\|_{\psi_2} \vee (2\sigma_\xi).$$

Recall the notation $f_A(\cdot) := \langle A, \cdot \rangle$. Let us view functions from $L_2(\Pi)$ as random variables defined on the space $\mathbb{H}_m(\mathbb{C})$ with probability measure $\Pi$ and let $\mathscr{L} \subset L_2(\Pi)$ be a subspace of subgaussian random variables such that $f_* \in \mathscr{L}$, for all $A \in \mathbb{H}_m(\mathbb{C})$, $\langle A, \cdot \rangle \in \mathscr{L}$, and, for some constant $\tau_1 > 0$,

$$\|f\|_{\psi_2} \leq \tau_1 \|f\|_{L_2(\Pi)}, \ f \in \mathscr{L} \tag{9.35}$$

(compare with (9.29)). We assume also that condition (9.30) holds.

Recall the notation $\beta^{(b)}(\mathbb{D}; L; \Pi)$ of Sect. 9.1 and define

$$\beta(S) := \beta^{(5)}(\mathbb{D}; \mathrm{supp}(S); \Pi).$$

Finally, denote

$$q(\varepsilon) := q(\mathbb{D}; \varepsilon) := \inf_{S \in \mathbb{D}} \left[ \|f_S - f_*\|_{L_2(\Pi)}^2 + \varepsilon \|S\|_1 \right].$$

Observe that

$$q(\varepsilon) \leq \|f_*\|_{L_2(\Pi)}^2$$

(take $S = 0$ in the expression after the infimum defining $q(\varepsilon)$). Note also that if, for some $S \in \mathbb{D}$,

$$\|S\|_1 \leq \frac{1}{2} \frac{q(\varepsilon)}{\varepsilon},$$

then

$$q(\varepsilon) \leq 2\|f_S - f_*\|_{L_2(\Pi)}^2. \tag{9.36}$$

Indeed,

$$\frac{q(\varepsilon)}{\varepsilon} \leq \frac{\|f_S - f_*\|_{L_2(\Pi)}^2}{\varepsilon} + \|S\|_1 \leq \frac{\|f_S - f_*\|_{L_2(\Pi)}^2}{\varepsilon} + \frac{1}{2} \frac{q(\varepsilon)}{\varepsilon},$$

implying (9.36).

Given $t > 0$ and $\varepsilon > 0$, denote

$$\kappa := \log\left( \log n \vee \log m \vee |\log \varepsilon| \vee \log \|f_*\|_{L_2(\Pi)} \vee 2 \right),$$

$$t_{n,m} := (t + \kappa) \log n + \log(2m).$$

**Theorem 9.3.** *There exist constants* $c, C, D > 0$ *with the following property.* *Suppose that* $t \geq 1$,

$$\varepsilon \geq D\left[\sigma_\xi \sqrt{\frac{m(t + \log(2m))}{n}} \vee U_\xi^{(2)} \log\left(\frac{U_\xi^{(2)}}{\sigma_\xi}\right) \frac{m^{1/2}(t + \log(2m))}{n}\right] \quad (9.37)$$

*and* $t_{n,m} \leq cn$. *Then, the following bound holds with probability at least* $1 - e^{-t}$ :

$$\|f_{\hat{A}^\varepsilon} - f_*\|^2_{L_2(\Pi)} \leq \inf_{S \in \mathbb{D}}\left[2\|f_S - f_*\|^2_{L_2(\Pi)}\right.$$

$$\left. +C\left(\beta^2(S)\mathrm{rank}(S)\varepsilon^2 + \left(\|S\|_1^2 \vee \frac{q^2(\varepsilon)}{\varepsilon^2}\right)\frac{m t_{n,m}}{n} + n^{-1}\right)\right]. \quad (9.38)$$

*Proof.* Throughout the proof, $C, c, c_1, \ldots$ denote constants (typically, numerical or dependent only on irrelevant parameters) whose values might be different in different parts of the proof. Recall the definitions and notations used in the proof of Theorem 9.1, in particular, the definitions of $\Xi$ and $\Delta$ :

$$\Xi = n^{-1}\sum_{j=1}^{n}\xi_j X_j, \quad \Delta = \|\Xi\|.$$

*Step 1. Bounding the norm* $\|\hat{A}^\varepsilon\|_1$. We start with the following lemma:

**Lemma 9.5.** *There exists a constant* $C > 0$ *such that, for all* $\varepsilon \geq 4\Delta$, *with probability at least* $1 - e^{-n}$

$$\|\hat{A}^\varepsilon\|_1 \leq C\frac{q(\varepsilon)}{\varepsilon}.$$

*Proof.* We argue exactly as in the proof of Theorem 9.1 to get the following version of bound (9.11): for all $S \in \mathbb{D}$,

$$\|f_{\hat{A}^\varepsilon} - f_*\|^2_{L_2(\Pi_n)} + \varepsilon\|\hat{A}^\varepsilon\|_1 \leq \|f_S - f_*\|^2_{L_2(\Pi_n)} + 2\Delta\|\hat{A}^\varepsilon - S\|_1 + \varepsilon\|S\|_1, \quad (9.39)$$

which implies

$$\|f_{\hat{A}^\varepsilon} - f_*\|^2_{L_2(\Pi_n)} + \varepsilon\|\hat{A}^\varepsilon\|_1 \quad (9.40)$$

$$\leq \|f_S - f_*\|^2_{L_2(\Pi)} + (\Pi_n - \Pi)(f_S - f_*)^2 + 2\Delta\|\hat{A}^\varepsilon - S\|_1 + \varepsilon\|S\|_1.$$

Since $(f_S - f_*)^2 \in L_{\psi_1}(\Pi)$ and, moreover,

$$\|(f_S - f_*)^2\|_{\psi_1} \leq c\|f_S - f_*\|^2_{L_2(\Pi)}$$

for some constant $c > 0$, we can use a version of Bernstein's inequality for $L_{\psi_1}$ random variables (see Sect. A.2) to get that, with some constant $c_1 > 0$ and with probability at least $1 - e^{-t}$,

$$\left|(\Pi_n - \Pi)(f_S - f_*)^2\right| \le c_1 \|f_S - f_*\|^2_{L_2(\Pi)} \left(\sqrt{\frac{t}{n}} \vee \frac{t}{n}\right). \tag{9.41}$$

As a consequence, for $t \le n$, we get from (9.40) that

$$\varepsilon \|\hat{A}^\varepsilon\|_1 \le \|f_S - f_*\|^2_{L_2(\Pi)} \left(1 + c_1 \sqrt{\frac{t}{n}}\right) + 2\Delta(\|\hat{A}^\varepsilon\|_1 + \|S\|_1) + \varepsilon\|S\|_1.$$

As soon as $\varepsilon \ge 4\Delta$, this implies

$$\frac{\varepsilon}{2}\|\hat{A}^\varepsilon\|_1 \le \|f_S - f_*\|^2_{L_2(\Pi)} \left(1 + c_1 \sqrt{\frac{t}{n}}\right) + \frac{3}{2}\varepsilon\|S\|_1.$$

For $t = n$, we get

$$\|\hat{A}^\varepsilon\|_1 \le \frac{2(1 + c_1)}{\varepsilon}\|f_S - f_*\|^2_{L_2(\Pi)} + 3\|S\|_1, \tag{9.42}$$

which holds with probability at least $1 - e^{-n}$ and under the assumption $\varepsilon \ge 4\Delta$. The bound of the lemma follows by applying (9.42) to the value of $S$ for which the infimum in the definition of $q(\varepsilon)$ is attained.    □

Since $q(\varepsilon) \le \|f_*\|^2_{L_2(\Pi)}$, under the assumptions of Lemma 9.5, we have

$$\|\hat{A}^\varepsilon\|_1 \le \frac{C}{\varepsilon}\|f_*\|^2_{L_2(\Pi)} \tag{9.43}$$

(again with probability at least $1 - e^{-n}$).

Next, observe that

$$\|f_{\hat{A}^\varepsilon} - f_*\|^2_{L_2(\Pi)} \le 2\left(\|f_{\hat{A}^\varepsilon}\|^2_{L_2(\Pi)} + \|f_*\|^2_{L_2(\Pi)}\right) \tag{9.44}$$

$$\le c\left(\|\hat{A}^\varepsilon\|^2_1 \mathbb{E}\|X\|^2 + \|f_*\|^2_{L_2(\Pi)}\right) \le c_1\left(m\|\hat{A}^\varepsilon\|^2_1 + \|f_*\|^2_{L_2(\Pi)}\right)$$

$$\le c_1\left(\frac{m\|f_*\|^4_{L_2(\Pi)}}{\varepsilon^2} \vee \|f_*\|^2_{L_2(\Pi)}\right),$$

where we used bound (9.43) and Proposition 9.1. Now it is easy to see that it will be enough to consider $S \in \mathbb{D}$ for which

$$\|f_S - f_*\|^2_{L_2(\Pi)} \le c_1 \left( \frac{m\|f_*\|^4_{L_2(\Pi)}}{\varepsilon^2} \vee \|f_*\|^2_{L_2(\Pi)} \right)$$

(otherwise bound (9.38) of the theorem trivially holds). This implies that, for some $c$,

$$\|f_{\hat{A}^\varepsilon} - f_S\|_{L_2(\Pi)} \le c \left( \frac{m^{1/2}\|f_*\|^2_{L_2(\Pi)}}{\varepsilon} \vee \|f_*\|_{L_2(\Pi)} \right). \tag{9.45}$$

*Step 2. Reduction to the bounds on empirical processes.* Arguing again as in the proof of Theorem 9.1, we get the following version of bound (9.16):

$$2\langle f_{\hat{A}^\varepsilon} - f_*, f_{\hat{A}^\varepsilon} - f_S \rangle_{L_2(\Pi_n)} + \varepsilon\|P_{L\perp}\hat{A}^\varepsilon P_{L\perp}\|_1$$
$$\le -\varepsilon\langle \mathrm{sign}(S), \hat{A}^\varepsilon - S \rangle + \langle 2\Xi, \hat{A}^\varepsilon - S \rangle. \tag{9.46}$$

If $\langle f_{\hat{A}^\varepsilon} - f_*, f_{\hat{A}^\varepsilon} - f_S \rangle_{L_2(\Pi_n)} \ge 0$ and $\varepsilon \ge 3\Delta$, then (9.46) implies that $\hat{A}^\varepsilon - S \in \mathscr{K}(\mathbb{D}, L, 5)$, where $L = \mathrm{supp}(S)$ (see the proof of Theorem 9.1). Because of this,

$$\|\mathscr{P}_L(\hat{A}^\varepsilon - S)\|_2 \le \beta(S)\|\hat{A}^\varepsilon - S\|_{L_2(\Pi)}. \tag{9.47}$$

Replacing in the left hand side of (9.46) $\Pi_n$ by $\Pi$ and using the identity

$$2\langle f_{\hat{A}^\varepsilon} - f_*, f_{\hat{A}^\varepsilon} - f_S \rangle_{L_2(\Pi)}$$
$$= \|f_{\hat{A}^\varepsilon} - f_*\|^2_{L_2(\Pi)} + \|f_{\hat{A}^\varepsilon} - f_S\|^2_{L_2(\Pi)} - \|f_S - f_*\|^2_{L_2(\Pi)}, \tag{9.48}$$

we get

$$\|f_{\hat{A}^\varepsilon} - f_*\|^2_{L_2(\Pi)} + \|f_{\hat{A}^\varepsilon} - f_S\|^2_{L_2(\Pi)} + \varepsilon\|P_{L\perp}\hat{A}^\varepsilon P_{L\perp}\|_1 \tag{9.49}$$
$$\le \|f_S - f_*\|^2_{L_2(\Pi)} - \varepsilon\langle \mathrm{sign}(S), \hat{A}^\varepsilon - S \rangle + \langle 2\Xi, \hat{A}^\varepsilon - S \rangle$$
$$+ 2(\Pi - \Pi_n)(f_S - f_*)(f_{\hat{A}^\varepsilon} - f_S) + 2(\Pi - \Pi_n)(f_{\hat{A}^\varepsilon} - f_S)^2.$$

This inequality will be used when $\langle f_{\hat{A}^\varepsilon} - f_*, f_{\hat{A}^\varepsilon} - f_S \rangle_{L_2(\Pi_n)} \ge 0$ (*case A*). Alternatively, when $\langle f_{\hat{A}^\varepsilon} - f_*, f_{\hat{A}^\varepsilon} - f_S \rangle_{L_2(\Pi_n)} < 0$ (*case B*), a simpler bound holds instead of (9.49):

$$\|f_{\hat{A}^\varepsilon} - f_*\|^2_{L_2(\Pi)} + \|f_{\hat{A}^\varepsilon} - f_S\|^2_{L_2(\Pi)} \tag{9.50}$$
$$\le \|f_S - f_*\|^2_{L_2(\Pi)} + 2(\Pi - \Pi_n)(f_S - f_*)(f_{\hat{A}^\varepsilon} - f_S) + 2(\Pi - \Pi_n)(f_{\hat{A}^\varepsilon} - f_S)^2.$$

It remains to bound the empirical processes in the right hand sides of (9.49) and (9.50) in each of these two cases:

$$\langle 2\Xi, \hat{A}^\varepsilon - S \rangle = \left\langle \frac{2}{n} \sum_{j=1}^{n} \xi_j X_j, \hat{A}^\varepsilon - S \right\rangle,$$

$2(\Pi - \Pi_n)(f_{\hat{A}^\varepsilon} - f_S)^2$ and $2(\Pi - \Pi_n)(f_S - f_*)(f_{\hat{A}^\varepsilon} - f_S)$.

*Step 3. Bounding $\langle \Xi, \hat{A}^\varepsilon - S \rangle$.* We use slightly modified bounds from the proof of Theorem 9.1 to control $\langle \Xi, \hat{A}^\varepsilon - S \rangle$, which is needed only in case A:

$$\langle \Xi, \hat{A}^\varepsilon - S \rangle = \langle \mathscr{P}_L(\Xi), \hat{A}^\varepsilon - S \rangle + \langle \mathscr{P}_L^\perp(\Xi), \hat{A}^\varepsilon - S \rangle \leq$$
$$\leq \Lambda \|\mathscr{P}_L(\hat{A}^\varepsilon - S)\|_2 + \Gamma \|\mathscr{P}_L^\perp(A^\varepsilon - S)\|_1,$$

where
$$\Lambda := \|\mathscr{P}_L(\Xi)\|_2, \quad \Gamma := \|\mathscr{P}_L^\perp(\Xi)\|.$$

Using the fact that, in case A, $\hat{A}^\varepsilon - S \in \mathscr{K}(\mathbb{D}, L, 5)$ and (9.47) holds, and arguing as in the proof of Theorem 9.1, we get

$$\langle \Xi, \hat{A}^\varepsilon - S \rangle \leq \Delta \|P_{L^\perp} \hat{A}^\varepsilon P_{L^\perp}\|_1 + \sqrt{2\mathrm{rank}(S)} \Delta \beta(S) \|\hat{A}^\varepsilon - S\|_{L_2(\Pi)}$$
$$\leq \Delta \|P_{L^\perp} \hat{A}^\varepsilon P_{L^\perp}\|_1 + 2\mathrm{rank}(S)\beta^2(S)\Delta^2 + \frac{1}{4} \|f_{\hat{A}^\varepsilon} - f_S\|_{L_2(\Pi)}^2. \qquad (9.51)$$

Since $X$ is a subgaussian random matrix, we can use Proposition 9.1 to get that with some constant $c$

$$\sigma_X := \|\mathbb{E}(X - \mathbb{E}X)^2\|^{1/2} \leq \|\mathbb{E}X^2\|^{1/2} \leq \mathbb{E}^{1/2}\|X\|^2 \leq c\sqrt{m}$$

and $U_X^{(2)} := \left\| \|X\| \right\|_{\psi_2} \leq c\sqrt{m}$. It easily follows that

$$\left\| \|X\| |\xi| \right\|_{\psi_1} \leq c_1 \left\| \|X\| \right\|_{\psi_2} \|\xi\|_{\psi_2} \leq c_2\sqrt{m} \|\xi\|_{\psi_2}.$$

Therefore, the second bound of Theorem 2.7 with $\alpha = 1$ implies that, with probability at least $1 - e^{-t}$,

$$\Delta \leq \bar{\Delta} := \qquad\qquad\qquad\qquad\qquad\qquad\qquad\qquad (9.52)$$
$$C\left[ \sigma_\xi \sqrt{\frac{m(t + \log(2m))}{n}} \bigvee U_\xi^{(2)} \log\left(\frac{U_\xi^{(2)}}{\sigma_\xi}\right) \frac{m^{1/2}(t + \log(2m))}{n} \right],$$

which will be used in combination with bound (9.51).

*Step 4. Bounding $(\Pi - \Pi_n)(f_S - f_*)(f_{\hat{A}^\varepsilon} - f_S)$.* Note that

$$\left|(\Pi - \Pi_n)(f_S - f_*)(f_{\hat{A}^\varepsilon} - f_S)\right| = |\langle \Upsilon, \hat{A}^\varepsilon - S \rangle|$$

$$\leq \|\Upsilon\| \|\hat{A}^\varepsilon - S\|_1 \leq \|\Upsilon\|(\|\hat{A}^\varepsilon\|_1 + \|S\|_1), \tag{9.53}$$

where

$$\Upsilon := n^{-1} \sum_{j=1}^n \left[(f_S(X_j) - f_*(X_j))X_j - \mathbb{E}(f_S(X_j) - f_*(X_j))X_j\right]. \tag{9.54}$$

Observe that with some constant $c > 0$

$$\left\|\mathbb{E}|f_S(X) - f_*(X)|^2 X^2\right\| \leq \mathbb{E}|f_S(X) - f_*(X)|^2 \|X\|^2$$

$$\leq \mathbb{E}^{1/2}|f_S(X) - f_*(X)|^4 \mathbb{E}^{1/2}\|X\|^4 \leq cm\|f_S - f_*\|^2_{L_2(\Pi)},$$

by the properties of subgaussian matrix $X$ and "subgaussian subspace" $\mathcal{L}$. Also, with some $c > 0$,

$$\left\|(f_S(X) - f_*(X))\|X\|\right\|_{\psi_1} \leq c\sqrt{m}\|f_S - f_*\|_{L_2(\Pi)}.$$

Therefore, we can use again exponential inequalities of Theorem 2.7 to bound $\Upsilon$ as follows: with probability at least $1 - e^{-t}$ and with some $c > 0$,

$$\|\Upsilon\| \leq c\|f_S - f_*\|_{L_2(\Pi)}\left[\sqrt{\frac{m(t + \log(2m))}{n}} \bigvee \frac{m^{1/2}(t + \log(2m))}{n}\right]. \tag{9.55}$$

Under the assumption $t + \log(2m) \leq n$, we now deduce from (9.53), (9.55) and Lemma 9.5 that, with probability at least $1 - 2e^{-t}$,

$$\left|(\Pi - \Pi_n)(f_S - f_*)(f_{\hat{A}^\varepsilon} - f_S)\right|$$

$$\leq C\left(\frac{q(\varepsilon)}{\varepsilon} \bigvee \|S\|_1\right)\|f_S - f_*\|_{L_2(\Pi)}\sqrt{\frac{m(t + \log(2m))}{n}}.$$

This leads to the bound

$$\left|(\Pi - \Pi_n)(f_S - f_*)(f_{\hat{A}^\varepsilon} - f_S)\right| \leq \frac{1}{4}\|f_S - f_*\|^2_{L_2(\Pi)}$$

$$+ C\left(\|S\|_1^2 \vee \frac{q^2(\varepsilon)}{\varepsilon^2}\right)\frac{m(t + \log(2m))}{n} \tag{9.56}$$

which holds with some constant $C > 0$ and with the same probability.

*Step 5. Bounding* $(\Pi - \Pi_n)(f_{\hat{A}^\varepsilon} - f_S)^2$. Here we will use the following lemma. Given $\delta > 0$ and $R > 0$, denote

$$
\Delta_n(\delta, R) := \sup \left\{ \left| n^{-1} \sum_{j=1}^{n} \langle S_1 - S_2, X_j \rangle^2 - \| S_1 - S_2 \|_{L_2(\Pi)}^2 \right| : \right.
$$

$$
\left. S_1, S_2 \in \mathbb{H}_m(\mathbb{C}), \| S_1 - S_2 \|_{L_2(\Pi)} \le \delta, \| S_1 \|_1 \le R, \| S_2 \|_1 \le R \right\}.
$$

**Lemma 9.6.** *Suppose* $X_1, \ldots, X_n$ *are i.i.d. copies of a subgaussian matrix* $X$ *satisfying conditions (9.29), (9.30). Let* $\delta > 0$ *and* $R > 0$. *There exists a constant* $C > 0$ *such that, for all* $t > 0$, *with probability at least* $1 - e^{-t}$

$$
\Delta_n(\delta, R) \le C \left[ \delta R \sqrt{\frac{m}{n}} \bigvee \frac{R^2 m}{n} \bigvee \delta^2 \sqrt{\frac{t}{n}} \bigvee \frac{R^2 m \log n}{n} \right]. \tag{9.57}
$$

*Moreover, if* $0 < \delta_- < \delta_+$, *then, with some constant* $C > 0$ *and with probability at least* $1 - e^{-t}$, *for all* $\delta \in [\delta_-, \delta_+]$,

$$
\Delta_n(\delta, R) \le C \left[ \delta R \sqrt{\frac{m}{n}} \bigvee \frac{R^2 m}{n} \bigvee \delta^2 \sqrt{\frac{t + \kappa}{n}} \bigvee \frac{R^2 m (t + \kappa) \log n}{n} \right], \tag{9.58}
$$

*where*

$$
\kappa := 2 \log \log_2 \left( \frac{2\delta_+}{\delta_-} \right).
$$

*Proof.* Clearly, the following representation of the quantity $\Delta_n(\delta, R)$ holds:

$$
\Delta_n(\delta, R) := \sup_{f \in \mathscr{F}_{\delta, R}} \left| n^{-1} \sum_{j=1}^{n} (f^2(X_j) - P f^2) \right|,
$$

where

$$
\mathscr{F}_{\delta, R} := \{ \langle S_1 - S_2, \cdot \rangle : S_1, S_2 \in \mathbb{H}_m(\mathbb{C}), \| S_1 - S_2 \|_{L_2(\Pi)} \le \delta, \| S_1 \|_1 \le R, \| S_2 \|_1 \le R \}.
$$

To bound this empirical process, we use a powerful inequality of Mendelson (see Theorem 3.15). It implies that

$$
\mathbb{E}\Delta_n(\delta, R) \le c \left[ \sup_{f \in \mathscr{F}_{\delta, R}} \| f \|_{\psi_1} \frac{\gamma_2(\mathscr{F}_{\delta, R}; \psi_2)}{\sqrt{n}} \bigvee \frac{\gamma_2^2(\mathscr{F}_{\delta, R}; \psi_2)}{n} \right] \tag{9.59}
$$

where $c > 0$ is a constant. By assumption (9.29), the $\psi_1$ and $\psi_2$-norms of functions from the class $\mathscr{F}_{\delta, R}$ can be bounded by the $L_2(P)$-norm (up to a constant).

Therefore,

$$\sup_{f \in \mathscr{F}_{\delta,R}} \|f\|_{\psi_1} \leq c\delta. \tag{9.60}$$

The next aim is to bound Talagrand's generic chaining complexities. First note that

$$\gamma_2(\mathscr{F}_{\delta,R}; \psi_2) \leq \gamma_2(\mathscr{F}_{\delta,R}; c\| \cdot \|_{L_2(\Pi)}) \tag{9.61}$$

for some $c > 0$ (again, by the bound on the $\psi_2$-norm). Let $W_\Pi(f), f \in L_2(\Pi)$ denote the isonormal Gaussian process, that is, a centered Gaussian process with covariance

$$\mathbb{E}W_\Pi(f)W_\Pi(g) = \int_{\mathbb{H}_m(\mathbb{C})} fg \, d\Pi.$$

We can also write

$$W_\Pi(f) = \int_{\mathbb{H}_m(\mathbb{C})} f(x) W_\Pi(dx),$$

where $W_\Pi(B) := W_\Pi(I_B)$ for Borel subsets $B \subset \mathbb{H}_m(\mathbb{C})$. Clearly, by linearity of $W_\Pi$; $W_\Pi(\langle S, \cdot \rangle) = \langle S, G \rangle$, where $G$ is a random matrix with the entries

$$g_{ij} := \int_{\mathbb{H}_m(\mathbb{C})} x_{ij} W_\Pi(dx).$$

Note that $G$ is a Gaussian matrix and, as a consequence, it is subgaussian. Moreover, it satisfies condition (9.30) since, for $u, v \in \mathbb{C}^m$ with $|u| = |v| = 1$,

$$\mathbb{E}|\langle Gu, v \rangle|^2 = \mathbb{E}|W_\Pi(\langle v \otimes u, \cdot \rangle)|^2 = \|v \otimes u\|_{L_2(\Pi)}^2 \leq \tau_2.$$

It follows from Talagrand's generic chaining bound (see Theorem 3.3) that, for some constant $C > 0$,

$$\gamma_2(\mathscr{F}_{\delta,R}; c\| \cdot \|_{L_2(\Pi)}) \leq C\omega(G; \delta, R), \tag{9.62}$$

where

$$\omega(G; \delta, R) := \mathbb{E} \sup_{\|S_1 - S_2\|_{L_2(\Pi)} \leq \delta, \|S_1\|_1 \leq R, \|S_2\|_1 \leq R} |W_\Pi(\langle S_1 - S_2, \cdot \rangle)|.$$

Bounds (9.59), (9.60), (9.61) and (9.62) imply that

$$\mathbb{E}\Delta_n(\delta, R) \leq C\left[\delta \frac{\omega(G; \delta, R)}{\sqrt{n}} \bigvee \frac{\omega^2(G; \delta, R)}{n}\right]. \tag{9.63}$$

Observe that, under the assumption $\|S_1\|_1 \leq R, \|S_2\|_1 \leq R$,

$$\left|\langle S_1 - S_2, G \rangle\right| \leq \|S_1 - S_2\|_1 \|G\| \leq 2R\|G\|.$$

It follows from Proposition 9.1 that

$$\omega(G; \delta, R) \leq 2R\mathbb{E}\|G\| \leq cR\sqrt{m}.$$

The last bound can be substituted in (9.63) to give that, for some constant $C > 0$,

$$\mathbb{E}\Delta_n(\delta, R) \leq C\left[\delta R\sqrt{\frac{m}{n}} \bigvee \frac{R^2 m}{n}\right] \tag{9.64}$$

To complete the proof, we use Adamczak's version of Talagrand's concentration inequality for unbounded function classes (see Sect. 2.3). To apply this inequality, one has to bound the uniform variance and the envelope of the function class $\mathscr{F}_{\delta,R}^2$. The uniform variance is bounded as follows: with some constant $c > 0$,

$$\sup_{f \in \mathscr{F}_{\delta,R}} (Pf^4)^{1/2} = \sup_{\|S_1 - S_2\|_{L_2(\Pi)} \leq \delta, \|S_1\|_1 \leq R, \|S_2\|_1 \leq R} \mathbb{E}^{1/2}\langle S_1 - S_2, X\rangle^4 =$$

$$\sup_{\|f_{S_1} - f_{S_2}\|_{L_2(\Pi)} \leq \delta, \|S_1\|_1 \leq R, \|S_2\|_1 \leq R} \|f_{S_1} - f_{S_2}\|_{L_4(\Pi)}^2 \leq c\delta^2,$$

where we used the equivalence properties of the norms in Orlicz spaces. For the envelope, we have the following bound:

$$\sup_{f \in \mathscr{F}_{\delta,R}} f^2(X) = \sup_{\|S_1 - S_2\|_{L_2(\Pi)} \leq \delta, \|S_1\|_1 \leq R, \|S_2\|_1 \leq R} \langle S_1 - S_2, X\rangle^2 \leq 4R^2\|X\|^2$$

and

$$\left\|\max_{1 \leq i \leq n} \sup_{f \in \mathscr{F}_{\delta,R}} f^2(X_i)\right\|_{\psi_1}$$

$$\leq c_1 R^2 \left\|\|X\|^2\right\|_{\psi_1} \log n \leq c_2 R^2 \left\|\|X\|\right\|_{\psi_2}^2 \log n \leq c_3 R^2 m \log n,$$

for some constants $c_1, c_2, c_3 > 0$. Here we used well known inequalities for maxima of random variables in Orlicz spaces (see Sect. A.1). Adamczak's inequality now yields that, with some constant $C > 0$ and with probability at least $1 - e^{-t}$,

$$\Delta_n(\delta, R) \leq 2\mathbb{E}\Delta_n(\delta, R) + C\delta^2\sqrt{\frac{t}{n}} + C\frac{R^2 m t \log n}{n}. \tag{9.65}$$

It remains to combine (9.64) with (9.65) to get that with probability at least $1 - e^{-t}$

$$\Delta_n(\delta, R) \leq C\left[\delta R\sqrt{\frac{m}{n}} \bigvee \frac{R^2 m}{n} \bigvee \delta^2\sqrt{\frac{t}{n}} \bigvee \frac{R^2 m t \log n}{n}\right]. \tag{9.66}$$

The second claim is proved by a standard discretization argument based on the union bound and on the monotonicity of $\Delta_n(\delta; R)$ and its upper bound with respect to $\delta$ (see, e.g., Lemma 8.1 for a similar argument).                                                       □

We will use the second bound of Lemma 9.6 to control $(\Pi - \Pi_n)(f_{\hat{A}^\varepsilon} - f_S)^2$. By a simple algebra (in particular, using the inequality $ab \leq a^2/8 + 2b^2$ and using the fact that $t_{n,m} \leq cn$ for a sufficiently small constant $c$), this bound implies that with some sufficiently large constant $C > 0$

$$\Delta_n(\delta, R) \leq \frac{1}{4}\delta^2 + C\frac{R^2 m(t + \kappa)\log n}{n}. \tag{9.67}$$

Clearly, we also have

$$\left|(\Pi - \Pi_n)(f_{\hat{A}^\varepsilon} - f_S)^2\right| \leq \Delta_n\left(\|S\|_1 \vee \|\hat{A}^\varepsilon\|_1; \|f_{\hat{A}^\varepsilon} - f_S\|_{L_2(\Pi)}\right). \tag{9.68}$$

Due to Lemma 9.5 and (9.45), we now use bound (9.67) for

$$R := C\left(\frac{q(\varepsilon)}{\varepsilon} \vee \|S\|_1\right),$$

$\delta_- = n^{-1/2}$ and

$$\delta_+ := c\left(\frac{m^{1/2}\|f_*\|_{L_2(\Pi)}^2}{\varepsilon} \vee \|f_*\|_{L_2(\Pi)}\right) \vee n^{-1/2}.$$

We get from bounds (9.67) and (9.68) that with some constant $C > 0$ and with probability at least $1 - e^{-t}$

$$\left|(\Pi - \Pi_n)(f_{\hat{A}^\varepsilon} - f_S)^2\right| \leq \frac{1}{4}\|f_{\hat{A}^\varepsilon} - f_S\|_{L_2(\Pi)}^2$$
$$+ C\left(\|S\|_1^2 \vee \frac{q^2(\varepsilon)}{\varepsilon^2}\right)\frac{m(t + \kappa)\log n}{n}, \tag{9.69}$$

where

$$\kappa = \log\left(\log n \vee \log m \vee |\log \varepsilon| \vee \log\|f_*\|_{L_2(\Pi)} \vee 2\right).$$

Bound (9.69) holds provided that $\delta_- \leq \|f_{\hat{A}^\varepsilon} - f_S\|_{L_2(\Pi)} \leq \delta_+$. Note that we do have $\|f_{\hat{A}^\varepsilon} - f_S\|_{L_2(\Pi)} \leq \delta_+$ because of (9.45). In the case when $\|f_{\hat{A}^\varepsilon} - f_S\|_{L_2(\Pi)} \leq \delta_-$, the proof of (9.38) even simplifies, so, we consider only the main case when $\|f_{\hat{A}^\varepsilon} - f_S\|_{L_2(\Pi)} \geq \delta_-$.

*Step 6. Conclusion.* To complete the proof, it is enough to use bound (9.49) in case A and bound (9.50) in case B in combination with the resulting bounds on empirical processes obtained in steps 3–5 (namely, bounds (9.51), (9.52), (9.56) and (9.69)). By a simple algebra, we get the following bound

$$\|f_{\hat{A}^{\varepsilon}} - f_*\|_{L_2(\Pi)}^2 \leq 2\|f_S - f_*\|_{L_2(\Pi)}^2$$

$$+C\left(\beta^2(S)\mathrm{rank}(S)\varepsilon^2 + \left(\|S\|_1^2 \vee \frac{q^2(\varepsilon)}{\varepsilon^2}\right)\frac{mt_{n,m}}{n} + n^{-1}\right) \quad (9.70)$$

that holds with probability at least $1 - 4e^{-t}$. A simple adjustment of constant $C$ allows one to rewrite the probability bound as $1 - e^{-t}$, which establishes (9.38). $\square$

The following corollary clarifies the meaning of bound (9.38) of Theorem 9.3 and explains the role of quantity $q(\varepsilon)$ in this bound.

**Corollary 9.2.** *Under the assumptions of Theorem 9.3, for all $S \in \mathbb{D}$ with $\|S\|_1 \geq \frac{q(\varepsilon)}{2\varepsilon}$, the following bound holds with probability at least $1 - e^{-t}$ :*

$$\|f_{\hat{A}^{\varepsilon}} - f_*\|_{L_2(\Pi)}^2 \leq 2\|f_S - f_*\|_{L_2(\Pi)}^2$$

$$+C\left(\beta^2(S)\mathrm{rank}(S)\varepsilon^2 + \frac{\|S\|_1^2 mt_{n,m} + 1}{n}\right). \quad (9.71)$$

*On the other hand, for all $S \in \mathbb{D}$ with $\|S\|_1 \leq \frac{q(\varepsilon)}{2\varepsilon}$, with the same probability,*

$$\|f_{\hat{A}^{\varepsilon}} - f_*\|_{L_2(\Pi)}^2$$

$$\leq \left(2 + C\frac{q(\varepsilon)}{\varepsilon^2}\frac{mt_{n,m}}{n}\right)\|f_S - f_*\|_{L_2(\Pi)}^2 + C\left(\beta^2(S)\mathrm{rank}(S)\varepsilon^2 + n^{-1}\right). \quad (9.72)$$

*If*

$$C\frac{q(\varepsilon)}{\varepsilon^2}\frac{mt_{n,m}}{n} \leq 1, \quad (9.73)$$

*this implies that*

$$\|f_{\hat{A}^{\varepsilon}} - f_*\|_{L_2(\Pi)}^2 \leq 3\|f_S - f_*\|_{L_2(\Pi)}^2 + C\left(\beta^2(S)\mathrm{rank}(S)\varepsilon^2 + n^{-1}\right). \quad (9.74)$$

*Proof.* For all $S \in \mathbb{D}$ with $\|S\|_1 \geq \frac{q(\varepsilon)}{2\varepsilon}$, (9.38) immediately implies (9.71). Alternatively, if $\|S\|_1 \leq \frac{q(\varepsilon)}{2\varepsilon}$, we have $q(\varepsilon) \leq 2\|f_S - f_*\|_{L_2(\Pi)}^2$ (see (9.36)) and (9.38) implies (9.72). $\square$

**Remarks.** • Note that the leading constants 2 in oracle inequalities (9.38), (9.71) or 3 in (9.74) can be replaced by $1 + \delta$ (with constant $C$ becoming of the order $\frac{1}{\delta}$).

• Note also that, when $\|S\|_1 \leq \frac{q(\varepsilon)}{2\varepsilon}$ and, as a consequence, $q(\varepsilon) \leq 2\|f_S - f_*\|_{L_2(\Pi)}^2$, condition (9.73) is satisfied provided that

$$\varepsilon \geq D_1 \|f_S - f_*\|_{L_2(\Pi)} \sqrt{\frac{m t_{n,m}}{n}} \tag{9.75}$$

for a sufficiently large constant $D_1 > 0$. This yields the following corollary.

**Corollary 9.3.** *Suppose the assumptions of Theorem 9.3 hold. For all $S \in \mathbb{D}$ satisfying condition (9.75) with sufficiently large constant $D_1 > 0$, the following bound holds with probability at least $1 - e^{-t}$ :*

$$\|f_{\hat{A}^\varepsilon} - f_*\|^2_{L_2(\Pi)} \leq 2\|f_S - f_*\|^2_{L_2(\Pi)}$$

$$+ C\left(\beta^2(S)\mathrm{rank}(S)\varepsilon^2 + \frac{\|S\|^2_1 m t_{n,m} + 1}{n}\right). \tag{9.76}$$

- In the case when the set $\mathbb{D}$ is bounded, the following version of Theorem 9.3 holds.

**Theorem 9.4.** *Suppose that $\mathbb{D} \subset \mathbb{H}_m(\mathbb{C})$ is a bounded closed convex set and*

$$R_{\mathbb{D}} := \sup_{S \in \mathbb{D}} \|S\|_1.$$

*There exist constants $c, C, D > 0$ with the following property. Suppose that $t \geq 1$ and*

$$\varepsilon \geq D\left[\sigma_\xi \sqrt{\frac{m(t + \log(2m))}{n}} \bigvee U_\xi^{(2)} \log\left(\frac{U_\xi^{(2)}}{\sigma_\xi}\right) \frac{m^{1/2}(t + \log(2m))}{n}\right]. \tag{9.77}$$

*Denote*

$$\kappa := \log\log_2(m R_{\mathbb{D}}) \quad \text{and} \quad t_{n,m} := (t + \kappa)\log n + \log(2m)$$

*and suppose that $t_{n,m} \leq cn$. Then, the following bound holds with probability at least $1 - e^{-t}$ :*

$$\|f_{\hat{A}^\varepsilon} - f_*\|^2_{L_2(\Pi)} \leq \inf_{S \in \mathbb{D}}\left[2\|f_S - f_*\|^2_{L_2(\Pi)}\right.$$

$$\left. + C\left(\beta^2(S)\mathrm{rank}(S)\varepsilon^2 + R^2_{\mathbb{D}}\frac{m t_{n,m}}{n} + n^{-1}\right)\right]. \tag{9.78}$$

## 9.4   Other Types of Design in Matrix Regression

In this section, we study matrix regression problem (9.27) under somewhat different assumptions on the design variables $X_1, \ldots, X_n$. In particular, our goal is to cover an important case of *sampling from an orthonormal basis*, that is, the case when

$X_1, \ldots, X_n$ are i.i.d. random variables sampled from a distribution $\Pi$ (most often, uniform) in an orthonormal basis $E_1, \ldots, E_{m^2}$ of $\mathbb{M}_m(\mathbb{C})$ that consists of Hermitian matrices.

We will study nuclear norm penalized least squares estimator (9.28) with some value of regularization parameter $\varepsilon > 0$ and establish oracle inequalities of the same type as in Theorem 9.3. One of the challenges will be to replace the bound of Lemma 9.6 that relied on the assumption that $X_1, \ldots, X_n$ were i.i.d. subgaussian matrices with another bound on empirical processes indexed by functions $\langle S, \cdot \rangle^2$. To this end, we use an approach based on $L_\infty(P_n)$-covering numbers that was developed in a different context by Rudelson [128] and, in high-dimensional learning problems, by Mendelson and Neeman [116] (see also [18, 98]). It is based on Theorem 3.16.

Assume that, for some constant $U_X > 0$, $\|X\| \leq U_X$ (the case when $\|X\|$ has a bounded $\psi_1$-norm can be handled similarly). In this section, we also use the notation

$$\sigma_X^2 = \|\mathbb{E}X^2\|.$$

We will need below the quantity

$$\gamma_n := \Gamma_{n,\infty}(\mathscr{F}) = \mathbb{E}\gamma_2^2(\mathscr{F}; L_\infty(P_n)),$$

based on generic chaining complexities with respect to $L_\infty(\Pi_n)$-distance, introduced and used earlier in Theorem 3.16. It will be used for the class

$$\mathscr{F} := \{\langle S, \cdot \rangle : S \in \text{l.s.}(\mathbb{D}), \|S\|_1 \leq 1\}.$$

It will be shown below (see Proposition 9.2) that in typical situations $\gamma_n$ grows as a power of $\log n$. In particular, we will see that, for $\mathbb{D} = \mathbb{H}_m(\mathbb{C})$,

$$\gamma_n \leq K\mathbb{E} \max_{1 \leq j \leq n} \|X_j\|_2^2 \log^2 n.$$

For $t \geq 1$ and $\varepsilon \geq 0$, denote

$$\kappa := \log\left(\log n \vee \log U_X \vee \log \varepsilon \vee \log \|f_*\|_{L_\infty(\Pi)} \vee 2\right),$$

$$t_{n,m} := \gamma_n + U_X^2(t + \kappa + \log(2m)). \qquad (9.79)$$

We will also use a modified version of function $q(\varepsilon)$ from the previous section. Given $t \geq 1$, it is defined as follows:

$$q(\varepsilon) := q_t(\mathbb{D}; \varepsilon) := \inf_{S \in \mathbb{D}}\left[\|f_S - f_*\|_{L_2(\Pi)}^2 + \varepsilon\|S\|_1 + \|f_S - f_*\|_{L_\infty(\Pi)}^2 \frac{t}{n}\right].$$

Clearly,

$$q(\varepsilon) \leq \|f_*\|^2_{L_2(\Pi)} + \|f_*\|^2_{L_\infty(\Pi)} \frac{t}{n}. \tag{9.80}$$

Moreover, if for some $S \in \mathbb{D}$,

$$\|S\|_1 \leq \frac{1}{2} \frac{q(\varepsilon)}{\varepsilon},$$

then

$$q(\varepsilon) \leq 2\|f_S - f_*\|^2_{L_2(\Pi)} + 2\|f_S - f_*\|^2_{L_\infty(\Pi_n)} \frac{t}{n}. \tag{9.81}$$

As in the previous section,

$$\beta(S) := \beta^{(5)}(\mathbb{D}; \operatorname{supp}(S); \Pi).$$

Assume that, for some $\alpha \geq 1$, $\xi \in L_{\psi_\alpha}(\mathbb{P})$ and denote

$$\sigma^2_\xi := \mathbb{E}\xi^2, \quad U^{(\alpha)}_\xi := \|\xi\|_{\psi_\alpha} \vee (2\sigma_\xi).$$

The following theorem is the main result of this section.

**Theorem 9.5.** *There exist constants $c, C, D > 0$ with the following property. Suppose that $t \geq 1$, $t_{n,m} \leq cn$ and also that*

$$\varepsilon \geq D\left[\sigma_\xi \sigma_X \sqrt{\frac{t + \log(2m)}{n}} \vee U^{(\alpha)}_\xi U_X \log^{1/\alpha}\left(\frac{U^{(\alpha)}_\xi}{\sigma_\xi} \frac{U_X}{\sigma_X}\right) \frac{t + \log(2m)}{n}\right]. \tag{9.82}$$

*Then the following bound holds with probability at least $1 - e^{-t}$ :*

$$\|f_{\hat{A}^\varepsilon} - f_*\|^2_{L_2(\Pi)} \leq \inf_{S \in \mathbb{D}}\left[2\|f_S - f_*\|^2_{L_2(\Pi)} + C\left(\beta^2(S)\operatorname{rank}(S)\varepsilon^2\right.\right. \tag{9.83}$$

$$\left.\left. + \left(\|S\|^2_1 \vee \frac{q^2(\varepsilon)}{\varepsilon^2}\right)\frac{t_{n,m}}{n} + \|f_S - f_*\|^2_{L_\infty(\Pi)} \frac{t + \log(2m)}{n} + n^{-1}\right)\right].$$

*Proof.* We follow the proof of Theorem 9.3 with some modifications and use the notations of this proof as well as the preceding proof of Theorem 9.1.

*Step 1. Bounding the norm $\|\hat{A}^\varepsilon\|_1$.* In this case, we need the following version of Lemma 9.5.

**Lemma 9.7.** *There exists a constant $C > 0$ such that, for all $\varepsilon \geq 4\Delta$ on an event of probability at least $1 - e^{-t}$,*

$$\|\hat{A}^\varepsilon\|_1 \leq C \frac{q(\varepsilon)}{\varepsilon}. \tag{9.84}$$

*Proof.* We repeat the proof of Lemma 9.5 using instead of (9.41) the bound

$$
\begin{aligned}
&\left| (\Pi_n - \Pi)(f_S - f_*)^2 \right| \\
&\leq c_1 \left( \|f_S - f_*\|_{L_2(\Pi)} \|f_S - f_*\|_{L_\infty(\Pi)} \sqrt{\frac{t}{n}} \bigvee \|f_S - f_*\|_{L_\infty(\Pi)}^2 \frac{t}{n} \right) \quad (9.85)
\end{aligned}
$$

that easily follows from the usual Bernstein inequality and that holds with probability at least $1 - e^{-t}$. Repeating the argument that follows in the proof of Lemma 9.5, it is easy to conclude that, for $\varepsilon \geq 4\Delta$, bound (9.84) holds with some constant $C > 0$ and on an event of probability at least $1 - e^{-t}$. □

Note that, bounds (9.84) and (9.80) yield

$$
\|\hat{A}^\varepsilon\|_1 \leq \frac{C}{\varepsilon} \left( \|f_*\|_{L_2(\Pi)}^2 + \|f_*\|_{L_\infty(\Pi)}^2 \frac{t}{n} \right) \quad (9.86)
$$

that holds with probability at least $1 - e^{-t}$ and with some $C > 0$. Using this bound for $t \leq n$ and arguing as in the proof of Theorem 9.3, we get that with some constant $c_1 > 0$

$$
\|f_{\hat{A}^\varepsilon} - f_*\|_{L_2(\Pi)}^2 \leq 2 \left( \|f_{\hat{A}^\varepsilon}\|_{L_2(\Pi)}^2 + \|f_*\|_{L_2(\Pi)}^2 \right) \quad (9.87)
$$

$$
\leq 2 \left( \|\hat{A}^\varepsilon\|_1^2 \mathbb{E}\|X\|^2 + \|f_*\|_{L_2(\Pi)}^2 \right) \leq 2 \left( U_X^2 \|\hat{A}^\varepsilon\|_1^2 + \|f_*\|_{L_2(\Pi)}^2 \right)
$$

$$
\leq c_1 \left( \frac{U_X^2 \|f_*\|_{L_\infty(\Pi)}^4}{\varepsilon^2} \bigvee \|f_*\|_{L_2(\Pi)}^2 \right).
$$

Hence, it is enough to consider only the oracles $S \in \mathbb{D}$ for which

$$
\|f_S - f_*\|_{L_2(\Pi)}^2 \leq c_1 \left( \frac{U_X^2 \|f_*\|_{L_\infty(\Pi)}^4}{\varepsilon^2} \bigvee \|f_*\|_{L_2(\Pi)}^2 \right),
$$

otherwise, the bound of the theorem trivially holds. This implies that, for some $c$,

$$
\|f_{\hat{A}^\varepsilon} - f_S\|_{L_2(\Pi)} \leq c \left( \frac{U_X \|f_*\|_{L_\infty(\Pi)}^2}{\varepsilon} \bigvee \|f_*\|_{L_2(\Pi)} \right). \quad (9.88)
$$

The last bound holds with probability at least $1 - e^{-t}$ and on the same event where (9.86) holds.

*Step 2. Reduction to the bounds on empirical processes.* This step is the same as in the proof of Theorem 9.3 and it results in bounds (9.49), (9.50) that have to be used in cases A and B, respectively.

*Step 3. Bounding* $\langle \Xi, \hat{A}^{\varepsilon} - S \rangle$. The changes in this step are minor. We still derive bound (9.51), but the bound on $\Delta$ that follows from Theorem 2.7 is slightly different: with probability at least $1 - e^{-t}$,

$$\Delta \leq \bar{\Delta} := \tag{9.89}$$

$$C \left[ \sigma_{\xi} \sigma_X \sqrt{\frac{t + \log(2m)}{n}} \bigvee U_{\xi}^{(\alpha)} U_X \log^{1/\alpha} \left( \frac{U_{\xi}^{(\alpha)}}{\sigma_{\xi}} \frac{U_X}{\sigma_X} \right) \frac{t + \log(2m)}{n} \right].$$

This bound will be used in combination with (9.51).

*Step 4. Bounding* $(\Pi - \Pi_n)(f_S - f_*)(f_{\hat{A}^{\varepsilon}} - f_S)$. The changes in this step are also minor: the bound on $\|\Upsilon\|$ becomes

$$\|\Upsilon\| \leq c U_X \left[ \|f_S - f_*\|_{L_2(\Pi)} \sqrt{\frac{t + \log(2m)}{n}} \bigvee \|f_S - f_*\|_{L_\infty(\Pi)} \frac{t + \log(2m)}{n} \right] \tag{9.90}$$

and it holds with probability at least $1 - e^{-t}$ and with some $c > 0$. It follows from (9.53), (9.90) and Lemma 9.7 that, with probability at least $1 - 2e^{-t}$,

$$\left| (\Pi - \Pi_n)(f_S - f_*)(f_{\hat{A}^{\varepsilon}} - f_S) \right|$$

$$\leq C U_X \left( \|S\|_1 \vee \frac{q(\varepsilon)}{\varepsilon} \right) \left( \|f_S - f_*\|_{L_2(\Pi)} \sqrt{\frac{m(t + \log(2m))}{n}} \bigvee \right.$$

$$\|f_S - f_*\|_{L_\infty(\Pi)} \frac{t + \log(2m)}{n} \right).$$

As a result, we easily get the bound

$$\left| (\Pi - \Pi_n)(f_S - f_*)(f_{\hat{A}^{\varepsilon}} - f_S) \right| \leq \frac{1}{4} \|f_S - f_*\|_{L_2(\Pi)}^2 \tag{9.91}$$

$$+ C U_X^2 \left( \|S\|_1^2 \vee \frac{q^2(\varepsilon)}{\varepsilon^2} \right) \frac{t + \log(2m)}{n} + \frac{1}{4} \|f_S - f_*\|_{L_\infty(\Pi)}^2 \frac{t + \log(2m)}{n}$$

which holds with some constant $C > 0$ and with the same probability.

*Step 5. Bounding* $(\Pi - \Pi_n)(f_{\hat{A}^{\varepsilon}} - f_S)^2$. We have to control $(\Pi - \Pi_n)(f_{\hat{A}^{\varepsilon}} - f_S)^2$ which would allow us to complete the proof of the theorem. Recall the notation

$$\Delta_n(\delta, R) := \sup \left\{ \left| n^{-1} \sum_{j=1}^{n} \langle S_1 - S_2, X_j \rangle^2 - \|S_1 - S_2\|_{L_2(\Pi)}^2 \right| : \right.$$

$$S_1, S_2 \in \mathbb{H}_m(\mathbb{C}), \|S_1 - S_2\|_{L_2(\Pi)} \leq \delta, \|S_1\|_1 \leq R, \|S_2\|_1 \leq R \right\}.$$

We are now in a position to prove a version of Lemma 9.6.

**Lemma 9.8.** *Let* $X_1, \ldots, X_n$ *be i.i.d. copies of a random Hermitian* $m \times m$ *matrix* $X$. *Let* $\delta > 0$ *and* $R > 0$. *There exists a constant* $C > 0$ *such that, for all* $t > 0$, *with probability at least* $1 - e^{-t}$

$$\Delta_n(\delta, R) \leq C \left[ \delta R \sqrt{\frac{\gamma_n}{n}} \bigvee \frac{R^2 \gamma_n}{n} \bigvee \delta R U_X \sqrt{\frac{t}{n}} \bigvee \frac{R^2 U_X^2 t}{n} \right]. \tag{9.92}$$

*Moreover, if* $0 < \delta_- < \delta_+$, *then, with some constant* $C > 0$ *and with probability at least* $1 - e^{-t}$, *for all* $\delta \in [\delta_-, \delta_+]$,

$$\Delta_n(\delta, R) \leq C \left[ \delta R \sqrt{\frac{\gamma_n}{n}} \bigvee \frac{R^2 \gamma_n}{n} \bigvee \delta R U_X \sqrt{\frac{t + \kappa}{n}} \bigvee \frac{R^2 U_X^2 (t + \kappa)}{n} \right]. \tag{9.93}$$

*where*

$$\kappa := 2 \log \log_2 \left( \frac{2\delta_+}{\delta_-} \right).$$

*Proof.* Applying bound (3.37) to the class

$$\mathscr{F} := \{ \langle S, \cdot \rangle : S \in \mathrm{l.s.}(\mathbb{D}), \|S\|_1 \leq 1, \|S\|_{L_2(\Pi)} \leq \delta \},$$

we get

$$\mathbb{E} \sup_{\|S\|_1 \leq 1, \|S\|_{L_2(\Pi)} \leq \delta} \left| (\Pi_n - \Pi) \langle S, \cdot \rangle^2 \right| \leq C \left[ \delta \sqrt{\frac{\gamma_n}{n}} \bigvee \frac{\gamma_n}{n} \right].$$

This bound easily implies that

$$\mathbb{E} \Delta_n(\delta, R) = 4 R^2 \mathbb{E} \Delta_n \left( \frac{\delta}{2R}; \frac{1}{2} \right) \leq \tag{9.94}$$

$$4 R^2 \mathbb{E} \sup_{\|S\|_1 \leq 1, \|S\|_{L_2(\Pi)} \leq \delta/(2R)} \left| (\Pi_n - \Pi) \langle S, \cdot \rangle^2 \right| \leq 4 C \left[ \delta R \sqrt{\frac{\gamma_n}{n}} \bigvee \frac{R^2 \gamma_n}{n} \right].$$

The rest of the proof is based on repeating the concentration argument of Lemma 9.6 with minor modifications. □

We are now ready to provide an upper bound on $(\Pi - \Pi_n)(f_{\hat{A}^\varepsilon} - f_S)^2$. In view of Lemma 9.7 and (9.88), we will use Lemma 9.8 with $\delta_- := n^{-1/2}$,

$$\delta_+ := c \left( \frac{U_X \|f_*\|_{L_\infty(\Pi)}^2}{\varepsilon} \bigvee \|f_*\|_{L_2(\Pi)} \right) \vee n^{-1/2}$$

and

$$R := C\left(\|S\|_1 \vee \frac{q(\varepsilon)}{\varepsilon}\right).$$

It follows from (9.88) that $\|f_{\hat{A}^\varepsilon} - f_S\|_{L_2(\Pi)} \leq \delta_+$ and the second statement of Lemma 9.8 implies that with probability at least $1 - e^{-t}$,

$$(\Pi - \Pi_n)(f_{\hat{A}^\varepsilon} - f_S)^2 \leq \Delta_n\left(\|f_{\hat{A}^\varepsilon} - f_S\|_{L_2(\Pi)}; R\right)$$

$$\leq C\left[\|f_{\hat{A}^\varepsilon} - f_S\|_{L_2(\Pi)} R \sqrt{\frac{\gamma_n}{n}} \vee \right.$$

$$\left. \frac{R^2\gamma_n}{n} \vee R U_X \|f_{\hat{A}^\varepsilon} - f_S\|_{L_2(\Pi)} \sqrt{\frac{t+\kappa}{n}} \vee \frac{R^2 U_X^2(t+\kappa)}{n}\right]$$

provided that

$$\|f_{\hat{A}^\varepsilon} - f_S\|_{L_2(\Pi)} \geq \delta_- = n^{-1/2}$$

(we are not going to consider the case when $\|f_{\hat{A}^\varepsilon} - f_S\|_{L_2(\Pi)}$ is smaller than $n^{-1/2}$, but it only simplifies the proof). Therefore, we can easily conclude that, with some $C' > 0$,

$$(\Pi - \Pi_n)(f_{\hat{A}^\varepsilon} - f_S)^2 \leq \frac{1}{8}\|f_{\hat{A}^\varepsilon} - f_S\|_{L_2(\Pi)}^2 + C'R^2\frac{U_X^2(t+\kappa) + \gamma_n}{n}.$$

It remains to substitute in the last bound the expression for $R$.

*Step 6. Conclusion.* To complete the proof, it is enough to combine the bounds of Steps 1–5 (as it was also done in the proof of Theorem 9.3). A simple inspection of probability bounds involved in the above arguments shows that the bound of the theorem holds with probability at least $1 - 5e^{-t}$, which can be rewritten as $1 - e^{-t}$ with a proper adjustment of the constants. $\square$

Arguing as in the proofs of Corollaries 9.2 and 9.3, one can show the following statement.

**Corollary 9.4.** *Suppose that all the notations and assumptions of Theorem 9.5, including (9.82), hold. Then, for all $S \in \mathbb{D}$, such that*

$$\varepsilon \geq D\left[\|f_S - f_*\|_{L_2(\Pi)} \vee \|f_S - f_*\|_{L_\infty(\Pi)}\sqrt{\frac{t}{n}}\right]\sqrt{\frac{t_{n,m}}{n}},$$

*with probability at least $1 - e^{-t}$,*

$$\|f_{\hat{A}^\varepsilon} - f_*\|_{L_2(\Pi)}^2 \leq 2\|f_S - f_*\|_{L_2(\Pi)}^2 + C\left(\beta^2(S)\mathrm{rank}(S)\varepsilon^2 \right. \qquad (9.95)$$

$$\left. + \frac{\|S\|_1^2 t_{n,m}}{n} + \|f_S - f_*\|_{L_\infty(\Pi)}^2\frac{t + \log(2m)}{n} + n^{-1}\right).$$

Note that in the case of sampling from an orthonormal basis $E_1, \ldots, E_{m^2}$ (that is, when $\Pi$ is a probability distribution supported in the basis), the $L_\infty(\Pi)$-norm involved in the bounds of Theorem 9.5 coincides with the $\ell_\infty$-norm:

$$\|f\|_{L_\infty(\Pi)} = \max_{1 \le j \le m^2} |f(E_j)|, \ f : \mathbb{H}_m(\mathbb{C}) \mapsto \mathbb{R}.$$

We now turn to the problem of bounding the quantity $\gamma_n := \Gamma_{n,\infty}(\mathscr{F})$, where

$$\mathscr{F} := \{ \langle S, \cdot \rangle : S \in \mathbb{H}_m(\mathbb{C}), \|S\|_1 \le 1 \}.$$

**Proposition 9.2.** *With some numerical constant $K > 0$, the following bound holds:*

$$\Gamma_{n,\infty}(\mathscr{F}) \le K \log^2 n \, \mathbb{E} \max_{1 \le j \le n} \|X_j\|_2^2.$$

*Proof.* In fact, we will even prove that

$$\Gamma_{n,\infty}(\mathscr{G}) \le K \log^2 n \, \mathbb{E} \max_{1 \le j \le n} \|X_j\|_2^2,$$

where $\mathscr{G} := \{ \langle S, \cdot \rangle : \|S\|_2 \le 1 \} \supset \mathscr{F}$. Note that

$$\max_{1 \le j \le n} |\langle S, X_j \rangle| = \sup_{A \in \mathscr{K}} \langle S, A \rangle = \|S\|_{\mathscr{K}},$$

where $\mathscr{K} := \mathrm{conv}\{X_j, -X_j, j = 1, \ldots, n\}$. Since $B_2 = \{S : \|S\|_2 \le 1\}$ is the unit ball in a Hilbert space, we can use bound (3.3) to control $\gamma_2(\mathscr{G}; L_\infty(P_n))$ as follows:

$$\gamma_2(\mathscr{G}; L_\infty(P_n)) = \gamma_2(B_2; \|\cdot\|_{\mathscr{K}}) \le C \left( \int_0^\infty \varepsilon H(B_2; \|\cdot\|_{\mathscr{K}}; \varepsilon) d\varepsilon \right)^{1/2}, \quad (9.96)$$

where $C > 0$ is a constant. Note that $|f(X_j)| \le \|X_j\|_2$, $f \in \mathscr{G}$, which implies

$$\{ (f(X_1), \ldots, f(X_n)) : f \in \mathscr{G} \} \subset [-V_n, V_n]^n,$$

where $V_n := \max_{1 \le j \le n} \|X_j\|_2$. Bounding the $\ell_\infty^n$-covering numbers of the cube $[-V_n, V_n]^n$, we get

$$N(B_2; \|\cdot\|_{\mathscr{K}}; \varepsilon) = N(\mathscr{G}; L_\infty(P_n); \varepsilon) \le \left( \frac{V_n}{\varepsilon} + 1 \right)^n, \ \varepsilon \le V_n. \quad (9.97)$$

This bound will be used for small values of $\varepsilon$, but, for larger values, we need a bound with logarithmic dependence on $n$ that can be derived from dual Sudakov's bound (3.2). Take an orthonormal basis $E_1, \ldots, E_N$, $N = m(m+1)/2$ of the space $\mathbb{H}_m(\mathbb{C})$ and use the isometry

$$\mathbb{H}_m(\mathbb{C}) \ni S \mapsto \left( \langle S, E_1 \rangle, \ldots, \langle S, E_N \rangle \right) \in \ell_2^N$$

to identify the unit ball $B_2$ in $\mathbb{H}_m(\mathbb{C})$ with the unit ball $B_2^N$ in $\ell_2^N$. It follows from dual Sudakov's bound (3.2) that, conditionally on $X_1, \ldots, X_n$,

$$\varepsilon H^{1/2}(B_2; \|\cdot\|_{\mathscr{X}}; \varepsilon) = \varepsilon H^{1/2}(B_2^N; \|\cdot\|_{\mathscr{X}}; \varepsilon)$$

$$\leq C' \mathbb{E}_Z \sup_{t \in \mathscr{X}} \langle Z, t \rangle_{\ell_2^N} = C' \mathbb{E}_Z \max_{1 \leq j \leq n} \langle Z, X_j \rangle_{\ell_2^N} \leq C_1 V_n \sqrt{\log n},$$

where $Z$ is a standard normal vector in $\mathbb{R}^N$ $(= \ell_2^N)$ and $C_1, C' > 0$ are numerical constants (with a minor abuse of notation, we identify matrices in $\mathbb{H}_m(\mathbb{C})$ with vectors in $\ell_2^N$ in the above relationships). As a result, the following bound holds:

$$H(B_2; \|\cdot\|_{\mathscr{X}}; \varepsilon) \leq \frac{C_1^2}{\varepsilon^2} V_n^2 \log n. \tag{9.98}$$

Let $\delta_n := n^{-1/2}$. We will use (9.97) for $\varepsilon \leq \delta_n V_n$ and (9.98) for $\varepsilon \in (\delta_n V_n, V_n]$. In view of (9.96), we get the following bound:

$$\Gamma_{n,\infty}(\mathscr{G}) = \mathbb{E}\gamma_2^2(\mathscr{G}; L_\infty(P_n))$$

$$\leq C \left( n\mathbb{E} \int_0^{\delta_n V_n} \log\left(\frac{V_n}{\varepsilon} + 1\right) \varepsilon d\varepsilon + C_1^2 \log n \, \mathbb{E}V_n^2 \int_{\delta_n V_n}^{V_n} \frac{1}{\varepsilon} d\varepsilon \right)$$

$$= C \left( n\mathbb{E}V_n^2 \int_0^{\delta_n} \log\left(\frac{1}{\varepsilon} + 1\right) \varepsilon d\varepsilon + C_1^2 \log n \, \mathbb{E}V_n^2 \log\frac{1}{\delta_n} \right)$$

$$\leq K' \left( n\delta_n^2 \log\frac{1}{\delta_n} + \log n \log\frac{1}{\delta_n} \right) \mathbb{E}V_n^2 \leq K \log^2 n \, \mathbb{E} \max_{1 \leq j \leq n} \|X_j\|_2^2$$

with some constant $K > 0$.                                                                        $\square$

In the case when the set $\mathbb{D}$ is bounded with $R_{\mathbb{D}} := \sup_{S \in \mathbb{D}} \|S\|_1$, it is easy to derive a version of Theorem 9.5 with control of the error in terms of $R_{\mathbb{D}}$. To this end, for $t \geq 1$, define

$$\kappa := \log\left(\log n \vee \log U_X \vee \log R_{\mathbb{D}} \vee \log \|f_*\|_{L_2(\Pi)} \vee 2\right),$$

$$t_{n,m} := \gamma_n + U_X^2(t + \kappa + \log(2m)). \tag{9.99}$$

**Theorem 9.6.** *There exist constants $c, C, D > 0$ with the following property. Suppose that $t_{n,m} \leq cn$ and also that (9.82) holds. Then the following bound holds* *with probability at least $1 - e^{-t}$ :*

$$\|f_{\hat{A}^\varepsilon} - f_*\|^2_{L_2(\Pi)} \leq \inf_{S \in \mathbb{D}} \left[ 2\|f_S - f_*\|^2_{L_2(\Pi)} + C\left( \beta^2(S)\mathrm{rank}(S)\varepsilon^2 \right. \right. \qquad (9.100)$$

$$\left. \left. + \frac{R^2_{\mathbb{D}} t_{n,m}}{n} + \|f_S - f_*\|^2_{L_\infty(\Pi)} \frac{t + \log(2m)}{n} + n^{-1} \right) \right].$$

Suppose now that $f_*(\cdot) := \langle A, \cdot \rangle$ for some matrix $A \in \mathbb{H}_m(\mathbb{C})$, that $\mathbb{D}$ is a closed convex subset of $\mathbb{H}_m(\mathbb{C})$ and that $A \in \mathbb{D}$. In particular, it includes the case when $\mathbb{D} = \mathbb{H}_m(\mathbb{C})$. Then, one can use $S = A$ as an oracle in Corollary 9.4 to get the following result.

**Corollary 9.5.** *Under the notations of Theorem 9.5 and under the conditions $t_{n,m} \leq cn$ and (9.82), the following bound holds with probability at least $1 - e^{-t}$ :*

$$\|\hat{A}^\varepsilon - A\|^2_{L_2(\Pi)} \leq C\left( \beta^2(A)\mathrm{rank}(A)\varepsilon^2 + \frac{\|A\|^2_1 t_{n,m} + 1}{n} \right). \qquad (9.101)$$

Next we turn to a couple of examples in which $X, X_1, \ldots, X_n$ are i.i.d. random matrices sampled from an orthonormal basis in the space $\mathbb{M}_m(\mathbb{C})$ of all $m \times m$ matrices or in the space $\mathbb{H}_m(\mathbb{C})$ of $m \times m$ Hermitian matrices. We will not discuss similar problems for rectangular matrices, but it has been already shown at the end of Sect. 9.2 how these problems can be reduced to the Hermitian case using the Paulsen dilation.

**Matrix Completion.** As a first example, consider an orthonormal basis $\{E_{kj} : k, j = 1, \ldots, m\}$ of the space $\mathbb{H}_m(\mathbb{C})$, where

$$E_{kk} := e_k \otimes e_k, \ k = 1, \ldots, m,$$

$$E_{kj} := \frac{1}{\sqrt{2}}(e_k \otimes e_j + e_j \otimes e_k), \ 1 \leq k < j \leq m \text{ and}$$

$$E_{jk} := \frac{i}{\sqrt{2}}(e_k \otimes e_j - e_j \otimes e_k), \ 1 \leq k < j \leq m.$$

Note that, for all $A \in \mathbb{H}_m(\mathbb{C})$ and for all $k$, $\langle A, E_{kk} \rangle = A_{kk}$, and for $k < j$, $\langle A, E_{kj} \rangle = \sqrt{2}\mathrm{Re}(A_{jk})$, $\langle A, E_{jk} \rangle = \sqrt{2}\mathrm{Im}(A_{jk})$. Let $\Pi$ denote the uniform distribution in the basis $\{E_{kj} : k, j = 1, \ldots, m\}$. Then, for all matrices $B$, $\|B\|^2_{L_2(\Pi)} = m^{-2}\|B\|^2_2$, which implies that $\beta(B) \leq m$. Sampling from the distribution $\Pi$ is equivalent to sampling the real parts and the imaginary parts of the entries of matrix $A$ at random with replacement.

Note that

$$\sigma^2_X = \|\mathbb{E}X^2\| = \sup_{v \in \mathbb{C}^m, |v|=1} \mathbb{E}\langle X^2 v, v \rangle = \sup_{v \in \mathbb{C}^m, |v|=1} \mathbb{E}|Xv|^2.$$

For $X = E_{kk}$, we have $|Xv|^2 = |\langle v, e_k \rangle|^2$, and, for $X = E_{kj}, k < j$,

$$|Xv|^2 = \frac{1}{2}\left|\langle e_k, v \rangle e_j + \langle e_j, v \rangle e_k\right|^2 = \frac{1}{2}|\langle e_k, v \rangle|^2 + \frac{1}{2}|\langle e_j, v \rangle|^2.$$

This easily implies that $\sigma_X^2 \leq \frac{2}{m}$. We also have $\|X\| \leq 1$, so, we can take $U_X = 1$.

In the case of i.i.d. Gaussian noise with mean 0 and variance $\sigma_\xi^2$, condition (9.82) with $\alpha = 2$ can be written as

$$\varepsilon \geq D\sigma_\xi \left[ \sqrt{\frac{t + \log(2m)}{mn}} \bigvee \log^{1/2} m \frac{t + \log(2m)}{n} \right].$$

with a large enough constant $D > 0$. Assuming for simplicity that

$$\frac{m(t + \log(2m)) \log m}{n} \leq 1, \tag{9.102}$$

one can take

$$\varepsilon = D\sigma_\xi \sqrt{\frac{t + \log(2m)}{mn}}. \tag{9.103}$$

With this choice of regularization parameter $\varepsilon$ and under the assumption that $A \in \mathbb{D}$, Corollary 9.5 implies the following.

**Corollary 9.6.** *Under the conditions* $t_{n,m} \leq cn$, *(9.102) and (9.103), the following bound holds with probability at least* $1 - e^{-t}$ :

$$\|\hat{A}^\varepsilon - A\|_{L_2(\Pi)}^2 \leq C \left( \sigma_\xi^2 \frac{m \, \mathrm{rank}(A)(t + \log(2m))}{n} + \frac{\|A\|_1^2 t_{n,m}}{n} + n^{-1} \right). \tag{9.104}$$

*Sampling from the Pauli basis.* Another example is sampling from the Pauli basis already discussed in Sect. 1.7. Recall that the *Pauli basis* in the space of $2 \times 2$ matrices $\mathbb{M}_2(\mathbb{C})$ was defined as $W_i := \frac{1}{\sqrt{2}}\sigma_i$, $i = 1, 2, 3, 4$, where

$$\sigma_1 := \begin{pmatrix} 0 & 1 \\ 1 & 0 \end{pmatrix}, \quad \sigma_2 := \begin{pmatrix} 0 & -i \\ i & 0 \end{pmatrix}, \quad \sigma_3 := \begin{pmatrix} 1 & 0 \\ 0 & -1 \end{pmatrix} \quad \text{and} \quad \sigma_4 := \begin{pmatrix} 1 & 0 \\ 0 & 1 \end{pmatrix}$$

are the Pauli matrices.. Let $m = 2^k$, $k \geq 1$. The Pauli basis in the space $\mathbb{M}_m(\mathbb{C})$ consists of all tensor products $W_{i_1} \otimes \cdots \otimes W_{i_k}$, $(i_1, \ldots, i_k) \in \{1, 2, 3, 4\}^k$. Assume that $\Pi$ is the uniform distribution in this basis and $X, X_1, \ldots, X_n$ are i.i.d. random matrices sampled from $\Pi$. Then, we have $\|B\|_{L_2(\Pi)}^2 = m^{-2}\|B\|_2^2$, implying that $\beta(B) \leq m$. It is also easy to see that $\|X\| \leq 2^{-k/2} = m^{-1/2}$. Therefore, one can take $U_X = m^{-1/2}$ and we have

$$\sigma_X^2 \leq \mathbb{E}\|X\|^2 \leq m^{-1}.$$

In condition (9.82), $\sigma_X$ can be replaced by an upper bound, say, $U_X = m^{-1/2}$. In the case of centered Gaussian noise, condition (9.82) can be rewritten as

$$\varepsilon \geq D\sigma_\xi \left[ \sqrt{\frac{t + \log(2m)}{nm}} \vee \frac{t + \log(2m)}{n\sqrt{m}} \right].$$

Under the assumption $t + \log(2m) \leq n$, one can use the value of regularization parameter

$$\varepsilon = D\sigma_\xi \sqrt{\frac{t + \log(2m)}{nm}} \tag{9.105}$$

with a sufficiently large constant $D$.

As before, assume that $\mathbb{D} \subset \mathbb{H}_m(\mathbb{C})$ is a closed convex subset that contains $A$.

**Corollary 9.7.** *Under the conditions* $t_{n,m} \leq cn$ *and (9.105), the following bound holds with probability at least* $1 - e^{-t}$ :

$$\|\hat{A}^\varepsilon - A\|_{L_2(\Pi)}^2 \leq C \left( \sigma_\xi^2 \frac{m \operatorname{rank}(A)(t + \log(2m))}{n} + \frac{\|A\|_1^2 t_{n,m}}{n} + n^{-1} \right). \tag{9.106}$$

An interesting special case is when $\mathbb{D}$ is the set of all *density matrices*, that is, Hermitian nonnegatively definite matrices of trace 1. Such matrices describe the states of a quantum system and $Y_j, j = 1, \ldots, n$ can be viewed as measurements of observables $X_j, j = 1, \ldots, n$, provided that the system has been prepared each time in the same state. The problem of estimation of the unknown state (density matrix) based on measurement $(X_1, Y_1), \ldots, (X_n, Y_n)$ is very basic in *quantum state tomography*, see [70, 71, 120]. In this case, for all $\varepsilon \geq 0$, the estimator $\hat{A}^\varepsilon$ coincides with the unpenalized least squares estimator $\hat{A}$,

$$\hat{A} := \operatorname{argmin}_{S \in \mathbb{D}} n^{-1} \sum_{j=1}^{n} \left( Y_j - \langle S, X_j \rangle \right)^2.$$

Indeed, for all $S \in \mathbb{D}$, $\|S\|_1 = \operatorname{tr}(S) = 1$, so, the penalty term in the definition of $\hat{A}^\varepsilon$ is a constant on the set $\mathbb{D}$. However, the bound of Corollary 9.7 and other bounds of this type hold for the estimator $\hat{A}$. For instance, the oracle inequality of Theorem 9.6 takes in this case the following form

$$\|f_{\hat{A}} - f_*\|_{L_2(\Pi)}^2 \leq 2\|f_S - f_*\|_{L_2(\Pi)}^2$$

$$+ C \left( \beta^2(S)\operatorname{rank}(S)\varepsilon^2 + \|f_S - f_*\|_{L_\infty(\Pi)}^2 \frac{t_{n,m}}{n} + \frac{t_{n,m}}{n} \right)$$

and it holds for all density matrices $S$.

Another interesting special case is when $\mathbb{D} \subset \mathbb{H}_m(\mathbb{C})$ is a closed convex set of diagonal $m \times m$ matrices that can be identified with $m$-dimensional vectors, so that $\mathbb{D}$ can be viewed as a convex subset of $\mathbb{R}^m$. As we have already pointed out in Sect. 9.2, this special case of low rank recovery is equivalent to the usual sparse recovery discussed in Chaps. 7–8 and the results on nuclear norm penalized least squares estimators, such as Theorem 9.5, easily imply oracle inequalities for the LASSO. This reduction has been already discussed in Sect. 9.2, so, we will only formulate here a corollary of Theorem 9.5. We will use the notations of Chaps. 7–8 and of Sect. 9.2. In particular, denote $\beta(\lambda) := \beta_2^{(5)}(J_\lambda; \Pi)$, where $J_\lambda := \text{supp}(\lambda)$. We will also assume that the functions in the dictionary $\{h_1, \ldots, h_m\}$ are uniformly bounded and denote

$$\sigma_X^2 := \max_{1 \leq k \leq m} \Pi h_k^2, \quad U_X := \max_{1 \leq k \leq m} \|h_k\|_{L_\infty(\Pi)}.$$

We will use the quantity $t_{n,m}$ defined by (9.79) with $\gamma_n = \Gamma_{n,\infty}(\mathscr{F})$, where

$$\mathscr{F} := \left\{ f_\lambda : \lambda \in \text{l.s.}(\mathbb{D}), \|\lambda\|_{\ell_1} \leq 1 \right\}$$

(which, in the case under consideration, is equivalent to the general definition used in Theorem 9.5). We also use the notation

$$q(\varepsilon) := q_t(\mathbb{D}; \varepsilon) := \inf_{\lambda \in \mathbb{D}} \left[ \|f_\lambda - f_*\|_{L_2(\Pi)}^2 + \varepsilon \|\lambda\|_{\ell_1} + \|f_\lambda - f_*\|_{L_\infty(\Pi)}^2 \frac{t}{n} \right].$$

**Corollary 9.8.** *There exist constants $c, C, D > 0$ with the following property. Suppose that $t_{n,m} \leq cn$ and that*

$$\varepsilon \geq D\left[ \sigma_\xi \sigma_X \sqrt{\frac{t + \log(2m)}{n}} \bigvee U_\xi^{(\alpha)} U_X \log^{1/\alpha}\left( \frac{U_\xi^{(\alpha)}}{\sigma_\xi} \frac{U_X}{\sigma_X} \right) \frac{t + \log(2m)}{n} \right].$$
(9.107)

*Then the following bound holds with probability at least $1 - e^{-t}$ :*

$$\|f_{\hat{\lambda}^\varepsilon} - f_*\|_{L_2(\Pi)}^2 \leq \inf_{\lambda \in \mathbb{D}} \left[ 2\|f_\lambda - f_*\|_{L_2(\Pi)}^2 + C\left( \beta^2(\lambda)\text{card}(J_\lambda)\varepsilon^2 \right. \right. \quad (9.108)$$

$$\left. \left. + \left( \|\lambda\|_{\ell_1}^2 \vee \frac{q^2(\varepsilon)}{\varepsilon^2} \right) \frac{t_{n,m}}{n} + \|f_\lambda - f_*\|_{L_\infty(\Pi)}^2 \frac{t + \log(2m)}{n} + n^{-1} \right) \right].$$

In the case when

$$\mathbb{D} := \left\{ (\lambda_1, \ldots, \lambda_m) : \lambda_j \geq 0, \sum_{j=1}^m \lambda_j = 1 \right\},$$

the estimators $\hat{\lambda}^\varepsilon, \varepsilon \geq 0$ coincide with the least squares estimator over the convex hull of the dictionary

$$\hat{\lambda} := \mathrm{argmin}_{\lambda \in \mathbb{D}} n^{-1} \sum_{j=1}^{n} (Y_j - f_\lambda(X_j))^2,$$

so, the oracle inequality of Corollary 9.8 applies to the least squares estimator $\hat{\lambda}$.

The following proposition can be used to bound the quantity $\gamma_n$ involved in the definition of $t_{n,m}$ and in oracle inequality (9.108).

**Proposition 9.3.** *There exists a constant $K > 0$ such that*

$$\gamma_n \le K U_X^2 \log^3 n \log m. \tag{9.109}$$

*Proof.* The proof is based on a version of well known Maurey's argument (see, for instance, Lemma 2.6.11 in [148]). We will start with bounding the $L_\infty(\Pi_n)$-covering numbers of the convex hull of the dictionary

$$\mathscr{G} := \mathrm{conv}(\{h_1, \dots, h_m\}) = \left\{ f_\lambda : \lambda_j \ge 0, \sum_{j=1}^{m} \lambda_j = 1 \right\}.$$

Any vector $\lambda$ with $\lambda_j \ge 0$ and $\sum_{j=1}^{m} \lambda_j = 1$ can be viewed as a probability distribution on the dictionary $\{h_1, \dots, h_m\}$. Let $\xi, \xi_1, \dots, \xi_N$ be an i.i.d. sample from this distribution (that is, $\xi_j$ takes value $h_k$ with probability $\lambda_k$) defined on a probability space $(\Omega', \Sigma', \mathbb{P}')$. Clearly, $\mathbb{E}'\xi = f_\lambda$. We will apply symmetrization inequality followed by Theorem 3.5 to an empirical process based on $(\xi_1, \dots, \xi_N)$ to get

$$\mathbb{E}' \left\| N^{-1} \sum_{j=1}^{N} \xi_j - f_\lambda \right\|_{L_\infty(\Pi_n)} \le C U_X \sqrt{\frac{\log n}{N}}.$$

It follows that there exists $\omega' \in \Omega'$ such that

$$\left\| N^{-1} \sum_{j=1}^{N} \xi_j(\omega') - f_\lambda \right\|_{L_\infty(\Pi_n)} \le C U_X \sqrt{\frac{\log n}{N}}.$$

Note that the number of possible choices of $(\xi_1, \dots, \xi_N)$, where $\xi_j \in \{h_1, \dots, h_m\}$, is at most $m^N$. Let $N$ be the smallest number for which $C U_X \sqrt{\frac{\log n}{N}} \le \varepsilon$, which implies

$$N \le C_1 U_X^2 \frac{\log n}{\varepsilon^2}. \tag{9.110}$$

Then $N(\mathscr{G}; L_\infty(\Pi_n); \varepsilon) \le m^N$. Note also that

$$\mathscr{F} = \{f_\lambda : \|\lambda\|_{\ell_1} \le 1\} = \mathrm{conv}(\{0, h_1, -h_1, \dots, h_m, -h_m\}).$$

Hence, we have

$$N(\mathscr{F}; L_\infty(\Pi_n); \varepsilon) \leq (2m + 1)^N. \tag{9.111}$$

On the other hand, we have an obvious bound (see also the proof of Proposition 9.2)

$$N(\mathscr{F}; L_\infty(P_n); \varepsilon) \leq \left(\frac{U_X}{\varepsilon} + 1\right)^n, \ \varepsilon \leq U_X \tag{9.112}$$

and we will use (9.111) for $\varepsilon \geq U_X n^{-1/2}$ and (9.112) for $\varepsilon \leq U_X n^{-1/2}$. We can bound the generic chaining complexity $\gamma_2(\mathscr{F}; L_\infty(\Pi_n))$ in terms of Dudley's entropy integral, use (9.112) and (9.111) in the respective intervals of values of $\varepsilon$ and also use (9.110) to get

$$\gamma_2(\mathscr{F}; L_\infty(\Pi_n)) \leq C_2 \Bigg[ \int_0^{U_X n^{-1/2}} \sqrt{n \log\left(\frac{U_X}{\varepsilon} + 1\right)} \, d\varepsilon$$

$$+ \int_{U_X n^{-1/2}}^{U_X} \sqrt{\log(2m + 1)} U_X \sqrt{\log n} \frac{d\varepsilon}{\varepsilon} \Bigg]$$

$$= C_2 \Bigg[ \sqrt{n} U_X \int_0^{n^{-1/2}} \sqrt{\log\left(\frac{1}{\varepsilon} + 1\right)} \, d\varepsilon + U_X \sqrt{\log(2m + 1) \log n} \log(n^{1/2}) \Bigg]$$

$$\leq C_3 U_X \sqrt{\log m} \log^{3/2} n. \tag{9.113}$$

It immediately follows that

$$\gamma_n = \Gamma_{n,\infty}(\mathscr{F}) = \mathbb{E}\gamma_2^2(\mathscr{F}; L_\infty(\Pi_n)) \leq K U_X^2 \log^3 n \log m. \qquad \square$$

## 9.5 Further Comments

Nuclear norm minimization has been used for a while as a heuristic approach to low rank matrix recovery. Theoretical understanding of this method in the case of noiseless matrix completion started with the papers by Recht et al. [125] and Candes and Recht [41]. Generally speaking, matrix completion with no error is impossible unless almost all the entries of the matrix are observed (it is enough to consider a matrix with only one non-zero entry: for such a matrix, the probability to miss the non-zero entry is close to 1 unless the number of the observed entries is comparable with the total number of entries in the matrix). However, under a reasonable assumption that the row and column spaces of the target matrix have "low coherence", these authors were able to prove that it is possible to recover the matrix based on a much smaller number of measurements and that this number depends on the rank of the target matrix. Candes and Tao [45] obtained the first tight bound on the number of randomly sampled entries needed for precise matrix completion. For an $m \times m$ matrix, this number is equal to $mr$ (up to a logarithmic factor), where $r$ is the rank of the matrix. The proof of this result was rather involved and relied heavily on noncommutative probability and random matrix theory. Sharp bounds for precise matrix completion were also obtained by Keshavan et al. [75]. Gross et al. [71] and Gross [70] developed a different approach to the analysis of low rank matrix recovery problems based on a noncommutative version of Bernstein's inequality developed earlier by Ahlswede and Winter [4]. Using this inequality, they simplified the argument of Candes and Tao, improved the logarithmic factor in their bound and extended the result to a broader class of matrix recovery problems based on sampling of Fourier coefficients in a given basis in the space of matrices, including important problems in quantum state tomography. Tropp [142] provided a review of exponential inequalities for sums of independent random matrices and matrix valued martingales extending the initial result of Ahlswede and Winter [4] (see also [88] for a $\psi_\alpha$-version of such exponential bounds).

Candes and Plan [40], Rohde and Tsybakov [127], Koltchinskii [88], Negahban and Wainwright [118], Koltchinskii et al. [90], Lecué and Gaiffas [98] started developing error bounds in low rank matrix recovery in the presence of noise. Our approach to the matrix regression with subgaussian isotropic design in the first part of Sect. 9.3 (reduction to the fixed design case using a matrix version of restricted isometry condition) is akin to the approach in [40] and the proof of Theorem 9.1 in Sect. 9.2 resembles the proof of the main result in [90]. In addition to the approach based on nuclear norm penalization, Rohde and Tsybakov [127] obtained error bounds for some other methods of low rank recovery based on the penalization with Schatten "$p$-norms" with $p < 1$ and studied a number of examples including multi-task learning. Koltchinskii [88] obtained "low rank oracle inequalities" for estimators of a density matrix (in quantum state tomography) based on penalized least squares method with a complexity penalty based on von Neumann entropy. Lecué and Gaiffas [98] studied a number of complexity penalties including a matrix version of "elastic nets" for which they proved oracle inequalities with "slow rates". Koltchinskii et al. [90] studied a modification of nuclear norm penalized least squares method suitable in the case of random design matrix regression problems with known design distribution (in the case of fixed design regression, this method coincides with the penalized least squares). They also obtained minimax lower bounds for low rank matrix recovery problems.

Propositions 9.2 and 9.3 are similar to what has been already used in sparse and low rank recovery (see Bartlett et al. [18] and Lecué and Gaiffas [98]).

The literature on low rank matrix recovery is vast, it keeps growing and it is not our goal here to provide its comprehensive review. Some further references can be found in the papers cited above.

# Appendix A
# Auxiliary Material

## A.1 Orlicz Norms

We frequently use Orlicz norms $\| \cdot \|_\psi$ of random variables. Given a convex nondecreasing function $\psi : \mathbb{R}_+ \mapsto \mathbb{R}_+$ with $\psi(0) = 0$ and a random variable $\eta$ on a probability space $(\Omega, \Sigma, \mathbb{P})$, define

$$\|\eta\|_\psi := \inf\left\{ C > 0 : \mathbb{E}\psi\left(\frac{|\eta|}{C}\right) \leq 1 \right\}$$

(see Ledoux and Talagrand [101], van der Vaart and Wellner [148], de la Pena and Giné [50]). If we want to emphasize the dependence of the Orlicz norms on the probability measure, we write $\| \cdot \|_{L_\psi(\mathbb{P})}$ (similarly, $\| \cdot \|_{L_\psi(P)}$, $\| \cdot \|_{L_\psi(\Pi)}$, etc).

If $\psi(u) = u^p$ for some $p \geq 1$, then the $\psi$-norm coincides with the usual $L_p$-norm. Some other useful choices of function $\psi$ correspond to Orlicz norms in spaces of random variables with subgaussian or subexponential tails. For $\alpha > 0$, define

$$\psi_\alpha(u) := e^{u^\alpha} - 1, \; u \geq 0.$$

Most often, $\psi_2$- and $\psi_1$-norms are used (the first one being the "subgaussian norm" and the second one being the "subexponential norm"). Note that, for $\alpha < 1$, the function $\psi_\alpha$ is not convex and, as a result, $\| \cdot \|_{\psi_\alpha}$ is not a norm. However, to overcome this difficulty, it is enough to modify $\psi_\alpha$ in a neighborhood of 0. As it is common in the literature, we ignore this minor inconvenience and use $\| \cdot \|_{\psi_\alpha}$ as a norm even for $\alpha < 1$. Moreover, usually, we need the $\psi_\alpha$-norms for $\alpha \geq 1$. The following bounds are well known (see [148], p. 95):

$$\|\eta\|_{\psi_{\alpha_1}} \leq (\log 2)^{\alpha_1/\alpha_2} \|\eta\|_{\psi_{\alpha_2}}, \; 1 \leq \alpha_1 \leq \alpha_2$$

and, for all $p \in (m-1, m]$, $m = 2, 3, \ldots$ $\|\eta\|_{L_p} \leq m! \|\eta\|_{\psi_1}$.

It easily follows from the definition of $\psi_\alpha$-norms that

V. Koltchinskii, *Oracle Inequalities in Empirical Risk Minimization and Sparse Recovery Problems*, Lecture Notes in Mathematics 2033, DOI 10.1007/978-3-642-22147-7,

$$\mathbb{P}\{|\eta| \geq t\} \leq 2\exp\left\{-\left(\frac{t}{\|\eta\|_{\psi_\alpha}}\right)^\alpha\right\}.$$

Another well known fact is that, for many convex nondecreasing functions $\psi$, including $\psi(u) = u^p$ with $p \geq 1$ and $\psi_\alpha$ with $\alpha \geq 1$, for all $N \geq 1$ and for all random variables $\eta_1, \ldots, \eta_N$

$$\left\|\max_{1\leq k\leq N}\eta_k\right\|_\psi \leq K\max_{1\leq k\leq N}\|\eta_k\|_\psi\psi^{-1}(N),$$

where $K$ is a constant depending on $\psi$ (see, e.g., [148], Lemma 2.2.2).

## A.2   Classical Exponential Inequalities

Let $X_1, \ldots, X_n$ be independent random variables with $\mathbb{E}X_j = 0$, $j = 1, \ldots, n$. We state below several classical exponential bounds for the sum

$$S_n := X_1 + \cdots + X_n.$$

Denote $B_n^2 := \mathbb{E}X_1^2 + \cdots + \mathbb{E}X_n^2$.

- *Bernstein's inequality.* Suppose $|X_j| \leq U$, $j = 1, \ldots, n$. Then,

$$\mathbb{P}\{S_n \geq t\} \leq \exp\left\{-\frac{t^2}{2B_n^2\left(1 + \frac{tU}{3B_n^2}\right)}\right\}.$$

- *Bennett's inequality.* Suppose $|X_j| \leq U$, $j = 1, \ldots, n$. Then,

$$\mathbb{P}\{S_n \geq t\} \leq \exp\left\{-\frac{B_n^2}{U^2}h\left(\frac{tU}{B_n^2}\right)\right\},$$

where $h(u) := (1 + u)\log(1 + u) - u$.
- *Hoeffding's inequality.* Suppose $a_j < b_j$, $j = 1, \ldots, n$, $X_j \in [a_j, b_j]$, $\mathbb{E}X_j = 0$, $j = 1, \ldots, n$. Then,

$$\mathbb{P}\left\{S_n \geq t\right\} \leq \exp\left\{-\frac{2t^2}{\sum_{j=1}^n(b_j - a_j)^2}\right\}, \quad t \geq 0.$$

- *Bernstein's type inequality for $\psi_1$-random variables.* Suppose $\|X_j\|_{\psi_1} \leq V$. Then,

$$\mathbb{P}\{S_n \geq t\} \leq \exp\left\{-c\left(\frac{t^2}{nV^2} \bigwedge \frac{t}{V}\right)\right\}$$

with some universal constant $c > 0$.

Bernstein's inequality easily implies that, for all $t > 0$, with probability at least $1 - e^{-t}$

$$|S_n| \leq C(B_n \sqrt{t} \vee Ut),$$

where $C$ is a numerical constant. We frequently use this form of Bernstein's inequality and other inequalities of similar type.

## A.3  Properties of ♯- and b-Transforms

Here we provide some properties of ♯- and b-transforms introduced in Sect. 4.1 and used in the construction of excess risk bounds. The proofs of these properties are rather elementary. We are mainly interested in ♯-transform.

1. If $\psi(u) = o(u)$ as $u \to \infty$, then the function $\psi^\sharp$ is defined on $(0, +\infty)$ and is a nonincreasing function on this interval.
2. If $\psi_1 \leq \psi_2$, then $\psi_1^\sharp \leq \psi_2^\sharp$. Moreover, if $\psi_1(\delta) \leq \psi_2(\delta)$ either for all $\delta \geq \psi_2^\sharp(\varepsilon)$, or for all $\delta \geq \psi_1^\sharp(\varepsilon) - \tau$ with an arbitrary $\tau > 0$, then also $\psi_1^\sharp(\varepsilon) \leq \psi_2^\sharp(\varepsilon)$.
3. For all $a > 0$,
$$(a\psi)^\sharp(\varepsilon) = \psi^\sharp(\varepsilon/a).$$

4. If $\varepsilon = \varepsilon_1 + \cdots + \varepsilon_m$, then

$$\psi_1^\sharp(\varepsilon) \bigvee \cdots \bigvee \psi_m^\sharp(\varepsilon) \leq (\psi_1 + \cdots + \psi_m)^\sharp(\varepsilon) \leq \psi_1^\sharp(\varepsilon_1) \bigvee \cdots \bigvee \psi_m^\sharp(\varepsilon_m).$$

5. If $\psi(u) \equiv c$, then
$$\psi^\sharp(\varepsilon) = c/\varepsilon.$$

6. If $\psi(u) := u^\alpha$ with $\alpha \leq 1$, then

$$\psi^\sharp(\varepsilon) := \varepsilon^{-1/(1-\alpha)}.$$

7. For $c > 0$, denote $\psi_c(\delta) := \psi(c\delta)$. Then

$$\psi_c^\sharp(\varepsilon) = \frac{1}{c}\psi^\sharp(\varepsilon/c).$$

If $\psi$ is nondecreasing and $c \geq 1$, then

$$c\psi^\sharp(u) \leq \psi^\sharp(u/c).$$

8. For $c > 0$, denote $\psi_c(\delta) := \psi(\delta + c)$. Then for all $u > 0, \varepsilon \in (0, 1]$

$$\psi_c^\sharp(u) \leq (\psi^\sharp(\varepsilon u/2) - c) \vee c\varepsilon.$$

Recall the definitions of functions of concave type and strictly concave type from Sect. 4.1.

9. If $\psi$ is of concave type, then $\psi^\sharp$ is the inverse of the function

$$\delta \mapsto \frac{\psi(\delta)}{\delta}.$$

In this case,

$$\psi^\sharp(cu) \geq \psi^\sharp(u)/c$$

for $c \leq 1$ and

$$\psi^\sharp(cu) \leq \psi^\sharp(u)/c$$

for $c \geq 1$.

10. If $\psi$ is of strictly concave type with exponent $\gamma$, then for $c \leq 1$

$$\psi^\sharp(cu) \leq \psi^\sharp(u)c^{-\frac{1}{1-\gamma}}.$$

## A.4   Some Notations and Facts in Linear Algebra

Let $L$ be a linear space. The following notations are frequently used: l.s.$(B)$ for a linear span of a subset $B \subset L$,

$$\text{l.s.}(B) := \left\{ \sum_{j=1}^{n} \lambda_j x_j : n \geq 1, \lambda_j \in \mathbb{R}, x_j \in B \right\};$$

conv$(B)$ for its convex hull,

$$\text{conv}(B) := \left\{ \sum_{j=1}^{n} \lambda_j x_j : n \geq 1, \lambda_j \geq 0, \sum_{j=1}^{n} \lambda_j = 1, x_j \in B \right\};$$

and conv$_s(B)$ for its symmetric convex hull,

$$\text{conv}_s(B) := \left\{ \sum_{j=1}^{n} \lambda_j x_j : n \geq 1, \lambda_j \in \mathbb{R}, \sum_{j=1}^{n} |\lambda_j| \leq 1, x_j \in B \right\}.$$

For vectors $u, v \in \mathbb{C}^m$ or $u, v \in \mathbb{R}^m$, $\langle u, v \rangle$ denotes the standard Euclidean inner product of $u$ and $v$; $|u|$ denotes the corresponding norm of $u$. Notations $\|u\|_{\ell_2}$ or $\|u\|_{\ell_2^m}$ are also used for the same purpose.

For vectors $u, v$ in $\mathbb{C}^m$ (or other real and complex Euclidean spaces), $u \otimes v$ denotes their tensor product, that is, the linear transformation defined by

$$(u \otimes v)x = \langle v, x \rangle u.$$

Given a subspace $L \subset \mathbb{C}^m$ (more generally, a subspace of any Euclidean space), $P_L$ denotes the orthogonal projection onto $L$ and $L^{\perp}$ denotes the orthogonal complement of $L$.

We use the notations $\mathbb{M}_{m_1, m_2}(\mathbb{R})$ and $\mathbb{M}_{m_1, m_2}(\mathbb{C})$ for the spaces of all $m_1 \times m_2$ matrices with real or complex entries, respectively. In the case when $m_1 = m_2 = m$, we use the notations $\mathbb{M}_m(\mathbb{R})$ and $\mathbb{M}_m(\mathbb{C})$. The space of all Hermitian $m \times m$ matrices is denoted by $\mathbb{H}_m(\mathbb{C})$. For $A, B \in \mathbb{H}_m(\mathbb{C})$, the notation $A \leq B$ means that $B - A$ is nonnegatively definite.

We denote by $\mathrm{rank}(A)$ the rank of a matrix $A$ and by $\mathrm{tr}(A)$ the trace of a square matrix $A$. Given $A \in \mathbb{M}_{m_1, m_2}(\mathbb{C})$, $A^*$ denotes its adjoint matrix. We use the notations $\langle \cdot, \cdot \rangle$ for the Hilbert–Schmidt inner product of two matrices of the same size,

$$\langle A, B \rangle = \mathrm{tr}(AB^*),$$

$\| \cdot \|$ for the operator norm of matrices and $\| \cdot \|_p$, $p \geq 1$ for their *Schatten p-norm*:

$$\|A\|_p := \left( \sum_k \sigma_k^p(A) \right)^{1/p},$$

where $\{\sigma_k(A)\}$ denote the singular values of the matrix $A$ (usually, arranged in a nonincreasing order). In particular, $\| \cdot \|_2$ is *the Hilbert–Schmidt* or *Frobenius norm* and $\| \cdot \|_1$ is *the nuclear norm*. The notation $\| \cdot \|$ is reserved for the operator norm. Given a probability distribution $\Pi$ in $\mathbb{H}_m(\mathbb{C})$, we also associate with a matrix $B \in \mathbb{H}_m(\mathbb{C})$ the linear functional $\langle B, \cdot \rangle$ and define the $L_2(\Pi)$ norm of $B$ as the $L_2(\Pi)$-norm of this functional:

$$\|B\|_{L_2(\Pi)}^2 := \int_{\mathbb{H}_m(\mathbb{C})} \langle B, x \rangle^2 \Pi(dx).$$

We use the corresponding inner product $\langle \cdot, \cdot \rangle_{L_2(\Pi)}$ in the space of matrices.

For a matrix $S \in \mathbb{H}_m(\mathbb{C})$ of rank $r$ with spectral decomposition

$$S = \sum_{j=1}^{r} \lambda_j (e_j \otimes e_j),$$

where $e_1, \ldots, e_r$ are the eigenvectors corresponding to the non-zero eigenvalues $\lambda_1, \ldots, \lambda_r$, define *the support* of $S$ as $\mathrm{supp}(S) := \mathrm{l.s.}(\{e_1, \ldots, e_r\})$. Also, define

$$|S| := \sqrt{S^2} = \sum_{j=1}^{r} |\lambda_j| (e_j \otimes e_j)$$

and

$$\text{sign}(S) := \sum_{j=1}^{r} \text{sign}(\lambda_j)(e_j \otimes e_j).$$

It is well known that the subdifferential of the convex function $\mathbb{H}_m(\mathbb{C}) \ni S \mapsto \|S\|_1$ has the following representation (see, e.g., [151]):

$$\partial\|S\|_1 = \left\{\text{sign}(S) + P_{L^\perp} W P_{L^\perp} : \|W\| \leq 1\right\},$$

where $L := \text{supp}(S)$.

Some other facts of linear algebra used in Chap. 9 can be found in [21].

# References

1. Adamczak, R. (2008) A tail inequality for suprema of unbounded empirical processes with applications to Markov chains. *Electronic Journal of Probability*, 13, 34, 1000-1034.
2. Adamczak, R., Litvak, A.E., Pajor, A. and Tomczak-Jaegermann, N. (2011) Restricted isometry property of matrices with independent columns and neighborly polytopes by random sampling. *Constructive Approximation*, to appear.
3. Affentranger, F. and Schneider, R. (1992) Random Projections of Regular Simplices. *Discrete Comput. Geom.*, 7(3), 219–226.
4. Ahlswede, R. and Winter, A. (2002) Strong converse for identification via quantum channels. *IEEE Transactions on Information Theory*, 48, 3, pp. 569–679.
5. Alexander, K.S. (1987) Rates of growth and sample moduli for weighted empirical processes indexed by sets. *Probability Theory and Related Fields*, 75, 379–423.
6. Anthony, M. and Bartlett, P. (1999) Neural Network Learning: Theoretical Foundations. Cambridge University Press.
7. Arlot, S. and Bartlett, P. (2011) Margin adaptive model selection in statistical learning. *Bernoulli*, 17, 2, 687–713.
8. Arlot, S. and Massart, P. (2009) Data-driven calibration of penalties for least-squares regression. *J. Machine Learning Research*, 10, 245–279.
9. Aubin, J.-P. and Ekeland, I. (1984) Applied Nonlinear Analysis. J. Wiley & Sons, New York.
10. Audibert, J.-Y. (2004) Une approche PAC-bayésienne de la théorie statistique de l'apprentissage. PhD Thesis, University of Paris 6.
11. Baraud, Y. (2002) Model selection for regression on a random design. *ESAIM: Probability and Statistics*, 6, 127–146.
12. Barron, A., Birgé, L. and Massart, P. (1999) Risk Bounds for Model Selection via Penalization. *Probability Theory and Related Fields*, 113, 301–413.
13. Bartlett, P. (2008) Fast rates for estimation error and oracle inequalities for model selection. *Econometric Theory*, 24, 2, 545–552.
14. Bartlett, P., Boucheron, S. and Lugosi, G. (2002) Model selection and error estimation. *Machine Learning*, 48, 85–113.
15. Bartlett, P., Bousquet, O. and Mendelson, S. (2005) Local Rademacher Complexities, *Annals of Statistics*, 33, 4, 1497-1537.
16. Bartlett, P., Jordan, M. and McAuliffe, J. (2006) Convexity, Classification and Risk Bounds. *Journal of American Statistical Association*, 101(473), 138–156.
17. Bartlett, P. and Mendelson, S. (2006) Empirical Minimization. *Probability Theory and Related Fields*, 135(3), 311–334.
18. Bartlett, P., Mendelson, S. and Neeman, J. (2010) $\ell_1$-regularized linear regression: persistence and oracle inequalities. *Probability Theory and Related Fields*, to appear.

19. Bartlett, P. and Williamson, R.C. (1996) Efficient agnostic learning of neural networks with bounded fan-in. *IEEE Transactions on Information Theory*, 42(6), 2118–2132.
20. Ben-Tal, A. and Nemirovski, A. (2001) Lectures on Modern Convex Optimization. Analysis, Algorithms and Engineering Applications. MPS/SIAM Series on Optimization, Philadelphia.
21. Bhatia, R. (1997) Matrix Analysis. Springer, New York.
22. Bickel, P., Ritov, Y. and Tsybakov, A. (2009) Simultaneous Analysis of LASSO and Dantzig Selector. *Annals of Statistics*, 37, 4, 1705–1732.
23. Birgé, L. and Massart, P. (1997) From Model Selection to Adaptive Estimation. In: Festschrift for L. Le Cam. Research Papers in Probability and Statistics. D. Pollard, E. Torgersen and G. Yang (Eds.), 55-87. Springer, New York.
24. Birgé, L. and Massart, P. (2007) Minimal Penalties for Gaussian Model Selection. *Probability Theory and Related Fields*.138, 33–73.
25. Blanchard, G., Bousquet, O. and Massart, P. (2008) Statistical performance of support vector machines. *Annals of Statistics*, 36, 2, 489–531.
26. Blanchard, G., Lugosi, G. and Vayatis, N. (2003) On the rate of convergence of regularized boosting classifiers. *Journal of Machine Learning Research*, 4, 861–894.
27. Bobkov, S. and Houdré, C. (1997) Isoperimetric constants for product probability measures. *Annals of Probability*, v. 25, no. 1, 184–205.
28. Borell, C. (1974) Convex measures on locally convex spaces. *Arkiv för Matematik*, 12, 239–252.
29. Boucheron, S., Bousquet, O. and Lugosi, G. (2005) Theory of Classification. A Survey of Some Recent Advances. *ESAIM Probability & Statistics*, 9, 323–371.
30. Boucheron, S., Bousquet, O., Lugosi, G. and Massart, P. (2005) Moment inequalities for functions of independent random variables. *Annals of Probability*, 33, 2, 514–560.
31. Boucheron, S., Lugosi, G. and Massart, P. (2000) A sharp concentration inequality with applications. *Random Structures and Algorithms*, 16, 277–292.
32. Boucheron, S. and Massart, P. (2010) A high dimensional Wilks phenomenon. *Probability Theory and Related Fields*, to appear.
33. Bousquet, O. (2002) A Bennett concentration inequality and its application to suprema of empirical processes. *C.R. Acad. Sci. Paris*, 334, 495–500.
34. Bousquet, O., Koltchinskii, V. and Panchenko, D. (2002) Some local measures of complexity of convex hulls and generalization bounds. In: COLT2002, Lecture Notes in Artificial Intelligence, 2375, Springer, 59 - 73.
35. Bunea, F., Tsybakov, A. and Wegkamp, M. (2007) Aggregation for Gaussian regression, *Annals of Statistics*, 35,4, 1674–1697.
36. Bunea, F., Tsybakov, A. and Wegkamp, M. (2007) Sparsity oracle inequalities for the LASSO. *Electronic Journal of Statistics*, 1, 169–194.
37. Bunea, F., Tsybakov, A. and Wegkamp, M. (2007) Sparse density estimation with $\ell_1$ penalties. In: *Proc. 20th Annual Conference on Learning Theory (COLT 2007)*, Lecture Notes in Artificial Intelligence, Springer, v. 4539, pp. 530–543.
38. Candes, E. (2008) The restricted isometry property and its implications for compressed sensing, *C. R. Acad. Sci. Paris, Ser. I*, 346, 589-592.
39. Candes, E. and Plan, Y. (2009) Near-ideal model selection by $\ell_1$-minimization. *Annals of Statistics*, 37, 5A, 2145-2177.
40. Candes, E. and Plan, Y. (2009) Tight Oracle Bounds for Low-Rank Matrix Recovery from a Minimal Number of Random Measurements. *IEEE Transactions on Information Theory*, to appear.
41. Candes, E. and Recht, B. (2009) Exact matrix completion via convex optimization. *Foundations of Computational Mathematics*, 9(6), 717–772.
42. Candes, E., Romberg, J. and Tao, T. (2006) Stable Signal Recovery from Incomplete and Inaccurate Measurements. *Communications on Pure and Applied Mathematics*, 59, 1207–1223.
43. Candes, E., Rudelson, M., Tao, T. and Vershynin, R. (2005) Error Correction via Linear Programming. *Proc. 46th Annual IEEE Symposium on Foundations of Computer Science (FOCS05)*, IEEE, 295–308.

44. Candes, E. and Tao, T. (2007) The Dantzig selector: statistical estimation when $p$ is much larger than $n$. *Annals of Statistics*, 35,6, 2313–2351.
45. Candes, E. and Tao, T. (2010) The power of convex relaxation: Near-optimal matrix completion. *IEEE Transactions on Information Theory*, 56, 2053–2080.
46. Catoni, O. (2004) Statistical Learning Theory and Stochastic Optimization. *Ecole d'Eté de Probabilités de Saint-Flour XXXI -2001*, Lecture Notes in Mathematics, **1851**, Springer, New York.
47. Cucker, F. and Smale, S. (2001) On the mathematical foundations of learning. *Bull. Amer. Math. Soc.*, 39, 1–49.
48. Dalalyan, A. and Tsybakov, A. (2008) Aggregation by exponential weighting, sharp PAC-Bayesian bounds and sparsity. *Machine Learning*, 72, 39–61.
49. DeVore, R., Kerkyacharian, G., Picard, D. and Temlyakov, V. (2006) Approximating Methods for Supervised Learning. *Foundations of Computational Mathematics*, 6, 3–58.
50. de la Pena, V. and Giné, E. (1998) Decoupling: From Dependence to Independence, Springer, New York.
51. Devroye, L., Györfi, G. and Lugosi, G. (1996) A Probabilistic Theory of Pattern Recognition. Springer, New York.
52. Donoho, D.L. (2004) Neighborly Polytopes and Sparse Solution of Underdetermined Linear Equations. Technical Report, Stanford.
53. Donoho, D.L. (2006) For Most Large Underdetermined Systems of Equations the Minimal $\ell^1$-norm Near-Solution Approximates the Sparsest Near-Solution. *Communications on Pure and Applied Mathematics*, 59, 7, 907–934.
54. Donoho, D.L. (2006) For Most Large Underdetermined Systems of Linear Equations the Minimal $\ell^1$-norm Solution is also the Sparsest Solution. *Communications on Pure and Applied Mathematics*, 59, 797–829.
55. Donoho, D.L. (2006) Compressed Sensing. *IEEE Transactions on Information Theory*, 52, 4, 1289–1306.
56. Donoho, D.L., Elad, M. and Temlyakov, V. (2006) Stable Recovery of Sparse Overcomplete Representations in the Presence of Noise. *IEEE Transactions on Information Theory*, 52, 1, 6–18.
57. Donoho, D.L. and Tanner, J. (2005) Neighborliness of Randomly-Projected Simplices in High Dimensions. *Proc. National Academy of Sciences*, 102(27), 9446–9451.
58. Dudley, R.M. (1978) Central Limit Theorems for Empirical Measures, *Annals of Probability*, 6,6, 899–929.
59. Dudley, R.M. (1999) Uniform Central Limit Theorems. Cambridge University Press.
60. Einmahl, U. and Mason, D. (2000) An empirical processes approach to the uniform consistency of kernel type function estimators. *J. Theoretical Probability*, 13, 1–37.
61. Fromont, M. (2007) Model selection by bootstrap penalization for classification. *Machine Learning*, 66, 2-3, 165–207.
62. van de Geer, S. (1999) Empirical Processes in M-estimation. Cambridge University Press. Cambridge.
63. van de Geer, S. (2008) High-dimensional generalized linear models and the Lasso, *Annals of Statistics*, 36, 2, 614–645.
64. Giné, E. and Guillou, A. (2001) A law of the iterated logarithm for kernel density estimators in the precense of censoring. *Annales Inst. H. Poincaré (B), Probabilités et Statistiques*, 37, 503–522.
65. Giné, E., Koltchinskii, V. and Wellner, J. (2003) Ratio Limit Theorems for Empirical Processes. In: *Stochastic Inequalities*, Birkhäuser, pp. 249–278.
66. Giné, E. and Koltchinskii, V. (2006) Concentration Inequalities and Asymptotic Results for Ratio Type Empirical Processes. *Annals of Probability*, 34, 3, 1143-1216.
67. Giné, E. and Nickl, R. (2010) Adaptive estimation of a distribution function and its density in sup-norm loss by wavelet and spline projections. *Bernoulli*, 16, 4, 1137–1163.
68. Giné, E. and Mason, D. (2007) On local $U$-statistic processes and the estimation of densities of functions of several variables. *Annals of Statistics*, 35, 1105-1145.

69. Giné, E. and Zinn, J. (1984) Some limit theorems for empirical processes. *Annals of Probability* 12, 929-989.

70. Gross, D. (2011) Recovering Low-Rank Matrices From Few Coefficients in Any Basis. *IEEE Transactions on Information Theory*, 57, 3, 1548–1566.

71. Gross, D., Liu, Y.-K., Flammia, S.T., Becker, S. and Eisert, J. (2010) Quantum state tomography via compressed sensing. *Phys. Rev. Lett.*, 105(15):150401, October 2010.

72. Györfi, L., Kohler, M., Krzyzak, A. and Walk, H. (2002) A Distribution-Free Theory of Nonparametric Regression. Springer.

73. Hanneke, S. (2011) Rates of convergence in active learning, *Annals of Statistics*, 39, 1, 333–361.

74. Johnstone, I.M. (1998) Oracle Inequalities and Nonparametric Function Estimation. In: Documenta Mathematica, *Journal der Deutschen Mathematiker Vereinigung*, Proc. of the International Congress of Mathematicians, Berlin, 1998, v.III, 267–278.

75. Keshavan, R.H., Montanari, A. and Oh, S. (2010) Matrix completion from a few entries. *IEEE Transactions on Information Theory*, 56, 2980–2998.

76. Klartag, B. and Mendelson, S. (2005) Empirical Processes and Random Projections. *Journal of Functional Analysis*, 225(1), 229–245.

77. Klein, T. (2002) Une inégalité de concentration à gauche pour les processus empiriques. *C.R. Acad. Sci. Paris*, Ser I, 334, 500–505.

78. Klein, T. and Rio, E. (2005) Concentration around the mean for maxima of empirical processes. *Annals of Probability*, 33,3, 1060–1077.

79. Kohler, M. (2000) Inequalities for uniform deviations of averages from expectations with applications to nonparametric regression. *J. of Statistical Planning and Inference*, 89, 1–23.

80. Koltchinskii, V. (1981) On the central limit theorem for empirical measures. *Theory of Probability and Mathematical Statistics*, 24, 71–82.

81. Koltchinskii, V. (2001) Rademacher penalties and structural risk minimization. *IEEE Transactions on Information Theory*, 47(5), 1902–1914.

82. Koltchinskii, V. (2005) Model selection and aggregation in sparse classification problems. *Oberwolfach Reports: Meeting on Statistical and Probabilistic Methods of Model Selection, October 2005.*

83. Koltchinskii, V. (2006) Local Rademacher Complexities and Oracle Inequalities in Risk Minimization. *Annals of Statistics*, 34, 6, 2593–2656.

84. Koltchinskii, V. (2009) Sparsity in penalized empirical risk minimization, *Annales Inst. H. Poincaré (B), Probabilités et Statistiques*, 45, 1, 7–57.

85. Koltchinskii, V. (2009) The Dantzig Selector and Sparsity Oracle Inequalities. *Bernoulli*, 15,3, 799-828.

86. Koltchinskii, V. (2009) Sparse Recovery in Convex Hulls via Entropy Penalization, *Annals of Statistics*, 37, 3, 1332-1359.

87. Koltchinskii, V. (2010) Rademacher Complexities and Bounding the Excess Risk in Active Learning, *Journal of Machine Learning Research*, 2010, 11, 2457–2485.

88. Koltchinskii, V. (2010) Von Neumann entropy penalization and low rank matrix estimation. Preprint.

89. Koltchinskii, V. (2010) A remark on low rank matrix recovery and noncommutative Bernstein type inequalities. Preprint.

90. Koltchinskii, V., Lounici, K. and Tsybakov, A. (2010) Nuclear norm penalization and optimal rates for noisy low rank matrix completion. *Annals of Statistics*, to appear.

91. Koltchinskii, V. and Minsker, S. (2010) Sparse recovery in convex hulls of infinite dictionaries. In: *COLT 2010, 23rd Conference on Learning Theory*, A. Kalai and M. Mohri (Eds), Haifa Israel, pp. 420–432.

92. Koltchinskii, V. and Panchenko, D. (2000) Rademacher processes and bounding the risk of function learning. In: Giné, E., Mason, D. and Wellner, J. *High Dimensional Probability II*, 443–459.

93. Koltchinskii, V. and Panchenko, D. (2002) Empirical margin distributions and bounding the generalization error of combined classifiers. *Annals of Statistics*, 30, 1, 1–50.

94. Koltchinskii, V., Panchenko, D. and Lozano, F. (2003) Bounding the generalization error of convex combinations of classifiers: balancing the dimensionality and the margins. *Ann. Appl. Probab.*, **13**, 1, 213–252.

95. Koltchinskii, V. and Panchenko, D. (2005) Complexities of convex combinations and bounding the generalization error in classification. *Annals of Statistics*, 33, 4, 1455–1496.

96. Koltchinskii, V. and Yuan, M. (2008) Sparse Recovery in Large Ensembles of Kernel Machines. In: *21st Annual Conference on Learning Theory COLT-2008*, Helsinki, 229–238, Omnipress.

97. Koltchinskii, V. and Yuan, M. (2010) Sparsity in Multiple Kernel Learning, *Annals of Statistics*, 38, 3660–3695.

98. Lecué, G. and Gaiffas, S. (2010) Sharp oracle inequalities for high-dimensional matrix prediction. *IEEE Transactions on Information Theory*, to appear.

99. Lecué, G. and Mendelson, S. (2009) Aggregation via empirical risk minimization. *Probability Theory and Related Fields*, 145(3–4), 591–613.

100. Ledoux, M. (2001) The Concentration of Measure Phenomenon. Mathematical Surveys and Monographs, Vol. 89, American Mathematical Society.

101. Ledoux, M. and Talagrand, M. (1991) Probability in Banach Spaces. Springer-Verlag, New York.

102. Lieb, E.H. (1973) Convex trace functions and the Wigner-Yanase-Dyson conjecture. *Adv. Math.*, 11, 267–288.

103. Lugosi, G. and Vayatis, N. (2004) On the Bayes-risk consistency of regularized boosting methods. *Annals of Statistics*, 32, 1, 30–55.

104. Lugosi, G. and Wegkamp, M. (2004) Complexity regularization via localized random penalties. *Annals of Statistics*, 32, 4, 1679–1697.

105. Mammen, E. and Tsybakov, A. (1999) Smooth discrimination analysis. *Annals of Statistics* 27, 1808–1829.

106. Massart, P. (2000) Some applications of concentration inequalities to statistics. *Ann. Fac. Sci. Toulouse (IX)*, 245–303.

107. Massart, P. (2007) Concentration Inequalities and Model Selection. *Ecole d'ete de Probabilités de Saint-Flour 2003*, Lecture Notes in Mathematics, Springer.

108. Massart, P. and Nedelec, E. (2006) Risk bounds for statistical learning. *Annals of Statistics*, 34, 5, 2326–2366.

109. Massart, P. and Meynet, C. (2010) An $\ell_1$-Oracle Inequality for the Lasso. *Electronic Journal of Statistics*, to appear.

110. McAllester, D.A. (1998) Some PAC-Bayesian theorems. In *Proc. 11th Annual Conference on Learning Theory*, pp. 230–234, ACM Press.

111. Meier, L., van de Geer, S. and Bühlmann, P. (2009) High-dimensional additive modeling, *Annals of Statistics*, 37, 3779–3821.

112. Mendelson, S. (2002) Improving the sample complexity using global data. *IEEE Transactions on Information Theory*, 48, 1977–1991.

113. Mendelson, S. (2002) Geometric parameters of kernel machines. In: COLT 2002, Lecture Notes in Artificial Intelligence, 2375, Springer, 29–43.

114. Mendelson, S. (2008) Lower bounds for the empirical minimization algorithm. *IEEE Transactions on Information Theory*, 54, 8, 3797–3803.

115. Mendelson, S. (2010) Empirical processes with a bounded $\psi_1$ diameter. *Geometric and Functional Analysis*, 20,4, 988–1027.

116. Mendelson, S. and Neeman, J. (2010) Regularization in kernel learning. *Annals of Statistics*, 38, 1, 526–565.

117. Mendelson, S., Pajor, A. and Tomczak-Jaegermann, N. (2007) Reconstruction and subgaussian operators in Asymptotic Geometric Analysis. *Geometric and Functional Analysis*, 17(4), 1248–1282.

118. Negahban, S. and Wainwright, M.J. (2010) Restricted strong convexity and weighted matrix completion: optimal bounds with noise. Preprint.

119. Nemirovski, A. (2000) Topics in non-parametric statistics, In: P. Bernard, editor, *Ecole d'Et'e de Probabilités de Saint-Flour XXVIII*, 1998, Lecture Notes in Mathematics, Springer, New York.

120. Nielsen, M.A. and Chuang, I.L. (2000) Quantum Computation and Quantum Information, Cambridge University Press.

121. Pajor, A. and Tomczak-Jaegermann, N. (1985) Remarques sur les nombres d'entropie d'un opérateur et de son transposé. *C.R. Acad. Sci. Paris Sér. I Math.*, 301(15), 743–746.

122. Pollard, D. (1982) A central limit theorem for empirical processes. *Journal of the Australian Mathematical Society*, A, 33, 235–248.

123. Pollard, D.(1984) Convergence of Stochastic Processes, Springer, New York.

124. Recht, B. (2009) A Simpler Approach to Matrix Completion. *Journal of Machine Learning Research*, to appear.

125. Recht, B., Fazel, M. and Parrilo, P. (2010) Guaranteed minimum rank solutions of matrix equations via nuclear norm minimization. *SIAM Review*, 52, 3, 471–501.

126. Rigollet, P. and Tsybakov, A. (2011) Exponential screening and optimal rates of sparse estimation. *Annals of Statistics*, 39, 731–771.

127. Rohde, A. and Tsybakov, A. (2011) Estimation of high-dimensional low rank matrices. *Annals of Statistics*, 39, 2, 887–930.

128. Rudelson, M. (1999) Random vectors in the isotropic position. *J. Funct. Analysis*, 164(1), 60–72.

129. Rudelson, M. and Vershynin, R. (2005) Geometric Approach to Error Correcting Codes and Reconstruction of Signals. *Int. Math. Res. Not.*, no 64, 4019-4041.

130. Rudelson, M. and Vershynin, R. (2010) Non-asymptotic theory of random matrices: extreme singular values. *Proceedings of the International Congress of Mathematicians*, Hyderabad, India.

131. Saumard, A. (2010) Estimation par Minimum de Contraste Régulier et Heuristique de Pente en Sélection de Modéles. PhD Thesis, Université de Rennes 1.

132. Shen, X. and Wong, W.H. (1994) Convergence rate of sieve estimates. *Annals of Statistics*, 22, 580-615.

133. Simon, B. (1979) Trace Ideals and their Applications. Cambridge University Press.

134. Steinwart, I. (2005) Consistency of support vector machines and other regularized kernel machines. *IEEE Transactions on Information Theory*, 51, 128–142.

135. Steinwart, I. and Christmann, A. (2008) Support Vector Machines. Springer.

136. Steinwart, I. and Scovel, C. (2007) Fast rates for support vector machines using Gaussian kernels. *Annals of Statistics*, 35, 575–607.

137. Talagrand, M. (1994) Sharper bounds for Gaussian and empirical processes. *Annals of Probability*, 22, 28-76.

138. Talagrand, M. (1996) A new look at independence. *Annals of Probability*, 24, 1-34.

139. Talagrand, M. (1996) New concentration inequalities in product spaces. *Invent. Math.*, 126, 505-563.

140. Talagrand, M. (2005) The Generic Chaining. Springer.

141. Tibshirani, R. (1996) Regression shrinkage and selection via Lasso, *J. Royal Statist. Soc., Ser B*, **58**, 267–288.

142. Tropp, J.A. (2010) User-friendly tail bounds for sums of random matrices. *Foundations of Computational Mathematics*, to appear.

143. Tsybakov, A. (2003) Optimal rates of aggregation. In: *Proc. 16th Annual Conference on Learning Theory (COLT) and 7th Annual Workshop on Kernel Machines*, Lecture Notes in Artificial Intelligence, **2777**, Springer, New York, 303–313.

144. Tsybakov, A. (2004) Optimal aggregation of classifiers in statistical learning. *Annals of Statistics*, 32, 135–166.

145. Tsybakov, A. and van de Geer, S. (2005) Square root penalty: adaptation to the margin in classification and in the edge estimation. *Annals of Statistics*, 33, 3, 1203–1224.

146. Vapnik, V. (1998) Statistical Learning Theory. John Wiley & Sons, New York.

147. Vapnik, V. and Chervonenkis, A. (1974) Theory of Pattern Recognition. Nauka, Moscow (in Russian).
148. van der Vaart, A.W. and Wellner, J.A. (1996) Weak Convergence and Empirical Processes. With Applications to Statistics. Springer-Verlag, New York.
149. van der Vaart, A.W. and Wellner, J.A. (2011) A Local Maximal Inequality under Uniform Entropy. *Electronic Journal of Statistics*, 5, 192–203.
150. Vershik, A.M. and Sporyshev, P.V. (1992) Asymptotic behavior of the number of faces of random polyhedra and the neighborliness problem. *Selecta Math. Soviet.*, 11(2), 181–201.
151. Watson, G. A. (1992) Characterization of the subdifferential of some matrix norms. *Linear Algebra Appl.*, 170, 33-45.
152. Yang, Y. (2000) Mixing strategies for density estimation. *Annals of Statistics*, **28**, 75–87.
153. Yang, Y. (2004) Aggregating regression procedures for a better performance. *Bernoulli*, **10**, 25–47.
154. Zhang, T. (2001) Regularized Winnow Method. In: *Advances in Neural Information Processing Systems 13 (NIPS2000)*, T.K. Leen, T.G. Dietrich and V. Tresp (Eds), MIT Press, pp. 703–709.
155. Zhang, T. (2004) Statistical behavior and consistency of classification methods based on convex risk minimization. *Annals of Statistics*, 32, 1, 56–134.
156. Zhang, T. (2006) From epsilon-entropy to KL-complexity: analysis of minimum information complexity density estimation. *Annals of Statistics*, 34, 2180–2210.
157. Zhang, T. (2006) Information Theoretical Upper and Lower Bounds for Statistical Estimation. *IEEE Transactions on Information Theory*, 52, 1307–1321.

# Index

V. Koltchinskii, *Oracle Inequalities in Empirical Risk Minimization and Sparse Recovery Problems*, Lecture Notes in Mathematics 2033, DOI 10.1007/978-3-642-22147-7,
© Springer-Verlag Berlin Heidelberg 2011

# Programme of the school

**Main lectures**

| | |
|---|---|
| Richard Kenyon | Lectures on dimers |
| Vladimir Koltchinskii | Oracle inequalities in empirical risk minimization and sparse recovery problems |
| Yves Le Jan | Markov paths, loops and fields |

**Short lectures**

| | |
|---|---|
| Michael Allman | Breaking the chain |
| Pierre Alquier | Lasso, iterative feature selection and other regression methods satisfying the "Dantzig constraint" |
| Jürgen Angst | Brownian motion and Lorentzian manifolds, the case of Robertson-Walker space-times |
| Witold Bednorz | Some comments on the Bernoulli conjecture |
| Charles Bordenave | Spectrum of large random graphs |
| Cédric Boutillier | The critical Ising model on isoradial graphs via dimers |
| Robert Cope | Modelling in birth-death and quasi-birth-death processes |
| Irmina Czarna | Two-dimensional dividend problems |
| Michel Émery | Geometric structure of Azéma martingales |
| Christophe Gomez | Time reversal of waves in random waveguides: a super resolution effect |
| Nastasiya Grinberg | Semimartingale decomposition of convex functions of continuous semimartingales |
| François d'Hautefeuille | Entropy and finance |
| Erwan Hillion | Ricci curvature bounds on graphs |
| Wilfried Huss | Internal diffusion limited aggregation and related growth models |
| Jean-Paul Ibrahim | Large deviations for directed percolation on a thin rectangle |
| Liza Jones | Infinite systems of non-colliding processes |
| Andreas Lagerås | General branching processes conditioned on extinction |

V. Koltchinskii, *Oracle Inequalities in Empirical Risk Minimization and Sparse Recovery Problems*, Lecture Notes in Mathematics 2033, DOI 10.1007/978-3-642-22147-7, © Springer-Verlag Berlin Heidelberg 2011

| Benjamin Laquerrière | Conditioning of Markov processes and applications |
| Krzysztof Łatuszynski | "If and only if" conditions (in terms of regeneration) for $\sqrt{n}$-CLTs for ergodic Markov chains |
| Thierry Lévy | The Poisson process indexed by loops |
| Wei Liu | Spectral gap and convex concentration inequalities for birth-death processes |
| Gregorio Moreno | Directed polymers on a disordered hierarchical lattice |
| Jonathan Novak | Deforming the increasing subsequence problem |
| Ecaterina Sava | The Poisson boundary of lamplighter random walks |
| Bruno Schapira | Windings of the SRW on $\mathbb{Z}^2$ or triangular lattice and averaged Dehn function |
| Laurent Tournier | Random walks in Dirichlet environment |
| Nicolas Verzelen | Kullback oracle inequalities for covariance estimation |
| Vincent Vigon | LU-factorization and probability |
| Guillaume Voisin | Pruning a Lévy continuum random tree |
| Peter Windridge | Blocking and pushing interactions for discrete interlaced processes |
| Olivier Wintenberger | Some Bernstein's type inequalities for dependent data |

# List of participants

## 38<sup>th</sup> Probability Summer School, Saint-Flour, France

**38**th **Probability Summer School, Saint-Flour, France**
July 6–19, 2008

**Lecturers**

| | |
|---|---|
| Richard KENYON | Brown University, Providence, USA |
| Vladimir KOLTCHINSKII | Georgia Inst. Technology, Atlanta, USA |
| Yves LE JAN | Université Paris-Sud, France |

**Participants**

| | |
|---|---|
| Michael ALLMAN | Warwick Univ., UK |
| Pierre ALQUIER | Univ. Paris Diderot, F |
| Jurgen ANGST | Univ. Strasbourg, F |
| Jean-Yves AUDIBERT | ENPC, Marne-la-Vallée, F |
| Witold BEDNORZ | Warsaw Univ., F |
| Charles BORDENAVE | Univ. Toulouse, F |
| Cédric BOUTILLIER | Univ. Pierre et Marie Curie, Paris, F |
| Pierre CONNAULT | Univ. Paris-Sud, Orsay, F |
| Robert COPE | Univ. Queensland, St Lucia, Australia |
| Irmina CZARNA | Univ. Wroclaw, Poland |
| Bassirou DIAGNE | Univ. Orléans, F |
| Roland DIEL | Univ. Orléans, F |
| Michel EMERY | Univ. Strasbourg, F |
| Mikael FALCONNET | Univ. Joseph Fourier, Grenoble, F |
| Benjamin FAVETTO | Univ. Paris Descartes, F |
| Jacques FRANCHI | Univ. Strasbourg, F |
| Laurent GOERGEN | ENS, Paris, F |
| Christophe GOMEZ | Univ. Paris Diderot, F |
| Nastasiya GRINBERG | Warwick Univ., UK |
| François HAUTEFEUILLE | London, UK |
| Erwan HILLION | Univ. Paul Sabatier, Toulouse, F |
| Wilfried HUSS | Graz Univ. Technology, Austria |
| Jean-Paul IBRAHIM | Univ. Paul Sabatier, Toulouse, F |

V. Koltchinskii, *Oracle Inequalities in Empirical Risk Minimization and Sparse Recovery*   253
*Problems*, Lecture Notes in Mathematics 2033, DOI 10.1007/978-3-642-22147-7,
© Springer-Verlag Berlin Heidelberg 2011

| | |
|---|---|
| Liza JONES | Univ. Oxford, UK |
| Adrien KASSEL | ENS, Paris, F |
| Andreas LAGERA S | Univ. Gothenburg & Chalmers Univ. Technology, Sweden |
| Benjamin LAQUERRIERE | Univ. La Rochelle, F |
| Krzysztof LATUSZYNSKI | Warsaw School Economics, Poland |
| Thierry LÉVY | ENS, Paris, F |
| Biao LI | Univ. Blaise Pascal, Clermont-Ferrand, F |
| Wei LIU | Univ. Blaise Pascal, Clermont-Ferrand, F |
| Karim LOUINICI | Univ. Paris Diderot, F |
| Sybille MICAN | Univ. München, Germany |
| Stanislav MINSKER | Georgia Inst. Technology, Atlanta, USA |
| Gregorio MORENO FLORES | Univ. Paris Diderot, F |
| Jonathan NOVAK | Queen's Univ., Kingston, Canada |
| Jean PICARD | Univ. Blaise Pascal, Clermont-Ferrand, F |
| Erwan SAINT LOUBERT BIÉ | Univ. Blaise Pascal, Clermont-Ferrand, F |
| Ecaterina SAVA | Technical Univ. Graz, Austria |
| Catherine SAVONA | Univ. Blaise Pascal, Clermont-Ferrand, F |
| Bruno SCHAPIRA | Univ. Paris-Sud, Orsay, F |
| Eckhard SCHLEMM | Univ. Toronto, Canada & Freie Univ. Berlin, Germany |
| Arseni SEREGIN | Univ. Washington, USA |
| Laurent SERLET | Univ. Blaise Pascal, Clermont-Ferrand, F |
| Frédéric SIMON | Univ. Claude Bernard, Lyon, F |
| Laurent TOURNIER | Univ. Claude Bernard, Lyon, F |
| Kiamars VAFAYI | Univ. Leiden, NL |
| Nicolas VERZELEN | Univ. Paris-Sud, Orsay, F |
| Vincent VIGON | Univ. Strasbourg, F |
| Guillaume VOISIN | Univ. Orléans, F |
| Peter WINDRIDGE | Univ. Warwick, UK |
| Olivier WINTERBERGER | Univ. Paris 1, F |
| Lorenzo ZAMBOTTI | Univ. Pierre et Marie Curie, Paris, F |
| Jean-Claude ZAMBRINI | GFMUL, Lisbon, Portugal |

# LECTURE NOTES IN MATHEMATICS    🐎 Springer

Edited by J.-M. Morel, B. Teissier; P.K. Maini

**Editorial Policy** (for the publication of monographs)

1. Lecture Notes aim to report new developments in all areas of mathematics and their applications - quickly, informally and at a high level. Mathematical texts analysing new developments in modelling and numerical simulation are welcome.

    Monograph manuscripts should be reasonably self-contained and rounded off. Thus they may, and often will, present not only results of the author but also related work by other people. They may be based on specialised lecture courses. Furthermore, the manuscripts should provide sufficient motivation, examples and applications. This clearly distinguishes Lecture Notes from journal articles or technical reports which normally are very concise. Articles intended for a journal but too long to be accepted by most journals, usually do not have this "lecture notes" character. For similar reasons it is unusual for doctoral theses to be accepted for the Lecture Notes series, though habilitation theses may be appropriate.

2. Manuscripts should be submitted either online at www.editorialmanager.com/lnm to Springer's mathematics editorial in Heidelberg, or to one of the series editors. In general, manuscripts will be sent out to 2 external referees for evaluation. If a decision cannot yet be reached on the basis of the first 2 reports, further referees may be contacted: The author will be informed of this. A final decision to publish can be made only on the basis of the complete manuscript, however a refereeing process leading to a preliminary decision can be based on a pre-final or incomplete manuscript. The strict minimum amount of material that will be considered should include a detailed outline describing the planned contents of each chapter, a bibliography and several sample chapters.

    Authors should be aware that incomplete or insufficiently close to final manuscripts almost always result in longer refereeing times and nevertheless unclear referees' recommendations, making further refereeing of a final draft necessary.

    Authors should also be aware that parallel submission of their manuscript to another publisher while under consideration for LNM will in general lead to immediate rejection.

3. Manuscripts should in general be submitted in English. Final manuscripts should contain at least 100 pages of mathematical text and should always include

    - a table of contents;
    - an informative introduction, with adequate motivation and perhaps some historical remarks: it should be accessible to a reader not intimately familiar with the topic treated;
    - a subject index: as a rule this is genuinely helpful for the reader.

    For evaluation purposes, manuscripts may be submitted in print or electronic form (print form is still preferred by most referees), in the latter case preferably as pdf- or zipped psfiles. Lecture Notes volumes are, as a rule, printed digitally from the authors' files. To ensure best results, authors are asked to use the LaTeX2e style files available from Springer's web-server at:

    ftp://ftp.springer.de/pub/tex/latex/svmonot1/ (for monographs) and
    ftp://ftp.springer.de/pub/tex/latex/svmultt1/ (for summer schools/tutorials).

Additional technical instructions, if necessary, are available on request from lnm@springer.com.

4. Careful preparation of the manuscripts will help keep production time short besides ensuring satisfactory appearance of the finished book in print and online. After acceptance of the manuscript authors will be asked to prepare the final LaTeX source files and also the corresponding dvi-, pdf- or zipped ps-file. The LaTeX source files are essential for producing the full-text online version of the book (see http://www.springerlink.com/openurl.asp?genre=journal&issn=0075-8434 for the existing online volumes of LNM). The actual production of a Lecture Notes volume takes approximately 12 weeks.

5. Authors receive a total of 50 free copies of their volume, but no royalties. They are entitled to a discount of 33.3 % on the price of Springer books purchased for their personal use, if ordering directly from Springer.

6. Commitment to publish is made by letter of intent rather than by signing a formal contract. Springer-Verlag secures the copyright for each volume. Authors are free to reuse material contained in their LNM volumes in later publications: a brief written (or e-mail) request for formal permission is sufficient.

**Addresses:**
Professor J.-M. Morel, CMLA,
École Normale Supérieure de Cachan,
61 Avenue du Président Wilson, 94235 Cachan Cedex, France
E-mail: morel@cmla.ens-cachan.fr

Professor B. Teissier, Institut Mathématique de Jussieu,
UMR 7586 du CNRS, Équipe "Géométrie et Dynamique",
175 rue du Chevaleret
75013 Paris, France
E-mail: teissier@math.jussieu.fr

*For the "Mathematical Biosciences Subseries" of LNM:*

Professor P. K. Maini, Center for Mathematical Biology,
Mathematical Institute, 24-29 St Giles,
Oxford OX1 3LP, UK
E-mail : maini@maths.ox.ac.uk

Springer, Mathematics Editorial, Tiergartenstr. 17,
69121 Heidelberg, Germany,
Tel.: +49 (6221) 487-8259

Fax: +49 (6221) 4876-8259
E-mail: lnm@springer.com